MODERN TRENDS IN PHYSICS RESEARCH

THIRD International Conference on Modern Trends in Physics Research

MODERN TRENDS IN PHYSICS RESEARCH

THIRD International Conference on Modern Trends in Physics Research

MTPR-08

Cairo, Egypt   6 – 10 April 2008

EDITOR

Lotfia M. El Nadi

Cairo University, Egypt

CONFERENCE PROCEEDINGS ■ VOLUME 998

 **World Scientific**

NEW JERSEY · LONDON · SINGAPORE · BEIJING · SHANGHAI · HONG KONG · TAIPEI · CHENNAI

Published by

World Scientific Publishing Co. Pte. Ltd.

5 Toh Tuck Link, Singapore 596224

USA office: 27 Warren Street, Suite 401-402, Hackensack, NJ 07601

UK office: 57 Shelton Street, Covent Garden, London WC2H 9HE

British Library Cataloguing-in-Publication Data
A catalogue record for this book is available from the British Library.

MODERN TRENDS IN PHYSICS RESEARCH
Proceedings of the Third International Conference on MTPR-08

ISBN-13 978-981-4317-50-4
ISBN-10 981-4317-50-0

Printed in Singapore by Mainland Press Pte Ltd.

FOREWORD

The Third International Conference "Modern Trends in Physics Research" MTPR-08 was held from 6–10 April 2008 at Cairo University, in collaboration with Misr University for Science and Technology (MUST), a newly-founded private university.

The organizing committees, with Professor Lotfia El Nadi as the leading program chair of the conference, had held whole day activities at the campuses of MUST. We considered this as a great honor to our university. It was an opportunity to familiarize the participants to the pioneer and initiator of our university, the late Professor Afaf Kafafi and innumerate her effort in order to establish MUST as one of the most advanced and modern Egyptian university. Our staff members and students were presented with up-to-date Keynote Talks, Plenary and Invited contributions from international experts participating in the activities of the conference. We were proud to receive the conference's Shield Award, presented to our university, by the program chair.

The opening ceremony at the Festival Main Hall of Cairo University was perfect. It was attended by the highest officials of higher education in Egypt, particularly the newly-appointed Professor Tarek Hussien, the President of the Academy of Scientific Research and Technology.

The Opening Keynote Lecture, delivered by Professor Jay Pasachoff, the director of Hopkins Observatory of USA, was fascinating and a perfect introduction to the world of space and its planetary systems.

The collaboration between the historical and well-established Cairo University with the young MUST had added a special flavor to the meeting.

I personally take this opportunity to thank the Cairo University authorities and the organizing committees for accomplishing this most successful event. I want to highly point and tag the importance of consistent cooperation between the universities to advance scientific knowledge and promote scientific research.

Professor Dr. Raafat Mahmoud

President of Misr University for Science and Technology (MUST)

PREFACE

The third of the international conference series, to be carried out biannually by the Physics Department, Faculty of Science, took place at Cairo University from April 6-10, 2008. The conference was carried out in collaboration with Misr University of Science and Technology (MUST), one of the newly-founded private universities in Egypt.

The year 2008 was an important year since celebrations of several important occasions were taking place:

1. **100** years anniversary of CAIRO UNIVERSITY
2. **70** years anniversary for The Egyptian Mathematical & Physical Society
3. **50** years anniversary for The Egyptian Physical Society EPS
4. **20** years anniversary for The Topical Society of Laser Sciences TSLS

The year **2008** was a showcase of creativity and innovation in Physics.

The GOAL & MOTIVATION of the meeting were to

Develop greater understanding of physics and its applications to new industries

Innovate knowledge about breakthroughs in PHYSICS fundamentals.

State the existing possibilities of the studies in the vast expanding fields PHYSICS

Consider methods for implementing local, regional and international cooperation

Outline the needs to upgrade the laboratories of PHYSICS research facilities

Verify & issue Memorandum of Understanding: PHYSICS: Path towards Appl.

Enable execution of projects forwarded by INDIVIDUALS or GROUPS

Reorganize plans for training graduate students

DISCOVER possible ways for our country to catch up with Modern Trends in Physics Research

The MTPR-08 technical program comprised 14 sessions 3.5 hours each within five days. The purpose of the conference was to paint an advanced picture of the activities in several branches of the physics domain in the last two years after the MTPR-06 conference. All papers were peer reviewed. This proceeding highlights the contributions presented at the conference. It provides some detailed accounts of the latest results in the fields of atomic, molecular, condensed matter, lasers, nuclear, particle and astrophysics. The papers and some of the review articles had been written by both international and national experts in these fields and should serve as reference materials to scientists wishing to get into the aspects of modern physics and technology of the twenty first century.

This conference proceedings has ten sectors. The first consists of the Honorary Keynote Lecture: "The Solar System in the Age of Space Exploration" by Professor Dr Jay M. Pasachoff of Williams College and Director of Hopkins Observatory in USA. His projects with NASA and MIT enriched the field of astrophysics with new information that is helping in the INTERSTELLAR JOURNEYS to OUTER SPACE. *He is providing means to deepen scientific knowledge to promote better life to HUMANITIES.*

The other chapters cover each of the topics which formed the theme of the conference. Each chapter opens with keynote, plenary, and invited presentations followed by the contributed papers.

I am grateful to all authors who took the effort to submit their excellent papers in a timely manner. Thanks in particular to all international participants for their continuous advice that contributed greatly to rank the conference as a highly successful meeting.

I wish to express my gratitude to both authorities of Cairo University and Misr University of Science and Technology for the generous finance given to MTPR-08 conference that helped greatly in shaping and accomplishing the meeting in perfect shape. Utmost thanks to my colleagues, Hussein A. Moniem, Galila A. Mehena and Magdy M. Omar for their sincere efforts before, during and after the conference. Thanks are also extended to some of the staff members of the Physics Department who directly or indirectly had contributed to the success of MTPR.

This book contains full reports of about 40 contributions presented at the MTPR-08 International Conference. Some of the contributions are article reviews. The book highlights the most important achievements and state-of-the-art research in modern physics. The book consequently contains the recent research in atomic and molecular spectroscopy, astrophysics, condensed matter, laser technology and applications, nanotechnology, nuclear and particle physics from both theoretical and experimental point of view.

The book covers the efforts of the physicists to contribute to the main branches of physics that could fulfill the curiosity of young researchers. The conference as well as the book gives the opportunity to young researchers to feel the several important fields of physics and could well help them, in a glance, to choose the most interesting topic to follow in their future physics research.

For the readers, I hope you will feel the excitement and enthusiasm and be motivated to join the next MTPR-010 (2010) conference which will be held from 12/12/2010 to 16/12/2010. The website for MTPR-010 is www.eun.eg/MTPR-010/home.htm that will appear on October 2010 and will continue till October 2011. Insha-Allah.

In conclusion, it is worthy to assure the readers that throughout my life, I will always be ready to serve in spreading knowledge, promoting international cooperation and caring to upgrade the wonderful world of physics research.

THE EDITOR
Professor Dr. Lotfia M. El Nadi
Program Chair of MTPR-08 (April 2008)
Professor of Nuclear and Laser Physics, Physics Department, Faculty of Science,
Vice Director of International Center of Scientific and Applied Studies of HDSP Lasers, NILES
Cairo University, Egypt

CONTENTS

OPENING HONORARY KEYNOTE PRESENTATION

I. ATOMIC, MOLECULAR & CONDENSED MATTER PHYSICS

I-1 KEYNOTE AND PLENARY PAPERS

I-2 INVITED LECTURE PAPERS

I-3 CONTRIBUTING PAPERS

II. LASERS, CHEMICAL PHYSICS & DEVICES

II-1 KEYNOTE AND PLENARY PAPERS

II-2 INVITED LECTURE PAPERS

II-3 CONTRIBUTING PAPERS

III. NUCLEAR, PARTICLE PHYSICS & ASTROPHYSICS

III-1 KEYNOTE AND PLENARY PAPERS

III-2 INVITED LECTURE PAPERS

III-3 CONTRIBUTING PAPERS

OPENING HONORARY KEYNOTE PRESENTATION

THE SOLAR SYSTEM IN THE AGE OF SPACE EXPLORATION

JAY M. PASACHOFF[*]

Williams College—Hopkins Observatory, Williamstown, Massachusetts 01267, USA

Abstract. We are celebrating the 50th anniversary of the launch of Sputnik, which began the space age. Though the manned exploration of the solar system has been limited to the Moon, in NASA's Apollo Program that ended over 35 years ago, robotic exploration of the solar system continues to be very successful. This paper explores the latest space mission and other observations of each planet and of each type of solar-system object, including dwarf planets, asteroids, and comets, as well as the sun.

4 October 1957 was an epochal day in the history of humanity. With the launch of Sputnik, we were launched into weightlessness in an important step toward escaping the Earth's gravity. The launch of Sputnik also marked the beginning of the space race, which spurred both the Soviet Union and the United States into heights that they would not have reached without the competition.

At the 50th anniversary of the space program, let us take stock of how we have explored our solar system. We have launched uncrewed rockets to all the newly limited number of 8 planets and received images and other data from them. We now even have a rocket en route to the former 9th planet, now the first dwarf planet, though we will never know if *New Horizons* or some other program would indeed have been financed had Pluto had its category changed before launch.

Further, objects between the planets are of increasing image. Spacecraft have visited asteroids and comets, and spectacular comets doing strange things randomly appear in our skies. Studies of asteroids have received surprising importance with the realization that our Earth may be targeted by one of them, and that we want notice in order to give ourselves a chance to divert any large object that might be heading for a collisions with us.

Studies of the sun are carried out with a different set of telescopes and spacecraft from studies of other solar-system objects, because the sun is so bright that it could overload systems optimized for fainter objects.

In this paper, I will survey the status of studies of each of the types of solar-system object.

Mercury

Mercury has long been one of the most ignored objects in the solar system. It is so close to the sun that from Earth it is rarely seen; even Copernicus is rumored never to have seen it. As the innermost planet, spacecraft to it have to be prevented from overheating.

Not since 1974-75, when NASA's Mariner 10 flew by Mercury three times, had Mercury been visited. But in 2008, NASA's MESSENGER spacecraft (an acronym for **ME**rcury Surface, Space ENvironment, GEochemistry, and Ranging) flew by and imaged the side that had been turned away when Mariner 10 made its imaging pass. The images then and now show that Mercury looks a lot like the Moon (Figure 1a) but that because of the different surface gravity, crater rims are different, debris is thrown out to different distances, and so on.

---

[*] *Jay M. Pasachoff is Field Memorial Professor of Astronomy at Williams College, Williamstown, Massachusetts, USA, and retiring president of the Commission on Astronomy Education and Development of the International Astronomical Union. He is the author of* A Field Guide to the Stars and Planets *and coauthor of textbooks* The Cosmos: Astronomy in the New Millennium *and* The Solar Corona.

CP 998, Modern Trends in Physics Research
Third International Conference MTPR-08
edited by L. El Nadi
Copyright @ 2011 by World Scientific Publishing Co. 978-981-4317-50-4 / 981-4317-50-0

4

Fig. 1(a). Mercury, as the MESSENGER spacecraft flew by in 2008. (Credit: MESSENGER, NASA, Johns Hopkins University Applied Physics Laboratory, Carnegie Institution of Washington).

Fig. 1(b). A closer view of Mercury's surface from MESSENGER in 2008, with a scarp—a line of cliffs—near the limb. . (Credit: MESSENGER, NASA, Johns Hopkins University Applied Physics Laboratory, Carnegie Institution of Washington).

A closer view (Figure 1b) shows one of the lines of cliffs known as scarps that are a sign that Mercury cooled and collapsed a bit billions of years ago.

MESSENGER, which carries a wide variety of instruments beyond cameras, will fly close to Mercury again additional times to slow it up so that it can go into orbit around Mercury in 2011.

Venus

Though Venus is not as close to the Sun as Mercury is, it winds up being even hotter at its surface. It was through Venus's high surface temperature, detected first by radio waves, that it was realized how the greenhouse effect has and will have an effect on the Earth's surface temperature as well. So Venus has played an important role in our understanding of global warming.

Venus's temperature is 750 K and its surface pressure is over 90 times that of Earth. It was thus an impressive feat when the Soviets, in a series of missions to Venus, succeeded not only on landing on Venus and sending measurements like temperatures, but also in sending back several images to Earth (Figure 2).

Fig. 2. Venus, as seen from the Soviet Venera 13 lander in 1982. Even in the 480°C temperature on Venus's surface and pressure over 90 times that at Earth's surface, the lander survived long enough to send back images. (Soviet Planetary Exploration Program, National Space Science Data Center).

Venus has been extensively mapped, especially by NASA's Magellan mission. It is composed of one large plate, showing none of the continental plates or drift we find on Earth. It does boast of several volcanoes, the largest named Maxwell after James Clerk Maxwell, the 19th-century unifier of electricity and magnetism.

Earth

Our home planet is being surveyed by a host of spacecraft. Indeed, the gains to us from weather prediction, Global Positioning System (GPS), mapping, and other measurements made using satellites in Earth orbit have more than paid back the investment in space satellite and launchers.

For example, we now know that warm or cool regions in the equatorial Pacific Ocean, known as El Niño (after its Christmastime appearance in some years) or La Niña for its opposite, can affect the weather thousands of miles away. (Figure 3a).

Space imaging was important for mapping the ozone hole, which opens when sunlight hits high-altitude particles each Antarctic springtime, and for monitoring its growth and recent shrinkage. The shrinkage wouldn't have happened without the space imaging that led worldwide agreement to limit production and use of the chlorofluorocarbons that helped break up the ozone layer.

To understand the Earth's atmosphere and weather better, we must continue our efforts in *comparative planetology*: studying phenomena on as many planets as possible in order to derive the underlying physical laws that govern planetary processes.

Of course, people went to the moon on six Apollo missions from NASA in 1969-1972.

6

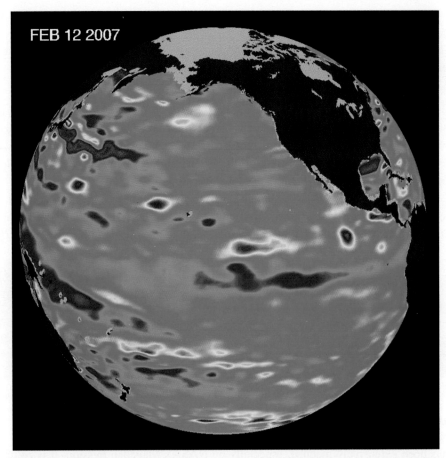

Fig. 3. Earth, showing sea-level heights to show the transition in 2007 from El Niño (warm water) to La Niña (cool water, shown in blue near the equator) in the tropical Pacific Ocean. El Niño and la Niña affect the weather thousands of kilometers away. The data are from NASA's Jason satellite. (NASA/JPL/Caltech Ocean Surface Topography Team).

Mars

Most people on Earth are interested in Mars, but not always for the right reasons. Nineteenth-century mistranslation of the Italian "canali" for the English word "canal," implying that lines seen on Mars were dug by Martians, led to an excessive believe in life on Mars. This belief carried through to the extensive studies of Mars in the twentieth century by Percival Lowell and others. The novel "War of the Worlds," and its American radio broadcast by Orson Wells that scared a huge number of people by invoking a Martian attack, remained in the public memory.

Yet there are astronomical reasons for hoping that life may be found on Mars, even if it is not intelligent life walking around. Life as we know it on Earth formed in association with water, and the existence of polar caps that include not only carbon-dioxide ice but also water ice, and the deduction from high-resolution images of Mars that many surface features were cut by rushing water, has led many astronomers to hope that life may be found on Mars.

NASA's Viking Missions landed on Mars in 1976 and carried small biology laboratories. They gave results that seemed to rule out the discovery of life, though that evaluation is doubted by some. Mars Phoenix landed on Mars in 2008, carrying a small laboratory that is digging down some centimeters through the soil at the edge of the northern polar cap and analyzing the ice that is lower down. (Figure 4).

Credit: NASA/JPL/University of Arizona/Texas A&M/James Canvin

Fig. 4. NASA's Mars Phoenix mission landed on a northern latitude in 2008, and tested ice just below the surface. (Phoenix Mission Team, NASA/JPL/Caltech, University of Arizona, Texas A&M University).

People talk about sending astronauts to Mars, but that is decades away and fantastically expensive. It would require astronauts to be in space for years, compared with the few days it takes to travel to the Moon, and questions about safety from solar flares are among the many problems. Recall that the Moon is about 1 ½ light seconds from Earth while Mars is 4 light minutes from Earth at its closest, much farther away.

Jupiter

Jupiter, by far the largest and most massive of the planets, dominates the solar system. We have misled by it to think for many years that huge planets like Jupiter would orbit their sunlike stars in periods measured in years, so it was a shock when scientists began a dozen years ago to detect "hot Jupiters" in orbit around other stars. They are as massive as Jupiter or more massive, yet orbit so close to their stars that their periods of revolution are only a few days.

The Jupiter in our solar system was observed by Galileo in 1609 and thereafter to be orbited by moons, an important step in verifying the Copernican system. With larger telescopes than that of Galileo, the 400th anniversary of whose telescope discoveries we are now celebrating in the International Year of Astronomy, a Great Red Spot has been a giant hurricane on Jupiter lasting hundreds of years. Applications to comparative planetology and the understanding of hurricanes and cyclones are obvious.

Jupiter has been the target of many spacecraft, from Pioneers 10 and 11 to Voyagers 1 and 2 and to the 1990's orbiting Galileo. Galileo dropped a probe into Jupiter's clouds, to measure water content (lower than expected, but it was only one path) and wind velocities. Galileo also sent back closeup images of many of Jupiter's moons, including the Galilean satellites Io, Europa, Ganymede (the largest moon in the solar system), and Callisto. These satellites are about as large as Mercury and have surface appearances and characters of their own. Io, for example, has over a dozen erupting volcanoes, and its surface is the youngest in the solar system as a result.

NASA's New Horizons mission en route to Pluto passed the Jupiter system in 2007 (Fig. 5).

Fig. 5. Jupiter and its moon Io, taken from NASA's New Horizons in 2007 as the spacecraft passed by en route to Pluto. The Great Red Spot appears as blue in this false-color infrared image of Jupiter. The image of Io, including the 330-km-high volcanic plume from Tvashtar, is in true color. (NASA/Johns Hopkins University Applied Physics Laboratory/Southwest Research Institute/Goddard Space Flight Center).

Saturn

Galileo could barely detect that Saturn didn't appear round, but he reported only "ears" there. It took decades before it could be discovered with a better telescope that Galileo is surrounded by a ring.

Saturn's rings are so magnificent that they can nowadays be seen with telescopes of almost any size. A joint American-European mission called Cassini, after the Italian-French astronomer who in the seventeenth century studied Saturn and discovered the main gap in its rings, is now in orbit in the Saturn system. The Cassini spacecraft itself, from NASA, arrived at Saturn in 2004. It takes closeup images of Saturn's disk, its rings, and its moons (Figure 6), measures magnetic fields, and records lots of additional data.

Fig. 6. Saturn, observed from NASA's Cassini spacecraft during 2008, with the rings illuminated from behind by the sun. (Cassini Imaging Team/Space Science Institute/JPL/ESA/NASA).

Carried from Earth to Saturn by Cassini was the European Space Agency's Huygens mission. In 2005, it plummeted through the atmosphere of Titan, Saturn's largest moon, which had been discovered by Christiaan Huygens in 1655, as a result of Huygens's own improvements in making a telescope. Titan has a thick atmosphere, thicker than Earth's and filled with photochemical smog. Huygens fantastically imaged riverbeds and lakeshores on Titan and its radar mapped Titan's surface precisely. The Huygens spacecraft survived on Titan's surface for over an hour, sending back images and other measurements.

Uranus

Uranus was, in 1781, the first planet to be discovered. The other planets up to then had always been known. But William Herschel spied it while sweeping the sky with his telescope in Bath, England. First thought to be a comet, its orbit showed it to be a new planet.

But Uranus is so far away, with an orbit 20 times across that of Earth, that little was known about it until NASA's Voyager 2 spacecraft visited it in 1986. Then its surface could be seen, though it turned out to be fairly uniformly bland. (Figure 7).

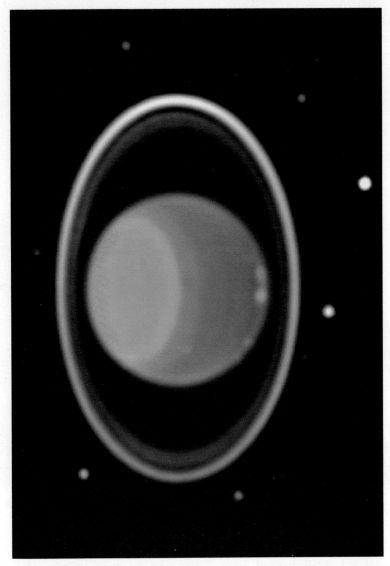

Fig. 7. Uranus and its rings, imaged in the infrared with the Near Infrared Camera/Multi-Object Spectrometer (NICMOS) on the Hubble Space Telescope). (NASA/E. Karkoschka of the University of Arizona).

Uranus had already been known to be tipped over almost on its side, with its moons orbiting in its equatorial plane. Also oddly, its magnetic field is skew, quite different from magnetic fields that had been known.

Uranus has rings, but for hundreds of years they were unknown. They showed up only during a stellar occultation, when Uranus was observed as it was about to hide (occult) a star). Because of the uncertainty in Uranus's orbit, the cameras on board NASA's aircraft observatory were turned on tens of minutes early, and they reported dips in the intensity of stellar radiation received. First reported to be a swarm of moons, they turned out to by symmetric before and after the occultation so they were actually rings. Occultation measurements remain very sensitive for studying the outer planets, as we shall see for Pluto. Today's infrared cameras are able to image Uranus's rings directly.

Neptune

Neptune, like Uranus, is an "ice giant," a fairly new term to distinguish these two giant planets from the even larger and more gaseous planets Jupiter and Saturn, which are known as "gas giants."

When Voyager 2 flew by Neptune in 1989, it found an atmosphere with more activity than that of Neptune. Advances in imaging techniques, both by sending the Hubble Space Telescope above the Earth's atmosphere and by using active optics for Earth-based telescopes ("active," in that the shape of the mirror is adjusted during observations by computer-controlled actuators pushing on the back of, usually, the secondary mirror), allow us to monitor the weather on Neptune even from Earth. (Figure 8)

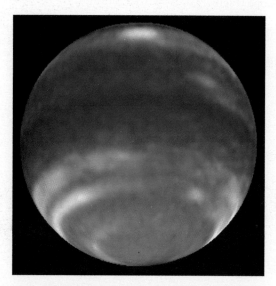

Fig. 8. Neptune, with its weather showing. This Hubble Space Telescope Image from 2002 shows increasing clouds. (L. Sromovsky and P. Fry of the University of Wisconsin-Madison, and others/NASA).

Occultation studies have allowed us to study the atmosphere of Triton, and to detect a slight global warming on that large but distant satellite.

Pluto

Pluto was founded by Clyde Tombaugh in 1930 at the Lowell Observatory, where Tombaugh had been hired to search for the "9th planet." But though he found the image after looking through hundreds of thousands of stellar images, comparing pairs of photographic plates taken at different times, the object was fainter and therefore smaller than expected. Also, its orbit was strangely out of round, even extending inside Neptune's. Still, it was thought for a while to have 90% the mass of Earth, and therefore to just be a type of terrestrial planet out beyond the giant planets.

But the discovery of a moon, soon named Charon, around Pluto in 1978 allowed Pluto's mass to be found accurately, using Newton's form of Kepler's third law. Pluto's mass turns out to be only 1/500 Earth's, making it a minuscule body. Requests escalated to demote Pluto from its planetary status.

Eventually, in 2003, Michael Brown of the California Institute of Technology discovered an object that was farther out than Pluto and that was arguably a little larger and a little more massive. He had also found a few objects almost that large and massive. The prospect is for dozens of objects at or beyond Pluto's orbit that are larger and more massive than Pluto. So it became difficult to maintain Pluto's planethood. Where school children to be expected to remember the names of dozens of planets?

As a result, the International Astronomical Union, at its triennial General Assembly held in Prague in 2006, set up a category known as "dwarf planets," with Pluto, Charon, the new object (which was then named Eris), and the asteroid Ceres as the first members. They were all large enough to be round but not massive enough to clear out their orbits of other objects. Some astronomers are trying to sell the renaming not as a demotion but as a promotion for Pluto to be the first in a new class of object. In 2009, the International Astronomical Union denoted "plutoid" as the name of a class of dwarf planet, so far populated only by Pluto and Eris.

Pluto has many characteristics of planets. It has three moons, two tiny ones having been discovered with the Hubble Space Telescope (Figure 9a).

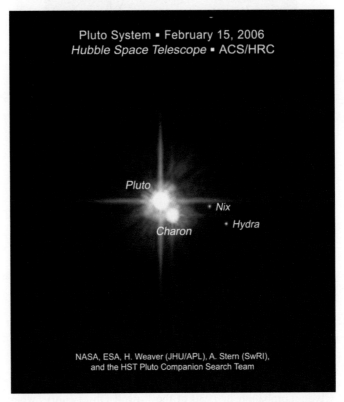

Fig. 9(a). Pluto, with a Hubble Space Telescope showing not only its relatively large moom, Charon, but also its tiny moons Nix and Hydra, only about 100 km across, which were discovered in 2005. (M. Mutchler/Space Telescope Science Institute, A. Stern/Southwest Research Institute, and the HST Pluto Companion Search Team/ESA/NASA).

It has an atmosphere, which since 2002 has been monitored through occultation studies, which revealed a major change in the atmosphere since the only previous observation in 1988 (Figure 9b).

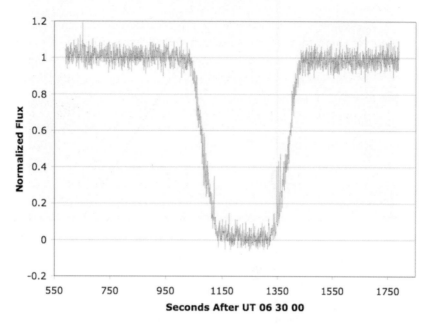

Fig. 9(b). A graph showing, over time, Pluto occulting (hiding) a star in 2002, with the slanted descent and ascent of brightness revealing Pluto's atmosphere as it distorts and refracts the starlight. The drop in intensity of the Pluto + star total would be vertical and abrupt in the absence of an atmosphere. Positive spikes in the light curve show focusing effects in atmospheric layers. (Jay M. Pasachoff, Bryce A. Babcock, Steven P. Souza, and David R. Ticehurst/Williams College, and the MIT-Williams Occultation Consortium).

Advances in telescopes, electronic imaging, and computer analysis of images that contain thousands of stellar and other images are leading to many discoveries in the outer solar system. We confidently expect that more plutoids will soon be discovered. My own group, which makes occultation observations of Pluto and Charon, are hoping to be able to observe and maybe even to discover atmospheres around some of them.

Comets

There are many objects between the planets and dwarf planets in our solar system. The most prominent are comets, since a brilliant comet appears in our skies every decade or so. The most famous comet is Halley's, officially numbered 1P, since it is reasonably bright every time it comes around, which it does every 76 years or so. Halley himself figured out in 1705 after he collected historical comet observations that one of the comet had come back a second and third time and predicted its return in 1758. Though he had died by then, when the recovery was successful, the comet became known after him and his idea that comets are in periodic orbits was verified. Newton's theory of gravity, which Halley had used in his calculations, was also verified by the discovery. Halley's Comet (1P/Halley) won't be back near us until 2061, though the Hubble Space Telescope is sufficiently powerful that it can be imaged throughout its orbit, even at its faintest.

A particularly bright and beautiful comet appeared in 2007, though it quickly moved from the northern to the southern hemisphere, where it gave its greatest display (Figure 10a).

14

Fig. 10(a). Comet McNaught, C2006/P1the Great Comet of 2007, viewed from the Siding Spring Observatory in Australia. At its peak, it was even visible in daylight. (Robert H. McNaught).

Later in 2007, a long known comet, 17P/Holmes (and therefore the 17th comet to have its periodic orbit recognized out of the approximately 200 now on that list) suddenly and abruptly brightened by a factor of a million. It became visible to the unaided eye, and remained so for weeks as it moved across the sky at a slightly different rate than the stars so moved through a set of constellations, as comets do. (Figure 10b)

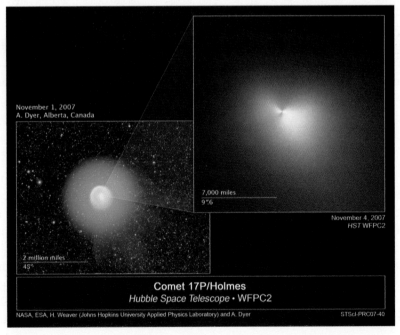

Fig. 10(b). Comet 17P/Holmes, which abruptly brightened by a factor of a million on 23 October 2007, presumably from an eruption of material from inside it. The ground-based image shows the huge comet head and a bit of tail, while the Hubble Space Telescope image shows the middle of the coma. (NASA/ESA/H. Weaver of the Johns Hopkins University Applied Physics Laboratory, and A. Dyer).

There are hundreds of thousands of small solar-system bodies known as asteroids. The bulk of them orbit between Mars and Jupiter, but some orbit closer to Earth or even cross Earth's orbit. We worry that one of them may hit us one day. Several spacecraft, some sent directly to asteroids and other en route elsewhere in the solar system, have provided us with closeup views of several asteroids.

The Sun

The sun is the most prominent object in the solar system, but since it is a star shining with its own energy it is often not discussed in articles and books on the solar system. Still, for completeness, let us include it here. The sun is millions of times brighter than the other stars; it is only 8 light minutes from us compared with 4.2 light years for the next nearest star.

Because the sun is so close to Earth, it shows a noticeable disk, though one can look at it only with safety precautions to maintain eye health. Many telescopes have been built especially to look at the sun and many spacecraft have been launched to study it.

The European Space Agency's Solar and Heliospheric Observatory (SoHO) has been aloft since 1995 with a dozen instruments to study the sun. It is in an orbit about a million kilometers toward the sun from Earth, moving in a small halo around the online point so as not to have the sun's strong radio radiation prevent receipt on Earth of radioed data. One instrument on SoHO, NASA's Goddard Space Flight Center's Extreme-ultraviolet Imaging Telescope, uses filters to pass only light from hot levels of the solar atmosphere, above the normal, everyday solar surface. Another instrument on SOHO, the U.S. Naval Research Laboratory's Large Angle Spectrometric Coronagraph, hides the everyday sun and a large region around it in which the everyday sun's light would scatter so much as to prevent observations. Those observations are valuable but nevertheless leave an unobserved doughnut-shaped region.

At a total eclipse of the sun, my group fills that region by providing coronal imaging of the low and middle corona. (Figure 11).

Fig. 11. The sun, with an image taken during the total solar eclipse of 29 March 2006 sandwiched between a view of the solar disk in the ultraviolet and a coronagraph view of the outer corona, both imaged with telescopes on board the European Space Agency's Solar and Heliospheric Observatory. (Eclipse: Jay M. Pasachoff, Bryce A. Babcock, Stephen P. Souza, Jesse S. Levitt, and the Williams College Eclipse Expedition, with National Science Foundation support/NASA composite by Steele Hill and C. Alex Young/EIT Team of NASA's Goddard Space Flight Center; LASCO Team at U.S. Naval Research Laboratory/NASA/ESA).

Such total eclipses occur about every year and a half somewhere in the world. The last total solar eclipse crossed Africa (including a region of Egypt near the Libyan border), a Greek island, Turkey, and Russia in 2006. A total eclipse will start in northern Canada and come down through Russia's Siberia and then western Mongolia and part of China. An even longer total eclipse, with totality lasting up to about 6 minutes, will cross from India to China, covering even Hangzhou and Shanghai with their tens of millions of people, in 2009. Another total eclipse will cross mainly the Pacific, including Easter Island, in 2010, ending low in the sky in southern Chile and Argentina.

Many scientific observations still come from eclipses, though such ground-based observations are now usually used in conjunction and collaboration with space-based research.

Final comments

The twentieth century has indeed been a golden era in the exploration of the solar system. We are the only people who were on Earth when these celestial bodies were seen closeup for the first time. We must treasure those memories and build on them.

I. ATOMIC, MOLECULAR & CONDENSED MATTER PHYSICS
I-1 KEYNOTE AND PLENARY PAPERS

The Iron Project: RADIATIVE ATOMIC PROCESSES IN ASTROPHYSICS

SULTANA N. NAHAR

Dept of Astronomy, The Ohio State University, Columbus, OH 43210, USA
E-mail: nahar@astronomy.ohio-state.edu

Astronomical objects, such as, stars, galaxies, blackhole environments, etc are studied through their spectra produced by various atomic processes in their plasmas. The positions, shifts, and strengths of the spectral lines provide information on physical processes with elements in all ionization states, and various diagnostics for temperature, density, distance, etc of these objects. With presence of a radiative source, such as a star, the astrophysical plasma is dominated by radiative atomic processes such as photoionization, electron-ion recombination, bound-bound transitions or photo-excitations and de-excitations. The relevant atomic parameters, such as photoionization cross sections, electron-ion recombination rate coefficients, oscillator strengths, radiative transition rates, rates for dielectronic satellite lines etc are needed to be highly accurate for precise diagnostics of physical conditions as well as accurate modeling, such as, for opacities of astrophysical plasmas.

This report illustrates detailed features of radiative atomic processes obtained from accurate ab initio methods of the latest developments in theoretical quantum mechanical calculations, especially under the international collaborations known as the Iron Project (IP) and the Opacity Project (OP). These projects aim in accurate study of radiative and collsional atomic processes of all astrophysically abundant atoms and ions, from hydrogen to nickel, and calculate stellar opacities and have resulted in a large number of atomic parameters for photoionization and radiative transition probabilities. The unified method, which is an extension of the OP and the IP, is a self-consistent treatment for the total electron-ion recombination and photoionization. It incorporates both the radiative and the dielectronic recombination processes and provides total recombination rates and level-specific recombination rates for hundreds of levels for a wide range of temperature of an ion. The recombination features are demonstrated. Calculations are carried out using the accurate and powerful R-matrix method in the close-coupling approximation. The relativistic fine structure effects are included in the Breit-Pauli approximation. The atomic data and opacities are available on-line from databases at CDS in France and at the Ohio Supercomputer Center in the USA. Some astrophysical applications of the results of the OP and IP from the Ohio State atomic-astrophysics group are also presented. These same studies, however with different elements, can be extended for bio-medical applications for treatments. This will also be explained with some preliminary findings.

Keywords: Photoionization; Electron-ion recombination; Oscillator strengths; Di-electronic satellite lines; Opacities

1. Introduction

The initial study of an astronomical object can be made via photometry which gives information on location, size, surroundings, etc alongwith low level spectroscopy. Figure 1 shows a photometric image of supernova remnant Cassiopia A[1] composed of infrared data from the Spitzer, visible data from the Hubble Space Telescope, and X-ray data from the Chandra observatory. A supernova remnant typically consists of an outer, shimmering shell of expelled material and a core skeleton of a once-massive star, called a neutron star. Heavy elements in universe are known to be formed during a supernova explosion.

However, details such as, temperature, density, speed, composition etc can be obtained only from precise spectroscopy. What we see in an astronomical object depends on how the nuclear energy (gamma rays) produced in the core of the star propagates to the surface. As the radiation travels outward it goes through repeated absorption and emission by atoms

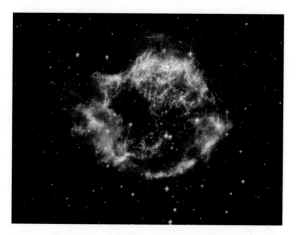

Fig. 1. Photometric picture of Cassiipia A made from observation by Spitzer (infrared), Hubble (visible), and Chandra (X-ray).[1]

and molecules and looses energy to become visible photons. This process depends on the opacity of the constituent materials. Stellar opacities give measures of radiation transport in the plasmas and are crucial in calculating various quantities and analyzing astrophysical spectra. Opacities depend mainly on oscillator strengths and photoionization cross sections of all transitions in the constituent elements.

About 25 years ago, it was realized that the existing calculated stellar opacities, obtained using atomic data from simple approximations, were incorrect by factors of 2 to 5 resulting in inaccurate stellar models. For example, pulsation periods of Cepheid stars, which are used in gauging the distances of astronomical objects, could not be modeled. A plea was made for accurate opacity using accorate atomic parameters. This initiated the international Opacity Project[2] for precise study of the radiative processes of all astrophysically abundant atoms and ions, from hydrogen to iron, and calculate the opacities. With the same theme, the Iron Project[3] was initiated later for both radiative and collisional process, however, for iron-peak elements and with inclusion of fine structure effect in the calculations. The OP and the IP include a very wide energy range, from infrared (IR), optical (O), ultraviolet (UV), extreme ultraviolet (EUV) to X-rays. The theoretical approach of R-matrix method using close coupling approximation enables consideration of large number of energy levels, and corresponding transitions, and photoionization.

2. Radiative Atomic Processes

Astrophysical plasmas are dominated by three radiative atomic processes. (1) Photo-excitation and de-excitation:

$$X^{+Z} + h\nu \rightleftharpoons X^{+Z*}$$

where X^{+Z} is the ion with charge Z. The emitted or absorbed photon ($h\nu$) is observed as a spectral line. The relevant atomic parameters for the direct and inverse processes are oscillator strength (f) and radiative decay rate (A-value).

The other two are the inverse processes of (2) photoionization (PI) and (3) electron-ion recombination. In direct photoionization and radiative recombination (RR), an electron is ejected with absorption of a photon or recombines to a ion with emission of a photon:

$$X^{+Z} + h\nu \rightleftharpoons X^{+Z+1} + \epsilon$$

Photoionization and recombination can also occur via an intermediate state as:

$$e + X^{+Z} \rightleftharpoons (X^{+Z-1})^{**} \rightleftharpoons \begin{cases} e + X^{+Z} & \text{AI} \\ X^{+Z-1} + h\nu & \text{DR} \end{cases}$$

where an electron colliding with a target of charge $+Z$ excites it as well as attaches itself to form a doubly excited quasi-bound state known as an autoionizing state. This state is short-lived and leads either to autoionization (AI) where the electron goes free and target drops to ground state or to dielectronic recombination (DR) where the electron gets bound by emission of a photon. The autoionizing state manifests as an enhancement or resonance in the process. Photoionization resonances can be seen in absorption spectra while recombination resonances can be seen in emission spectra. The relevant atomic parameters for these processes are photoionization cross sections (σ_{PI}), recombination cross sections (σ_{RC}) and rate coefficients (α_{RC}).

3. Theory

The relativistic Hamiltonian of a multi-electron system in Breit-Pauli R-matrix (BPRM) method is given by,

$$H_{N+1}^{\text{BP}} = H_{N+1}^{NR} + H_{N+1}^{\text{mass}} + H_{N+1}^{\text{Dar}} + H_{N+1}^{\text{so}}, \quad (1)$$

where the non-relativistic Hamiltonian is,

$$H_{N+1}^{NR} = \left[\sum_{i=1}^{N+1} \left\{ -\nabla_i^2 - \frac{2Z}{r_i} + \sum_{j>i}^{N+1} \frac{2}{r_{ij}} \right\} \right] \quad (2)$$

and the three one-body relativisitc terms are mass correction, $H^{mass} = -\frac{\alpha^2}{4} \sum_i p_i^4$, Darwin term, $H^{Dar} = \frac{\alpha^2}{4} \sum_i \nabla^2 \left(\frac{Z}{r_i} \right)$, and spin-orbit interation term, $H^{so} = \left[\frac{Ze^2\hbar^2}{2m^2c^2r^3} \right] \mathbf{L.S}$. However, the effects of higher order terms, especially the two-body terms, in the Breit-Pauli Hamiltonian,

$$\frac{1}{2} \sum_{i\neq j}^{N} [g_{ij}(so + so') + g_{ij}(ss') + g_{ij}(css') + g_{ij}(d) + g_{ij}(oo')]. \quad (3)$$

where the notation are c for contraction, d for Darwin, o for orbit, s for spin and a prime indicates 'other' can improve the energies and weak transitions. We include full Breit interaction,

$$H^B = \sum_{i>j} [g_{ij}(so + so') + g_{ij}(ss')] \quad (4)$$

for the weak forbidden transitions.

The energies and wavefunctions of the (N+1)-electron system can be obtained from solving

$$H_{N+1}^{BP}\Psi = E\Psi. \qquad (5)$$

The close coupling (CC) approximation for wavefunction describe an atomic process as a target or the core ion of N electrons interacting with the (N+1)th electron (e.g. Seaton[4]). Total wavefunction expansion is expressed as:

$$\Psi_E(e + ion) = A\sum_i^N \chi_i(ion)\theta_i + \sum_j c_j\Phi_j(e + ion)$$

where χ_i is the target ion or core wavefunction at state i, θ_i is the interacting electron wavefunction, which is a continuum state wavefunction Ψ_F when E ≥ 0 or bound state function Ψ_B when E < 0. Φ_j is a correlation function of (e+ion) compensating the short-range correlation and orthogonality condition. The complex resonant structures in the atomic processes are introduced through couplings of bound and continuum channels in the transition matrix. Substitution of $\Psi_E(e+ion)$ in Hamiltonian equation results in a set of coupled equations which are solved by R-matrix method.

R-matrix method divides the space in two regions, the inner and the outer regions, of a sphere of radius r_a with the ion at the center. r_a is large enough for electron-electon interaction potential to be zero outside the boundary; wavefunction at $r > r_a$ is Coulombic due to perturbation from the long-range multipole potentials. In the inner region, the partial wave function of the interacting electron is expanded in terns of a basis set, called the R-matrix basis,

$$F_i = \sum_k a_k u_k$$

which satisfies

$$\left[\frac{d^2}{dr^2} - \frac{l(l+1)}{r^2} + V(r) + \epsilon_{lk}\right]u_{lk} + \sum_n \lambda_{nlk}P_{nl}(r) = 0. \qquad (6)$$

and are made continuous at the boundary by matching with the Coulomb functions outside the boundary.

For radiative processes, the transition matrix elements $< \Psi_B||\mathbf{D}||\Psi_{B'} >$ for bound-bound transitions and $< \Psi_B||\mathbf{D}||\Psi_F >$ for photoionization and recombination are obtained with dipole operator, $\mathbf{D} = \sum_i \mathbf{r}_i$, where i is the number of electrons. The generalized line strength, S, can be obtained as

$$S = \left|\left\langle \Psi_f| \sum_{j=1}^{N+1} r_j|\Psi_i \right\rangle\right|^2 \qquad (7)$$

The oscillator strength (f_{ij}) and radiative decay rate (A_{ji}) for a dipole bound-bound transition are then obtained as

$$f_{ij} = \left[\frac{E_{ji}}{3g_i}\right] S, \; A_{ji}(sec^{-1}) = \left[0.8032 \times 10^{10}\frac{E_{ji}^3}{3g_j}\right] S \qquad (8)$$

The photoionization cross section, σ_{PI}, is related to S as

$$\sigma_{PI} = \left[\frac{4\pi}{3c}\frac{1}{g_i}\right]\omega S, \qquad (9)$$

where ω is incident photon energy in Rydberg units and g_i is the statistical weight factor of the initial state.

Recombination cross section, σ_{RC}, for any level can be obtained from σ_{PI} using principle of detailed balance (Milne relation) as

$$\sigma_{RC} = \sigma_{PI}\frac{g_i}{g_j}\frac{h^2\omega^2}{4\pi^2m^2c^2v^2}. \qquad (10)$$

These can be summed for the total σ_{RC}. Recombination rate coefficient in terms of photoelectron energy is given by

$$\alpha_{\mathbf{RC}}(\mathbf{E}) = \mathbf{v}\sigma_{\mathbf{RC}}$$

The recombination rate coefficient, $\alpha_{RC}(T)$, at various temperatures is obtained as

$$\alpha_{RC}(T) = \int_0^\infty vf(v)\sigma_{RC}dv, \qquad (11)$$

where $f(v,T) = \frac{4}{\sqrt{\pi}}(\frac{m}{2kT})^{3/2}v^2e^{-\frac{mv^2}{2kT}}$ is the Maxwellian velocity distribution function. The total α_{RC} is obtained from summed contributions of infinite number of recombined states.

The unified theory for electron-ion recombination considers all infinite recombining levels and subsumes RR and DR in unified manner[5-7] in contrast to existing methods that treat RR and DR separately and the total rate is obtained from $\alpha_{RC} = \alpha_{RR} + \alpha_{DR}$ neglecting the interference between RR and DR. The unified method divides the recombined states in to two groups. The contributions from states with $n \leq 10$ (group A) are obtained from σ_{PI} using principle of detailed balance, while the contributions from states with $10 < n \leq \infty$

22

(group B), which are dominated by narrow dense resonances, are obtained from an extension of the DR theory:[6,8]

$$\Omega(DR) = \sum_{J\pi}\sum_n (1/2)(2J+1)P_n^{J\pi}(DR). \quad (12)$$

where the DR probability $P_n^{J\pi}$ in entrance channel n is, $P_n^{J\pi}(DR) = (1 - \boldsymbol{\mathcal{S}}_{ee}^\dagger \boldsymbol{\mathcal{S}}_{ee})_n$, $\mathcal{S}_{ee}$ is the matrix for electron scattering *including* radiation damping. The recombination cross section, σ_{RC} in Megabarns (Mb), is related to the collision strength, Ω_{RC}, as

$$\sigma_{RC}(i \to j)(Mb) = \frac{\pi\Omega_{RC}(i,j)}{(g_i k_i^2)}(a_o^2/1. \times 10^{-18}), \quad (13)$$

where k_i^2 is in Rydberg. The method provides self-consistent set of σ_{PI}, σ_{RC}, and α_{RC} as they obtained using the same wavefunction. It also provides level-specific recombination rate coefficients and photoionization cross sections for many bound levels.

Under the IP, f, S, and A-values for fine structure forbidden transitions, and allowed for some complex ions, are treated with relativistic configuration interaction atomic structure calculations using code SUPERSTRUCTURE (SS).[9] The allowed and forbidden transitions are as summarized as: Allowed electric dipole (E1) transtitions are of two types: (i) same-spin multiplet (Δ j=0,±1, $\Delta L = 0$, ±1, ±2, $\Delta S = 0$, parity π changes), (ii) intercombination (Δ j=0,±1, $\Delta L = 0$, ±1, ±2, $\Delta S \neq 0$, π changes):

$$A_{ji}(sec^{-1}) = 0.8032 \times 10^{10}\frac{E_{ji}^3}{3g_j}S^{E1}, \; f_{ij} = \frac{E_{ji}}{3g_i}S^{E1}(ij) \quad (14)$$

where $S^{E1}(ij)$ is the line strength for E1 transition.

Forbidden transitions, which are usually much weaker than the allowed transitions, are of higher order magnetic and electric poles.
i) Electric quadrupole (E2) transitions (Δ J = 0,±1,±2, parity does not change):

$$A_{ji}^{E2} = 2.6733 \times 10^3\frac{E_{ij}^5}{g_j}S^{E2}(i,j) \; s^{-1}, \quad (15)$$

ii) Magnetic dipole (M1) transitions (Δ J = 0,±1, parity does not change):

$$A_{ji}^{M1} = 3.5644 \times 10^4\frac{E_{ij}^3}{g_j}S^{M1}(i,j) \; s^{-1}, \quad (16)$$

iii) Electric octupole (E3) transitions (Δ J= ±2, ±3, parity changes):

$$A_{ji}^{E3} = 1.2050 \times 10^{-3}\frac{E_{ij}^7}{g_j}S^{E3}(i,j) \; s^{-1}, \quad (17)$$

iv) Magnetic quadrupole (M2) transitions (Δ J = ±2, parity changes):

$$A_{ji}^{M2} = 2.3727 \times 10^{-2}s^{-1}\frac{E_{ij}^5}{g_j}S^{M2}(i,j). \quad (18)$$

The lifetime of a level can be calculated from the A-values, $\tau_k(s) = 1/\sum_i A_{ki}(s^{-1})$

4. Results

Results with various features on oscillator strengths and radiative decay rates for bound-bound transitions, photoionization cross sections and recombination rate coefficients are illustrated below.

4.1. *Radiative Transitions: f, S, A-values*

Under the IP, E1 transitions are calculated for levels going up to $n=10$ and $l \leq 9$ while weaker forbidden transitions are considered up to $n \leq 5$ in general. The advantage of BPRM method is that a large number of energy levels and transitions can be calculated, however, without spectroscopic identification. Theoretical spectroscopy for level identification, which is a major task, is carried out using quantum defect analysis, percentage of channels contributions, and angular momenta algebra.

For example, a recent BPRM calculations for Fe VX[11] has resulted in 507 fine structure levels and corresponding 27,812 transitions of type E1. The energies are found be within 1% in agreement with the observed values available in NIST[12] compiled table. Table 1 presents an example set of identified energy values grouped together as fine structure components of Fe XV.

Close coupling approximation with R-matrix method does not optimize individual transitions. Hence, consistent accuracies are expected in general. Table 2 presents comparison of BPRM transition probabilities of Fe XV[11] with the available values. Good agreement can be seen for most transitions. However, some differences are also exist. Table 2 also presents comparison of A-values for forbidden transitions in Fe XV,[11] obtained from relativistic Breit-Pauli approximation using SUPERSTRUCTURE,[9] with those available in NIST compilation.[12] Good agreement is found between SS and NIST for E2, M1 transitions. Variable agreement is noted between SS and NIST for M2 transitions.

Table 1. Sample set of fine structure energy levels of Fe XV grouped as sets of LS term components. C_t is the core configuration, ν is the effective quantum number.

$C_t(S_t L_t \pi_t)$	J_t	nl	$2J$	E(Ry)	ν	$SL\pi$

Eqv electron/unidentified levels, parity: e

| 2p63s2 | | | 0 | -3.35050E+01 | 0.00 | 1 S e |

Nlv(c)= 1 : set complete

Nlv= 3, $^3L^o$: P (2 1 0)

2p63s	(2Se)	1/2 3p	0	-3.13740E+01	2.68	3 P o
2p63s	(2Se)	1/2 3p	1	-3.13214E+01	2.68	3 P o
2p63s	(2Se)	1/2 3p	2	-3.11951E+01	2.69	3 P o

Nlv(c)= 3 : set complete

Nlv= 1, $^1L^o$: P (1)

| 2p63s | (2Se) | 1/2 3p | 1 | -3.02823E+01 | 2.73 | 1 P o |

Nlv(c)= 1 : set complete

Eqv electron/unidentified levels, parity: e

2p63p2			0	-2.84377E+01	0.00	3 P e
2p63p2			2	-2.83981E+01	0.00	3 P e
2p63p2			1	-2.83502E+01	0.00	3 P e

Nlv(c)= 3 : set complete

Eqv electron/unidentified levels, parity: e

| 2p63p2 | | | 2 | -2.81990E+01 | 0.00 | 1 D e |

Nlv(c)= 1 : set complete

Eqv electron/unidentified levels, parity: e

| 2p63p2 | | | 0 | -2.74705E+01 | 0.00 | 1 S e |

Nlv(c)= 1 : set complete

Nlv= 3, $^3L^e$: D (3 2 1)

2p63s	(2Se)	1/2 3d	1	-2.72920E+01	2.87	3 D e
2p63s	(2Se)	1/2 3d	2	-2.72822E+01	2.88	3 D e
2p63s	(2Se)	1/2 3d	3	-2.72669E+01	2.87	3 D e

Nlv(c)= 3 : set complete

Nlv= 1, $^1L^e$: D (2)

| 2p63s | (2Se) | 1/2 3d | 2 | -2.65180E+01 | 2.88 | 1 D e |

Nlv(c)= 1 : set complete

Nlv= 9, $^3L^o$: P (2 1 0) D (3 2 1) F (4 3 2)

2p63p	(2Po)	1/2 3d	2	-2.50182E+01	2.88	3 PDF o
2p63p	(2Po)	1/2 3d	3	-2.49282E+01	2.88	3 DF o
2p63p	(2Po)	1/2 3d	4	-2.48254E+01	2.86	3 F o
2p63p	(2Po)	1/2 3d	1	-2.45113E+01	2.89	3 PD o
2p63p	(2Po)	1/2 3d	2	-2.45045E+01	2.88	3 PDF o
2p63p	(2Po)	1/2 3d	3	-2.44046E+01	2.88	3 DF o
2p63p	(2Po)	1/2 3d	0	-2.43943E+01	2.88	3 P o
2p63p	(2Po)	1/2 3d	1	-2.43941E+01	2.89	3 PD o
2p63p	(2Po)	1/2 3d	2	-2.43929E+01	2.88	3 PDF o

Nlv(c)= 9 : set complete

Nlv= 3, $^1L^o$: P (1) D (2) F (3)

2p63p	(2Po)	1/2 3d	2	-2.48346E+01	2.88	1 D o
2p63p	(2Po)	1/2 3d	3	-2.37794E+01	2.84	1 F o
2p63p	(2Po)	3/2 3d	1	-2.36558E+01	2.85	1 P o

Nlv(c)= 3 : set complete

Table 2. Comparison of present A-values in unit of sec^{-1} for E1 transitions with those in NIST[12] compilation from references. The alphabetic letter is NIST accuracy rating.

λ Å	A:Ac NIST	A(Present) BPRM	SS	C i-j	SLπ i-j	g i-j
52.911	2.94e+11[1]: C	2.28e+11	2.38e+11	3s2 -3s4p	1S- 1P	1-3
59.404	3.4e+11[2] : C	2.40e+11	2.23e+11	3s3p -3s4d	1P-1D	3-5
307.73	4.91e+09[3]:C	4.78E+09	4.91E+09	3s3p -3p2	3P-3P	3-3
65.370	3.2e+10[2] : C	3.70e+10	3.52e+10	3s3p -3s4s	3P-3S	1-3
65.612	9.8e+10[2] : C	1.14e+11	1.06e+11	3s3p -3s4s	3P-3S	3-3
66.238	1.6e+11[2] : C	1.95e+11	1.82e+11	3s3p -3s4s	3P-3S	5-3
224.754	1.38e+04[4]:C	1.35e+10	1.40e+10	3s3p -3s3d	3P-3D	1-3
227.206	1.8e+04[4]:C	1.76e+10	1.83e+10	3s3p -3s3d	3P-3D	3-5
227.734	9.8e+09[4]:C	9.68e+09	1.0e+10	3s3p -3s3d	3P-3D	3-3
69.66	1.9e+11[5] :C	2.18e+11	2.16e+11	3s3p -3s4s	1P-1S	3-1
243.794	4.2e+04[4]:D	4.09e+10	4.25e+10	3s3p -3s3d	1P-1D	3-5
63.96	1.6e+11[2] : E	4.61e+10	1.97e+11	3p2 -3s4f	1D- 1F	5-7
194.067	3.8e+08[5]:E	1.04e+08	3.85e+08	3p2 -3p3d	1D-1P	1-3
231.68	1.5e+05[5]:E	1.43e+10	1.58e+10	3p2 -3p3d	3P- 3P	3-3
231.87	2.1e+05[5]:E	2.04e+10	2.13e+10	3p2 -3p3d	3P-3P	3-1
242.100	2.3e+05[5]:D	6.30e+09	2.56e+10	3p2 -3p3d	3P- 3D	5-7
68.849	9.2e+11[6] : C	9.15e+11	8.66e+11	3p3d -3p4f	3F-3G	9-11
69.945	7.4e+11[2] :C	7.25e+11	7.29e+11	3s3d -3s4f	3D-3F	3-5
69.987	7.9e+11[2] :C	7.67e+11	7.72e+11	3s3d -3s4f	3D-3F	5-7
70.054	8.8e+11[2] :C	8.63e+11	8.70e+11	3s3d -3s4f	3D-3F	7-9
70.224	4.13e+11[5]:C	4.23e+11	4.12e+11	3p3d -3p4f	3P-3D	1-3
73.199	8.8e+11[5] :C	6.89e+11	7.95e+11	3p3d -3p4f	1F-1G	7-9
73.473	6.2e+11[2] :C	6.01e+11	6.12e+11	3s3d -3s4f	1D-1F	5-7
312.556	1.1e+09[3] :E	3.51e+09	1.01E+09	3s3p -3p2	3P-1D	3-5
238.114	3.2e+08[4]:E	2.85e+08	2.81e+08	3s3p -3p2	3P-1S	3-1
191.408	3.5e+08[4]:E	3.53e+08	2.67e+08	3s3p -3s3d	3P-1D	3-5
196.741	1.6e+07[4]:E	1.45e+07	8.49e+06	3s3p -3s3d	3P-1D	5-5
304.998	3.0e+07[4]: E	2.20e+07	1.17e+07	3s3p -3s3d	1P-3D	3-5
305.889	2.6e+07[4]: E	2.63e+07	2.34e+07	3s3p -3s3d	1P-3D	3-3
38.95	1.69e+11[1]: C	7.4e+10	1.56e+11	3s2- 3s5p	1S- 1P	1-3

λ Å	A:Ac NIST	A(sec^{-1}) SS	C i-j	SLπ i-j	g i-j
		E2,M1,M2			
131.216	1.6e+06:D	1.73e+06	2p63s2 -3s3d	1S-1D	1-5 E2
171.913	4.3e+04[1]:E	3.81e+04	2p63s2 -3p2	1S-3P	1-5 E2
178.702	4.1e+05[1]:E	1.59e+05	2p63s2 -3p2	1S-1D	1-5 E2
20080	4.4e-01:E	4.13e-01	3p2 -3p2	1D-3P	5-3 M1
847.43	1.90e+02:E	1.70e+02	3s3p -3s3p	3P-1P	1-3 M1
975.84	3.0e+01:E	2.70e+01	3p2 -3s3d	1S- 1D	1-5 E2
999.63	2.70e+02:E	2.70e+02	3p2 -3p2	1D-1S	5-1 E2
1019.43	1.40e+02:E	1.24e+02	3s3p -3s3p	3P-1P	5-3 M1
1052.00	1.400e+03:E	1.25e+03	3p2 -3p2	3P- 1S	3-1 M1
1283.09	2.4e+01 :E	2.13e+01	3p2 -3p2	3P- 1S	5-1 E2
224.278	1.2e+00:D	31.9	3s3p -3s3d	3P-3D	1-5 M2
226.372	1.98e+00:C	26.3	3s3p -3s3d	3P-3D	3-7 M2
246.423	4.2e+00:D	40.3	3s3p -3p2	3P-1S 2	5-1 M2
303.494	6.4e+00:C	7.88e+00	3s3p -3s3d	1P-3D	3-7 M2
393.980	3.39e+00:C	3.38e+00	2p63s2 -3s3p	1S-3P	1-5 M2

Consideration of large number of transitions is needed for opacity calculations, spectral modeling of astrophysical plasmas. Figure 2 shows modeling of Fe I, Fe II, and Fe III lines observed in spectra of active galaxy 1 Zwicky 1. With large number oscillator strengths of Fe I-III in various modelings, agreement is good in general except in shorter wavelength region.

4.2. *Photoionization - Cross sections and Resonances*

Similar to oscillator strenghts, IP considers photoionization cross sections σ_{PI} for the ground and large number of excited bound states with $n \le 10$ and $l \le 9$. CC approximation enables inclusion of extensive resonances in σ_{PI} that are not possible with central field approximation. Resonances exist inher-

24

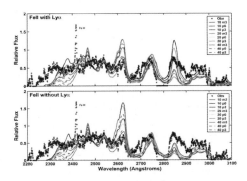

Fig. 2. Modeling of emission spectra of Fe I-III in active galaxy 1 Zwicky 1. Dots - observation; curves - various models with 1000 energy levels, millions of transitions. With (top) and without (bottom) Lyman-alpha fluorescent excitation of Fe II by recombining H-atoms. The models reproduce many of the observed features.

ently in photoionization and recombination processes for all atomic systems with more than one electron. Hence, with no core in hydrogen σ_{PI} is smooth, but it shows resonances in two electron helium atom. Fig. 3 presents σ_{PI} for the ground states of H[13] and He.[14]

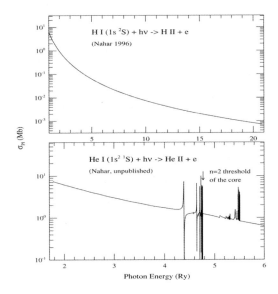

Fig. 3. Photoionization cross sections of ground state of H (top) and He (bottom). While σ_{PI} for hydrogen decays smoothly, helium shows resonances due to core excitations. Resonances converge to the excited core threshold; arrow points at n=2 threshold.

Resonant features in photoionization can be of

several types. i) The resonances due to Rydberg series of autoionizing states forming at energies E_p,

$$(E_t - E_p) = z^2/\nu^2$$

where E_t is an excited core state or threshold and ν is the effective quantum number of the state. Each excited core state corresponds to a Rydberg series although the resonances may not be prominent for the particular state. These resonances are usually narrow and more common. ii) The resonances due to photoexcitation-of-core, PEC, forming at excited states of the core that are dipole allowed by the core ground state. PEC or Seaton (who interpreted and named them PEC[16]) resonances exist in valence electron excited states only, that is, no PEC resonance for the ground or equivalent electron states, and often are enhanced and wider. iii) Overlapped resonant features forming from Rydberg series of states that belong to closely spaced core levels. iv) Resonances from quasibound equivalent electron states, usually broad, from Coulomb attraction and require atomic structure calculations for their identification. They occur not common. v) Fine structure effects can introduce narrow resonances which are allowed in fine structure, but not allowed in LS coupling.

For a complex system, such as neutral cromium Cr I, the resonant features in photoionization can be intriguing as seen in Fig. 4 which presents σ_{PI} of $3d^5 4s(^5P)$ state of Cr I.[15] The entire energy range is filled with extensive narrow resonances. Due to large number of core excitations, 39 in total, in the energy range overlapping of series of resonances is obvious. The background cross section has been enhanced by the first core excited state $3d^5 4s(^6D)$ (pointed by arrow). These structures will provide considerable contributions to quantities such as photoionization rates, recombination rates etc, from low to high temperatures.

The other prominent type of resonance, Seaton or the PEC (photo-excitation-of-core) resonance, forms at an excited core threshold as the core excites to an allowed state while the outer electron remains as a 'spectator' which is followed by photoionization and core dropping to the ground state. It is usually distinguishable because of its wider width and higher peak. A PEC or Seaton resonance is illustrated in Fig. 5 for excited $3d^5 \, ^6S4d(^7D)$ state of of Cr I.[15] It exhibits non-hydrogenic nature of an excited state cross sections in contrast to hydrogenic that is often

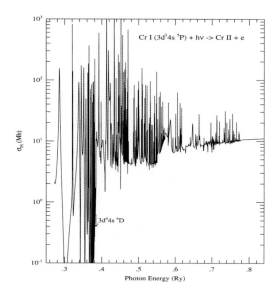

Fig. 4. Photoionization cross sections of $3d^54s(^5P)$ state of Cr I[15] showing complex resonant structures due to overlapping of Rydberg series of resonances. The calculation includes excitations of the core to 39 states.

assumed. The enhancement in the background due to a PEC can be orders of magnitude.

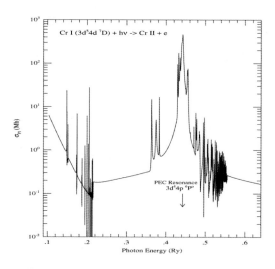

Fig. 5. Illustration of a PEC or Seaton resonance, due to core excitation $3d^5\ ^6S - 3d^44p\ 6P^o$ at energy 0.442 Ry (pointed by an arrow), in photoionization cross sections of excited $3d^5\ 6S4d^7D$ state of Cr I.[15] PEC resonance has enhanced the background by orders of magnitude.

4.3. Total and Level-Specific Electron-Ion Recombination

Electron-ion recombination in low temperature is usually dominated by radiative recombination when the electron combines with an ion by emission of a photon. The electron may not be energetic enough to excite the ion and go through dielectronic recombination. Hence typically the total recombination rate coefficient (α_R) starts high at low temperature due to dominance by RR, falls with temperature and then rises again at high temperature due to dominance of DR and forms a DR 'bump' which is followed by monotonic decay at very high temperature. An example showing the typical features of α_R for recombination to a Li-like ion, S XIV,[17] is presented in Fig. 6. The solid curve corresponds to the total unified recombination rate which is being compared with other existing separate RR and DR rates.

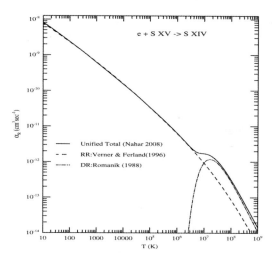

Fig. 6. Total recombination rate coefficients (solid curve) for (e+S XV → S XIV) from unified method that is valid for entire temperature range incorporating both RR and DR.[17] Existing rates are available individually for RR and DR.

The recombination feature often changes for complex atomic systems by forming two or multiple DR 'bump's. The positions of these bumps depend on the resonance positions in photoionzation. For example, one DR bump may form in the low temperature region, such as for recombination of Fe II.[27] In a rare case, almost four bumps may form, as seen in Fig. 7

26

for recombination to Ar XIII.[19]

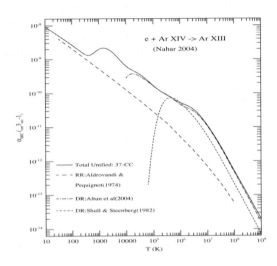

Fig. 7. Unified total recombination rate coefficients of Ar XIII showing multiple DR bumps.[19] Unified rates are compared with available RR and DR rates obtained separately by others.

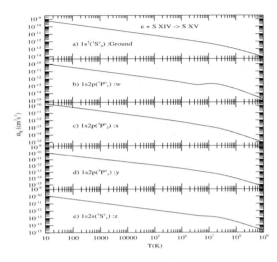

Fig. 8. Level-specific recombination rate coefficients, including RR and DR, of ground and four excited levels, corresponding to four diagositic w, x, y, z lines, of He-like S XV.[17]

One important advantage of the unified method is to have state-specific recombination rate coefficients $\alpha_R(nLS)$, including both RR and DR, for many bound levels. These are obtained from integration of state-specific photoionization cross sections as explained in the Theory section. Fig. 8 shows level-specific recombination rate coefficients for He-like S XV.[17] The rates correspond to (a) the ground level and four excited levels (b) $2s2p(^1P_1^o)$, (c) $2s2p(^3P_2^o)$, (d) $2s2p(^3P_1^o)$, (e) $2s2s(^3S_1)$ that form the well-known diagnostic allowed w, intercombination y, and forbidden x, z lines of He-like ions for high temperature plasmas. The hydrogenic decay of a level-specific recombination rate changes by a DR bump or shoulder in the high temperature. These rates are needed for the cascade matrix and determination of level populations.

4.4. Unified Method Extension for Dielectronic Satellite Lines

The unified method for total electron-ion recombination has been extended recently[20] to study the dielectronic satellite (DES) lines. These lines, formed from radiative decay of autoionizing states, are highly

sensitive temperature diagnostics of astrophysical and laboratory plasma sources. Most common DES lines in astrophysical spectra are formed by collision of (e+He-like ion) creating a 3-electron Li-like ion. Among DES lines, the 22 DES lines of KLL (1s2l2l') complex below the core excitation $1s^2(^1S_0) \rightarrow 1s2p(^1P_1^o)$ (w-line) are used most because of higher resolution or separation among them, especially for highly charged ions.

The earlier treatment of DES lines is based on isolated resonance approximation (IRA), e.g.,[21] where rate coefficient of a DES line is obtained as,

$$\alpha_R = a_o^3 \frac{g_i}{2g_f} \left[\frac{4\pi}{T} \right] e^{-\frac{\epsilon}{kT}} \frac{A_r A_a}{\sum_m A_a(m) + \sum_n A_r(n)} \quad (19)$$

ϵ is the DES energy. The single energy point can not provide the natural shape of the DES lines and hence a line profile (or cross section shape depending on one's perspective) is often assumed. IRA treats the resonances essentially as bound features, except for the inclusion of dielectronic capture into, and autoionization out of these levels.

In contrast, the unified method, by including the coupling between the autoionizing and continuum channels, gives the *intrinsic* spectrum of DES lines which includes not only the energies and line strengths as $S = \int_{\epsilon_i}^{\epsilon_f} \sigma_{RC} d\epsilon$ from the unified σ_{RC} that includes both RR background and DR, but also the natural line (or cross section) shapes. Since unified recombination cross sections (σ_{RC}) include both the resonant and the non-resonant background contributions. the DES spectra correspond directly to resonances in σ_{RC}. The recombination rate coefficients of DES lines can be calculated from direct integration over the product of the resonant cross sections and temperature-dependent factors.

Using the unified approach, the entire spectra of DES line intensities and recombination rates for the 22 satellite lines of KLL complexes can be produced. Fig. 9 presents the DES spectra of helium-like Fe XXV[20] which is applicable to analysis of $K\alpha$ complexes observed in high-temperature X-ray emission spectra. The top panel (a) presents total spectrum that can be observed and bottom three panels (b-d) present individual component states that contribute to the total spectrum. Individual spectra provides the identification of the DES lines. It may be noted that the natural widths and overlapping of lines are inherently included in the total spectrum. The method was benchmarked by comparing the recombination rates of the DES lines with the existing rates and very good agreement was found for the strong lines.

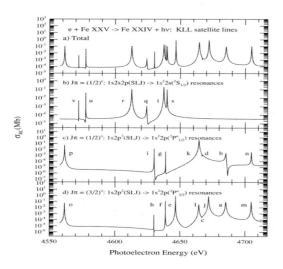

Fig. 9. Satellite lines of Fe XXV in the K_α complex.[20] The top panel (a) shows the total spectrum and the lower three panels show the resolved and identified lines belonging to final recombined level (b) $1s^2 2s(^2S_{1/2})$, (c) $1s^2 2p(^2P^o_{1/2})$, (d) $1s^2 2p(^2P^o_{3/2})$.

5. X-ray Spectroscopy from Astronomy to Bio-medical Science

Study of atomic astrophysics research is applicable to medical and nanotechnology research because of similarities. However, interest of atomic species is different from those in astrophysics. We have extended the current investigation to spectral models for X-ray absorption and transmission, and properties of chemical compounds of high-Z elements for possible bio-medical application. Heavy elements, such as gold, interact very efficiently with X-rays with large attenuation coefficients. Ionization and excitation of extended electronic shells have large photoabsorption cross sections up to very high energies. Gold nanoparticles, which are intoxic in living tissues, are being used in study of therapy and diagnostics for cancer treatment. Irradiation with gold nanoparticles in nice cancer has shown reduction of tumor growth.[22] Most of the irradiation methods use broad-band radiation that cause sufficient damages. Present work aim in narrow band irradiation using resonant energy ranges.

We have calculated the Auger resonant probabilities and cross sections of gold ions to obtain total mass attenuation coefficients using detailed resonance structures for the K $\longrightarrow$ L,M,N,O,P shell transitions. The X-ray mass absorption is shown in Fig. 10 for gold nanoparticles with vacancies in 2s, 2p sub-shells. It is seen that X-ray absorption is considerably enhanced by factors of up to 1000 or more at energies of K-shell resonances from K-alpha excitation energy, ~ 67.7 to K-ionization edge at $\sim$82 keV.[?] This is potentially useful in the calculation of resonant plus non-resonant attenuation coefficients by high-Z elements in plasmas created with high-intensity lasers, monochromatic synchrotron light sources, and electron-beam-ion-traps.

A simulation was carried out for resonant K_α X-ray (68 keV) absorption by gold nanoparticles in tissues. A phantom is assumed where a thin film of Au, 1.0mm/g, is deposited in a tumor 10 cm inside from the skin. As the 68 keV X-rays pass through the phantom, the percentage depth deposition (relative to background) of the radiation due to partial $K\alpha$ attenuation is found to be in complete absorption within $<$ 1 cm of the Au-layer, as shown in Fig. 11. This relates to the development of novel monochromatic or narrow-band X-ray sources.

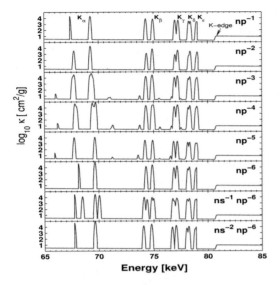

Fig. 10. X-ray mass absorption (κ) by nano gold particles with 2s, 2p-subshell vacancies via 1s-np K-shell transitions. The K-complexes of resonances, in E = 67.5 - 79 keV, show photo-absorption exceeding the background below the K-edge ionization at 82 keV by large factors.[?]

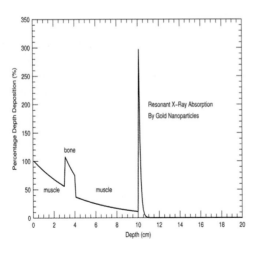

Fig. 11. Percentage depth deposition (relative to background) of 68 keV X-rays irradiated on body tissues where a film of gold nanoparticles of concentration 1.0mm/g is embedded in a tumor 10 cm inside the surface. Deposition due to partial $K\alpha$ attenuation by gold show complete absorption of X-rays within < 1 cm of gold layer.[?]

6. Conclusion

Dominant radiative atomic processes, such as photo-excitaions and de-excitations, photoionizatioon, and electron-ion recombination, in astrophyical plasmas are discussed. Features revealed from detailed study of the processes under the Iron Project are reported. Extensions to total and level-specific electron-ion recombination and di-electronic satellite lines provide self-consistent set of atomic data for the inverse processes vio unified theory.

The international collaborations of the Opacity Project and the Iron Project have resulted in large amount of atomic data for the astrophysically abundant atoms and ions from hydrogen to nickel in various ionization stages. They are available through databases, TOPbase and TIPbase at vizier.u-strasbg.fr/topbase/topbase.html and opacities.osc.edu. New and more updated results on photoionization cross sections and recombination cross sections and rates from Ohio State Atomic Astrophysics group (Nahar et al) is available from NORAD webpage at www.astronomy.ohio-state.edu/~nahar

Acknowledgments

I would like to thank Professors Fayaz Shahin, Lotfia El-Nadi, and organizers of MTPR08 conference for inviting me for a keynote speech and their kind hospitalities. The research is supported partially by NASA APRA program. The computations were carried out by various Cray computers at the Ohio Supercomputer Center.

References

1. Imag credit: NASA/JPL-Caltech/STScI/CXC/SAO at http://chandra.harvard.edu/photo/2005/casa/
2. The Opacity Project Team . *The Opacity Project*, Vol 1, 1995, Vol. 2, 1996, Institute of Physics, London UK 1995 and 1996
3. D.G. Hummer *et al.*, *Astron. Astrophys.* **279**, 298 (1993)
4. M. J. Seaton, *J. Phys.* B **20**, 6363 (1987)
5. S.N. Nahar & A.K. Pradhan, *Phys. Rev. Lett.* **68**, 1488 (1992)
6. S.N. Nahar & A.K. Pradhan, *Phys. Rev.* A **49**, 1816 (1994)
7. H.L. Zhang, S.N. Nahar, & A.K. Pradhan, *J. Phys* B **32**, 1459 (1999)
8. R.H. Bell and M.J. Seaton, *J. Phys.* B **18**, 1589 (1985)
9. W. Eissner, M, Jones, H. Nussbaumer, *Comput. Phys. Commun.* 8 **270**, 1974 (

10. spn T.A.A. Sigut, A.K. Pradhan, S.N. Nahar, *Astrophys. J.* **611**, 81 (2004)

11. S.N. Nahar (submitted, 2008)

12. National Institute for Standards and Technology (NIST), compilation atomic data are available at *http : //physics.nist.gov/cgi − bin/AtData/main_asd*

13. S.N. Nahar, *Phys. Rev. A* **53**, 2417 (1996)

14. S.N. Nahar (unpublished)

15. S.N. Nahar (in preparation 2008)

16. Yu Y. & M.J. Seaton, *J. Phys. B* **20**, 6409 (1987)

17. S.N. Nahar, *Open Astron. J.* **I**, 1 (2008)

18. S.N. Nahar, *Phys. Rev. A* **55**, 1980 (1997)

19. S.N. Nahar, *ApJS* **156**, 93 (2004)

20. S.N. Nahar & A.K. Pradhan, *Phys. Rev. A* **73**, 62718-1 (2006)

21. A.H. Gabriel, *MNRAS* **160**, 99 (1972)

22. Hainfeld J *et al. Phys. Med. Biol.* **49**, N309 (2004)

23. A.K. Pradhan *et al* (in preparation 2008)

RESOLVED FLUORESCENCE SPECTROSCOPY OF THE Cs₂ $3^3\Pi_g \rightarrow b^3\Pi_u$ TRANSITION

DAN LI, FENG XIE, AND LI LI[†]

Department of Physics and Key Laboratory of Atomic and Molecular Nanosciences,
Tsinghua University, Beijing, 100084, China

Perturbation facilitated Infrared-Infrared double resonance spectroscopy has been used to study the triplet states of Cs₂. The $3^3\Pi_g$ state has been observed and rotationally resolved fluorescence spectra into the $b^3\Pi_u$ state and the perturbed $A^1\Sigma_u^+$ levels have been recorded. Molecular constants of the $b^3\Pi_{0u}$ and $A^1\Sigma_u^+$ states were derived from the fluorescence spectra.

1. Introduction

Alkali dimers have been selected as *prototype* molecules for the study of the spectra and structure of diatomic molecules. More information can be obtained from triplet states than from singlet states alone: spin-orbit, spin-spin, and spin-rotation interactions. For study of triplet states of alkali diatomic molecules, Li and Field [1] developed perturbation facilitated optical-optical double resonance (PFOODR) spectroscopy.

While Li₂ and Na₂ have been well studied, the excited states of Cs₂, especially triplet states, are not well known experimentally due to experimental difficulties: small vibrational and rotational constants, very high density of energy levels, requirement of single mode infrared lasers and sub-Doppler resolution, etc. The ground state of the Cs₂ molecule, $X^1\Sigma_g^+$, has been well characterized with accurate molecular constants [2]. The lowest excited singlet state, $A^1\Sigma_u^+$, which is strongly perturbed by the $b^3\Pi_u$ $\Omega = 0^+$ state, has been observed by Fourier transform spectroscopy (FTS) [3]. The only experimental observation of the $b^3\Pi_u$ state was very low resolution $X^1\Sigma_g^+ \rightarrow b^3\Pi_u$ absorption in the region of 7700-8200 cm⁻¹[4]. The perturbed $A^1\Sigma_u^+$ levels can be used as the intermediate *window* levels in the two-step excitation into the triplet Rydberg states.

Perturbation facilitated infrared-infrared double resonance (PFIIDR) spectroscopy has been applied to study triplet states of Cs₂. Recently we performed PFIIDR excitation and resolved fluorescence spectroscopy of Cs₂ and observed the $3^3\Sigma_g^+$, $a^3\Sigma_u^+$, $2^3\Delta_g$ and $b^3\Pi_u$ states [5,6]. Here we report the resolved fluorescence spectroscopy of the $3^3\Pi_g \rightarrow b^3\Pi_u$ transition.

2. Experimental

The experimental setup was similar to our previous Cs₂ and K₂ experiments [5-7]. Cesium vapor was generated in a heatpipe oven with 1 Torr Argon buffer gas. A single mode tunable DL100 diode laser was used as the pump laser to selectively excite an $A^1\Sigma_u^+$ intermediate level from the ground state. Another DL100 diode laser was used as the probe laser to further excite the $3^3\Pi_g \leftarrow A^1\Sigma_u^+$ transition. The two laser beams counter-propagated and crossed at the heatpipe center. The laser frequencies were measured by WA-1600 wavemeters. When the pump laser was held fixed to excite an $A^1\Sigma_u^+ \leftarrow X^1\Sigma_g^+$ transition, the probe laser was scanned and double resonance signals were detected by monitoring the $3^3\Pi_g \rightarrow a^3\Sigma_u^+$ / $b^3\Pi_u$ fluorescence with interference filters and a photomultiplier tube. While the two laser frequencies were held fixed to excite a $3^3\Pi_g$ level, the $3^3\Pi_g \rightarrow b^3\Pi_u$ fluorescence was dispersed with a 0.85 m double grating Spex 1404 monochromator.

3. Results

Our earlier study reported the first observation of the Cs₂ $2^3\Delta_g \rightarrow b^3\Pi_u$ fluorescence spectra [6]. Resolved fluorescence from the $2^3\Delta_{1g}$ state into the $b^3\Pi_{0u}$ $v_b' = 0$-48, $J_b' = 12$-100 levels has been recorded. Recently we observed the Cs₂ $3^3\Pi_g$ state by PFIIDR spectroscopy, [8] and the rotationally resolved fluorescence from this $3^3\Pi_g$ state to the high rotational levels of the $b^3\Pi_{0u}$ state has been studied.

The $3^3\Pi_g$ $(v, J) \rightarrow b^3\Pi_u$ (v_b', J_b') fluorescence consists of a "P" $(J \rightarrow J_b'=J+1)$ rotational line and an "R" $(J \rightarrow J_b'=J-1)$ rotational line, as predicted for the

---

[†] To whom all correspondences should be addressed. E-mail address: lili@mail.tsinghua.edu.cn.

$^3\Pi_0$ (case a) J, e-symmetry $\rightarrow$ $^3\Pi_0$ (case a) transition. The $b^3\Pi_u$ $v_b' = 18\text{-}38$, $J_b' = 230\text{-}236$ levels have been observed. [8]

Fig. 1 shows the resolved fluorescence spectrum from the $3^3\Pi_g$ $v = v_x+3$, $J = 233$, $T = 22282.441$ cm^{-1} level. The fluorescence lines in the region of 795-825 nm are the transitions into the $b^3\Pi_{0u}$ $v_b' = 27\text{-}38$ levels, and those lines in the region of 820-850 nm are transitions into the $A^1\Sigma_u^+$ $v_A' = 0\text{-}9$ levels. Our previous study showed that the intermediate $A^1\Sigma_u^+$ levels are strongly perturbed by the $b^3\Pi_u$ $\Omega = 0^+$ state. [6] These $A^1\Sigma_u^+$ $v_A' = 0\text{-}9$ levels may have borrowed transition possibility from the $b^3\Pi_u$ $\Omega = 0^+$ state. The upper $3^3\Pi_g$ $v = v_x+3$, $J = 233$ e-symmetry level may be perturbed by nearby singlet states. Fluorescence transitions into each vibrational level of the $b^3\Pi_{0u}$ and $A^1\Sigma_u^+$ states consist of two lines separated by ~11 cm^{-1}. These two lines have been assigned to the $J \rightarrow J'=J+1$ "P" and $J \rightarrow J'=J-1$ "R" rotational transitions.

Fig. 2 is the resolved fluorescence spectrum from the $3^3\Pi_g$ $v = v_x+6$, $J = 235$, $T = 22326.210$ cm^{-1} level. The fluorescence lines in the region of 775-825 nm are transitions into the $b^3\Pi_{0u}$ $v_b' = 18\text{-}38$ levels, and the lines in the region of 820-845 nm are transitions into the $A^1\Sigma_u^+$ $v_A' = 0\text{-}9$ levels. The band peaking at 760 nm is collision-induced $B^1\Pi_u \rightarrow X^1\Sigma_g^+$ fluorescence, which appears as a background of the $3^3\Pi_g \rightarrow b^3\Pi_{0u}$ fluorescence in the spectrum.

The term values of the upper $3^3\Pi_g$ v, J levels have been determined from the PFIIDR excitation spectroscopy, [8] thus the term values of the lower $b^3\Pi_{0u}$ and $A^1\Sigma_u^+$ e-symmetry levels can be calculated from the term values of the upper levels and fluorescence frequencies. Table 1 gives the molecular constants of the $b^3\Pi_{0u}$ state from a Dunham fit of 96 e-symmetry levels with $v_b' = 18\text{-}38$, $J_b' = 230\text{-}236$. The uncertainty of the $T_{v=-1/2}$ is quite big due to several reasons: the levels observed in the fluorescence spectra have high vibrational and rotational quantum numbers, small data set, and the accuracy of the fluorescence lines is about ~1.5 cm^{-1}.

The theoretical calculated constants are also given in Table 1 for comparison. The theoretical constants of Ref. 6 are not for the $b^3\Pi_{0u}$ state but for the $b^3\Pi_u$ state (mostly the $b^3\Pi_{1u}$ state). Ref. 9 gives a spin-orbit constant of A ~200 cm^{-1} for the $b^3\Pi_u$ state. So the T_e for the $b^3\Pi_u$ $\Omega = 0$ state of Ref. 6 will be ~7962 cm^{-1}. Considering the e-symmetry levels observed are pushed

down due to strong perturbation of the higher-lying $A^1\Sigma_u^+$ state, the experimental results agree with the theoretical constants reasonably well.

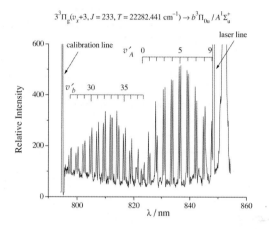

Figure 1. The resolved fluorescence spectrum from the $3^3\Pi_g$ $v = v_x+3$, $J = 233$, $T = 22282.441$ cm^{-1} level.

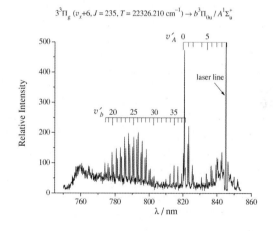

Figure 2. The resolved fluorescence spectrum from the $3^3\Pi_g$ $v = v_x+6$, $J = 235$, $T = 22326.210$ cm^{-1} level.

We also performed a Dunham fit with the $b^3\Pi_{0u}$ $v_b' = 0\text{-}30$ e-symmetry levels observed in this experiment and the levels observed in our earlier $2^3\Delta_g \rightarrow b^3\Pi_{0u}$ fluorescence spectroscopy [6]. The constants from the combined fit have better agreement with the theoretical results (Table 1). The $b^3\Pi_{0u}$ levels with $v_b' > 30$ are strongly perturbed by the $A^1\Sigma_u^+$ state and could not be introduced in the fit without destroying its quality.

Table 1. The Dunham constants of the $b^3\Pi_u$ $\Omega=0^+$ state. All constants are in cm^{-1} except R_e, which is in Å.

	This Work [a] $v_{b}' = 18\text{-}38, J_{b}' = 230\text{-}236$	Combined fit [b] $v_{b}' = 0\text{-}30, J_{b}' = 12\text{-}236$	Magnier [6]	Spies [9]
$T_{v=-1/2}$	7935.01 (17.00)	7959.31 (54)	8162 ($b^3\Pi_u$, not $b^3\Pi_{0u}$)	7955
Y_{10}	47.351 (480)	42.682 (78)	45.90	44.7
Y_{20}	−0.1730 (82)	−0.0516 (26)		
Y_{01}	0.01193 (27)	0.0129584 (630)	0.01271	0.01261
Y_{11}		−2.65 (18) × 10^{-5}		
Y_{02}	−3.029 (270) × 10^{-9}	−4.8 (1.1) × 10^{-9}		
R_e	4.611 (52)	4.42452 (1100)	4.465	4.483
y_{00}		0.001696 (1500)		

[a] Y_{02} was fixed as $-4*Y_{01}{}^3/Y_{10}{}^2$ at the fitting procedure.
[b] Combining our earlier data of the $b^3\Pi_u$ $v_{b}' = 0\text{-}30, J_{b}' = 12\text{-}100$ e-symmetry levels from the $2^3\Delta_g \rightarrow b^3\Pi_u$ fluorescence.

Table 2. The Dunham constants of the $A^1\Sigma_u{}^+$ state. All constants are in cm^{-1} except R_e, which is in Å.

	This Work [a] $v_{A}' = 0\text{-}9, J_{A}' = 230\text{-}236$	Combined fit [b] $v_{A}' = 0\text{-}15, J_{A}' = 8\text{-}237$	Verges and Amiot [3]	Magnier [10]	Spies [9]
$T_{v=-1/2}$	9619.14 (11.00)	9620.348 (360)	9627.06 (2)	9628	9631
Y_{10}	38.226 (240)	38.600 (71)	36.09 (2)	32.95	37.1
Y_{20}	0.182 (26)	0.0719 (35)			
Y_{01}	0.00908 (19)	0.0091695 (120)	0.009058 (2)	0.009140	0.009027
Y_{11}		1.314 (93) × 10^{-5}			
Y_{02}		−2.66 (16) × 10^{-9}			
R_e	5.29 (6)	5.2598 (34)	5.292 (1)	5.269	5.300
y_{00}		0.01798 (87)			

[a] $Y_{02} = -4*Y_{01}{}^3/Y_{10}{}^2 = -2.049 \times 10^{-9}$ cm^{-1}.
[b] Combining our intermediate levels of $A^1\Sigma_u{}^+$ $v_{A}' = 4\text{-}15, J_{A}' = 8\text{-}237$ levels confirmed in PFIIDR excitation experiment.

We have performed a Dunham fit for the $A^1\Sigma_u{}^+$ state with term values determined from these fluorescence spectroscopy and the term values determined in our PFIIDR excitation spectroscopy [8]. Table 2 gives the experimental and theoretical molecular constants of the $A^1\Sigma_u{}^+$ state. The experimental and theoretical results show very good agreement.

Rydberg-Klein-Rees (RKR) curves of the $A^1\Sigma_u{}^+$ and $b^3\Pi_{0u}$ states have been calculated using the molecular constants from the combined fits and compared with the theoretical potentials by Spies [9] and Magnier [10] in Fig. 3. The theoretical curve of Ref. 10 is for the $b^3\Pi_u$ state, whose T_e is about 200 cm^{-1} higher than the T_e of the $b^3\Pi_{0u}$ state. In Fig. 3, the theoretical potential curve of Ref. 10 is shifted down by 200 cm^{-1}. The RKR and theoretical potential curves are in good agreement.

Acknowledgments

This work was supported by NSFC (20773072) and NKBRSF of China.

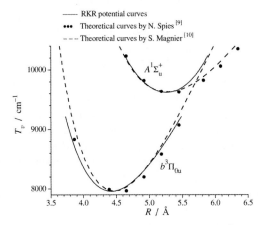

Figure 3. The RKR and theoretical potential curves of the $b^3\Pi_u$ $\Omega=0^+$ and $A^1\Sigma_u{}^+$ states.

References

1. Li Li and R. W. Field, *J. Phys. Chem.* **87**, 3020 (1983).
2. C. Amiot and O. Dulieu, *J. Chem. Phys.* **117**, 5155 (2002).

3. J. Verges and C. Amiot, *J. Mol. Spectrosc.* **126**, 393 (1987).
4. R. P. Benedict, D. L. Drummond, and L. A. Schlie, *J. Chem. Phys.* **66**, 4600 (1977).
5. D. Li, F. Xie, and Li Li, *Chem. Phys. Lett.* **in Press** (2008).
6. F. Xie, D. Li, L. Tyree, Li Li, V. B. Sovkov, V. S. Ivanov, S. Magnier, and A. M. Lyyra, *J. Chem. Phys.* **in Press** (2008).
7. F. Xie, D. Li, and Li Li, *MTPR-06, AIP Conference Proceedings*, **888**, 3 (2007).
8. D. Li, *Ph. D. Thesis, Department of Physics, Tsinghua University, China* (2008).
9. N. Spies, *Ph. D. Thesis, Fachbereich Chemie, University of Kaiserslautern, Germany* (1989).
10. S. Magnier, to be published.

EFFECTS OF RARE EARTH OXIDES ON SOME PHYSICAL PROPERTIES OF Li- Zn NANOPARTICLE FERRITES

M. A. AHMED[1], N. OKASHA[2], A. I. ALI[3]*, M. HAMMAM[3] and J. Y. SONG[4]

[1] *Materials Science Lab (1), Physics Department, Faculty of Science, Cairo University, Giza, Egypt*
[2] *Physics Department, Faculty of Girls, Ain Shams University, Cairo, Egypt*
[3] *Physics Department, Helwan University Ain Helwan, 11795, Cairo, Egypt*
[4] *Department of Materials Science and Engineering, Pohang University of Science and Technology (POSTECH), Pohang 790-784, Korea*

Abstract

The spinel ferrite $Li_{0.2}Zn_{0.6}La_yFe_{2.2-y}O_4$; $0.01 \leq y \leq 0.1$ were prepared by the usual ceramic sintering technique. XRD confirmed the formation of the samples in single phase spinel structure for all the samples. The lattice parameters decreased with increasing the La-content. The d c resistivity was measured as a function of temperature. The obtained results indicate the semiconductor like behavior, where more than straight lines indicating the presence of different conductions mechanism exist. The density of states near Fremi level as a function of La-content and discussed based on the variable range hoping model. The dielectric constant and dielectric loss were measured as a function of temperature and frequency. The dispersion peak at low frequency (10 kHz) was splitted by increasing La-content up to the suggested absorber splitting.

Keywords: Nanoparticle; Li-Zn ferrites; Rare earths; substitutions; XRD analyses; Dielectric behavior.

1. Introduction

Li-Zn-Ferrites have high potential for several electromagnetic devices in the radio frequency region, since they have frequency-dependent physical properties, such as permittivity and permeability. Polycrystalline ferrite has been extensively used in many electronic devices because of its high permeability in the radio frequency region, high electrical resistivity, mechanical toughness and chemical stability. There are many experimental and theoretical investigations on the frequency dispersion of complex permeability in polycrystalline ferrite [1- 6]. The complex permeability spectra of polycrystalline ferrite depend not only on the chemical composition of the ferrite but also on the post-sintering density and the microstructure such as grain size and porosity. These are attributed to the fact that the permeability of the polycrystalline ferrite is described as the superposition of two different magnetizing processes: spin rotation and domain wall motion [1, 6 and 7]. It is known that there are much applications of polycrystalline ferrite in the radio frequency devices. Lithium based ferrite is a pertinent magnetic material for applications because of its better properties of high frequency (high resistivity), high Curie temperature, low dielectric loss and lower densification temperature than NiZn ferrite. It is known that the preparation of Liz ferrite in dense polycrystalline form by conventional ceramic processing is difficult because of the Lithia evaporation implies a limitation of the sintering temperature.

Some authors [8-10] have been reported on the effect of the addition of divalent, trivalent and tetravalent ions on the electrical conductivity and dielectric properties. The electrical properties of mixed Li- Cd ferrites were reported by Ravinder and Radha [12] have studied the Li- Cd ferrites for their frequency and composition dependence of dielectric behavior.

Some researches have studied the influence of La-content on ferromagnetic oxides [15- 19]. D. Ravider et al [18, 19] studied the influence of the rare earth atoms with a large radius and stable valance (3^+) such as Gd_2O_3 and La_2O_3 in modification the ferrites structure and properties, where they measured at low frequency (1kHz or 1MHz). Moreover, they could not explain the effect of different amounts of rare earth on the dielectric properties.

In the present work, we aimed to study the effect of La substitution on the structure and the transport properties of the $Li_{0.2}Zn_{0.6}La_yFe_{2.2-y}O_4$; ($0.01 \leq y \leq 0.1$) system in the high temperature region as a function of frequency and temperature. Also one of our goals was to reach the critical concentration at which the physical properties reach the optimum values and the sample become more applicable.

2. Experimental procedures
2a- Sample preparation:

High purity oxides of Fe_2O_3, LiOH, ZnO and La_2O_3 were mixed together in molar ratio to prepare ferrites of composition $Li_{0.2}Zn_{0.6}La_yFe_{2.2-y}O_4$ ($0.01 \leq y \leq 0.1$) using the conventional double sintering technique. Final sintering of the specimens was carried out for 15h at 1150°C. The details of the method of preparation have been given in our earlier publication [20].

2b- Characterization and measurements:

X-ray powder diffraction was performed on the investigated samples using Rigaku Co- Miniflex X-ray diffractometer employing CuK_α radiation with λ=1.5418Å. The particle size (D) was calculated using Scherer's relationship, $D = 0.9\lambda /\beta cos\theta$; where λ is the X- ray wavelength, θ is the Bragg's diffraction angle and β is the half width of the (311) XRD peaks. The complex dielectric properties measurements were carried out using programmable automatic LCR-Meter (HP model 4284A) in the frequency range of 100 Hz ~ 1 MHz. The dielectric constants were measured for both heating and cooling runs. The cooling rate was about 0.6 K/min, and the sample temperature was monitored with a copper-constantan thermocouple connected to a Keithley 2000 digital multimeter. Electrical resistivity was measured both on cooling and heating runs in the temperature range from 10 K to 300 K by standard dc four-probe method using a CCR type refrigerator in NIT (Nanotechnology and Information Lab., Pusan National University, Busan, Korea).

3. Results and discussion
3a- The structural analysis:

XRD diffraction pattern of the $Li_{0.2}Zn_{0.6}La_yFe_{2.2-y}O_4$ ($0.01 \leq y \leq 0.1$) system was shown in Fig. (1a).

36

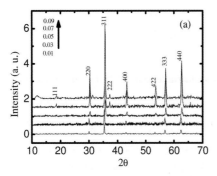

Fig. 1 (a) XRD patterns for $Li_{0.2}Zn_{0.6}La_yFe_{2.2-y}O_4$ system.

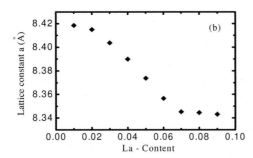

Fig. 1 (b) The change of lattice constant (a) with La content.

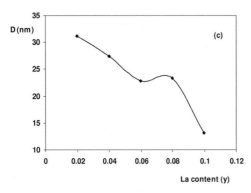

Fig. 1 (c) The change of particle size (D) with La content.

The spectra indicate that, there is nanocrystalline ferrites with no extra reflections and cubic structure belong to fcc spinel structure crystal symmetry was obtained. This result was good agreement with the previous studies on Li-Zn ferrite [12, 13, 17]. The lattice constant, (a) is plotted as a function of y in Fig. (1b) where the values of (a) is agree well with

JCPDS card [21], within experimental errors. It is clear that, the value of (a) decreases with increasing La-content up to y=0.07 then reach stable values. This behavior can be explained as follows: due to increase of the heat treatment, the magnetic domains will be increased leading to an increase in the crystallite size for all samples. When some Fe^{3+} ions in ferrite lattice were substituted by La ions, the lattice constant will be changed [22, 23]. The variation of lattice constant leads to an increase in the lattice strains which produce an internal stress [24, 25]. Such a stress hinders the growth of grains, so the particle sizes of the samples doped with La ions are smaller than that of Li- Zn ferrite nanoparticle. On the other hand, due to the larger bond of La^{3+}- O^{2-} as compared with that of Fe^{3+}- O^{2-}, it is obvious that more energy is needed to make La ions enter into lattice and form La^{3+}- O^{2-} bonds. All doped La^{3+} ferrite have higher thermal stability relative to Li-Zn ferrite nanocrystalline, and hence more energy is needed for the substituted samples to complete the crystallization and grow grains.

Moreover, after y=0.07, some La ions may be reside on the grain boundary as the ionic radii of La^{3+} ions is larger than that of Fe^{3+} ions. The average particle size (D) was calculated and shows in Fig. (1c).

It is reasonable to observe that, the particle sizes (D) for all concentrations were decreased with increasing La content where the presence of La ions on the grain boundary, cause pressure on the grains leads to hinder the growth of the grains as mentioned before.

3b- The electrical properties:

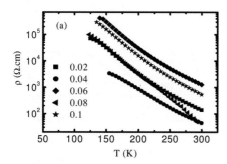

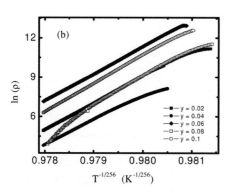

Fig. 2 (a) The electrical resistivity as a function of temperature, b- The logarithms of the electrical conductivity Vs. $1/T^{1/256}$ for the investigated samples.

The electrical resistivity as a function of temperature has been summarized in Fig. (2a). The resistivity of all compositions show insulating behavior, which gives an increasing resistance with decreasing temperature, from room temperature down to 100 K. To know the type of conduction mechanism, we checked the resistivity variation as a function of temperature, $1000/T$, and the hopping mechanism $(1/T)^{1/2}$.

The temperature dependent on resistivity for all compositions can be well described by the variable range hopping (VRH) model [14], depicted in Fig. (2b). In the VRH model, the resistivity of compositions is given by,

$$\rho = \rho_o \exp(\frac{T_o}{T})^{\frac{1}{4}} \qquad (1)$$

where T_o is the reduced temperature which is related to the localization length by,

$$T_o = \beta / k_B g(\mu) a_o^3 \qquad (2)$$

where β is the numerical factor, k_B is the Boltzmann's constant, $g(\mu)$ is the electronic density of states, and a_o is the Bohr radius [15]. From the plot of ln (ρ) versus $T^{1/256}$, we have estimated the fitting factor (T_o) as shown with La-content in Fig. (2b), to acceptable values which are around 6×10^9 K.

The density of state g (m) versus La concentration (y) is shown in Fig. (3b).

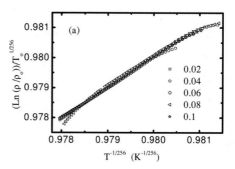

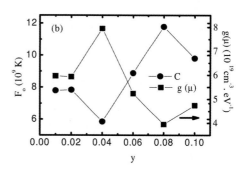

Fig. 3 (a) The master plot of the reduced electrical resistivity Vs reduced temperature, (b) Density of state g (m) and fitting temperature as a function of La content.

38

The data shows that for the low doping compositions (y = 0.01 and 0.02) the density of state has nearly constant values $5 \sim 6 \times 10^{19}$ $cm^{-3}.eV^{-1}$. However, it increased with an increasing in y from 0.03 to 0.04 to become 12×10^{19} $cm^{-3}.eV^{-1}$). It was decreased down to minimum values with the highest doping composition (y = 0.1).

To confirm the fitting for all compositions, we draw the master plot for $((ln~(\rho)~+~Ln~(\rho_o))/(To)^{1/4})$ Vs $T^{-1/4}$ as shown in Fig. (3a) providing an excellent fit to eq. (1) [14–17]. We can conclude that, since La^{3+} contents can change the Fe^{2+}/Fe^{3+} ratio, it is reasonable to assume that g(m) is nearly dependent on La-content. Therefore, the variation of to and g (m) with La-content reflects the change of localization length. This implies that the disorder in the lattice decreases with La-doping. As the extent of disorder decreases with La-doping, it is evident from Fig.(2) and Fig. (3), that the conduction mechanisms in this material are not thermally activated but are related to the hopping between Fe^{2+} and Fe^{3+} ions.

3c- The dielectric properties:

Figure (4:a-d) shows the dependence of the dielectric constant ($\acute{\epsilon}$) on the absolute temperature for $Li_{0.2}Zn_{0.6}La_yFe_{2.2-y}O_4$ system at different frequencies 10kHz, 100kHz, and 1MHz.

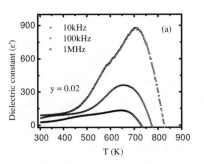

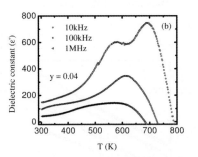

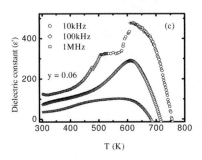

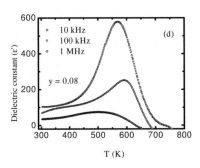

Fig. 4 (a-d) The dielectric constant as function of temperature for $Li_{0.2}Zn_{0.6}La_yFe_{2.2-y}O_4$ system at different frequencies 10kHz, 100kHz, and 1MHz with La-contents.

We presented here two samples depending on the La-contents at y = 0.02, 0.04. From the obtained data it is clear that the dielectric constant is kept constant from 300 K up to 450 K. With increasing the temperature around 500 K, the dielectric constant start to increase and a sharp peak appeared around 700 K at the applied frequency 10 kHz, while this peak is shifted to lower temperature (650 K) at frequency 100 kHz and decreased to the lowest temperature (625 K) when the applied frequency become 1MHz. This means that, at low temperature region (T < 450 K), the dipoles were frozen as well as the polarization is low, as a result, the dielectric constant gives no observable change in έ. With an increasing the temperature up to 500 K, the dipoles starts to thermally activated as a results of increasing the polarization and the dielectric constant increases. Further increasing temperature up to 700 K the thermal energy increases the lattice vibrations and the polarization as well as έ was to reach the smaller values.

One of the most important features here is that with applying the low electric field at (10 kHz) as in Fig. (5: a- c),

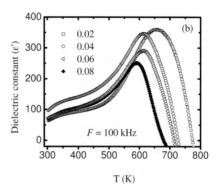

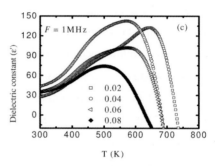

Fig. 5 (a-c) The dielectric constant as function of temperature with different La-contents and at different frequencies (a) 10 kHz (b) 100 kHz. and (c).

The dielectric constant gives the highest value while its value decreases with increasing the frequency. This can be explained as due to applying the low frequency the dipole moments rotates very fast and can not fallow the field variation with the result of the decreasing έ with increasing the frequency. At low frequency more dipoles become free and the field aligned them in its direction leading to an increase in polarization and έ. Moreover, the data in the Figure pronounced that the plateau region of (ε') extends up to T = 400 K. This relaxation is supposed to be arising from the electron hopping mechanism between the iron ions of different valances as reported for other ferrites containing Fe^{2+} ions [26- 28].

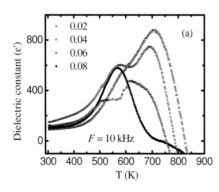

40

Therefore, Fe^{3+} produced from Fe^{2+} ion initiate vacancies which would form pairs with Fe^{2+} ions and give rise to electron hopping analogous to the general behavior of ferrites. Accordingly, the values of dielectric constant increase with increasing temperature for all La concentrations. This behavior of (ε') with temperature for the investigated samples agrees well with the known spinel ferrites [29] where (ε') increases with increasing temperature and decreases with increasing frequency. The increase of (ε') with temperature in this group can be ascribed to the cooperation of more than one type of polarization from room temperature up to 650 K. The electronic polarization is the most predominant one under the effect of both small thermal energy and applied electric field. With increasing frequency, the scattering processes between s–s and s–d interbands take place due to electron phonon interaction which decreases the intergranular spacing giving rise the same hopping length in both tetrahedral and octahedral sites. Rabkin et al. [30] suggested that, the processes of dielectric polarization in ferrites take place through a mechanism which is the same as that of conduction process because of the two processes are of the same origin. Due to electronic exchange $Fe^{2+} \rightarrow Fe^{3+} + e^-$, one can obtain local displacements of charge in the direction of the applied electric field, these displacements determine the polarization. Both types of charge carriers n and p contribute to polarization and they depend on temperature. Since the influence of temperature on the electronic exchange $Fe^{2+} \rightarrow Fe^{3+}$ is more pronounced than on the displacement of P-carriers, ε' will be increased rapidly with increasing temperature. Indeed, this behavior can be seen clearly for various La concentrations.

4. Conclusions

XRD patterns show that, the investigated ferrite has single spinel structure. The lattice constant are decreased with increasing La content up to 0.07 then stable where all doped La ions cannot enter into the lattice but reside on the grain boundary and this lead to decrease in the particle size. The decrease in dielectric constant with increasing frequency for all samples is attributed to the decrease in the polarization because of the dipoles cannot follow up the field variation. The dielectric losses are reflected on the conductivity measurements where the materials of high conductivity exhibiting high losses and vice versa. Finally, this study was an attempt to improve the quality of the classic Li- Zn ferrite by introducing different substitution of R_2O_3 instead of Fe_2O_3.

Acknowledgements

This work was supported by Korea Research Foundation Grant. (KRF-2006-005-C00045) from the Korea Research Foundation. A. I. Ali was supported by the Brain Korea 21 Project 2006.

References

1. G.T. Rado Rev. Mod. Phys. 25 (1953) 81.
2. D. Polder and J. Smit Rev. Mod. Phys. 25 (1953) 81.

3. J. Smit and H.M.J. Wijn Ferrites, Phillips Technical Library, Eindhoven, the Netherlands (1959).

4. E. Scloemann J. Appl. Phys. 41 (1970) 204.

5. A. Globus J. Phys. C 1 (1977) 1.

6. J.P. Bouchaud and P.G. Zerah J. Appl. Phys. 67 (1990) 5512.

7. Y. Naito, Proceedings of the First International Conference on Ferrites, (1970) 558.

8. S. A. Mazen, Phys. Stat. Sol. (a) 154 (1996)681.

9. A. Ahmed, J. Mater. Sci., 27 (1992) 4120.

10. S. A. Mazen, F. Metawe, S. F. Mansour, J. Phys. D: Appl. Phys., 30 (1997)1799.

11. D. Ravinder, T. S. Res, Crys. Res. Technol., (Est Germany) 25 (8) (1990)963.

12. K. Radha, D. Ravinder, Indian J. Pure & Appl. Phys., 33 (1995)74.

13. Yang Ying- Chang, Kong Lin- Shu, Sun Shu- he, Gu Dong- mei, J. Appl. Phys., 63 (1988)3702.

14. Li Hong- Shuo, Hu Bo- Ping, Coey J. M. D., Solid State Commun., 66 (1988)133.

15. N. Rezlescu, E. Rezlescu, C. Pasnicu, M. L. Craus, J. Phys., 6 (1994)5707.

16. N. Rezlescu, E. Rezlescu, C. Pasnicu, M. L. Craus, J. Mag. Mag. Mater., 136 (1994)319.

17. N. Rezlescu, E. Rezlescu, P. D. Popa,, L. Rezlescu, J. Alloys Compounds, 275 (1998)657.

18. D. Ravinder, B. Ravi Kumar, J. Materials letters 57 (2003)1738.

19. T. Nakamura, K. Hatakeyama IEEE Trans. Magn., 365 (2000)3415.

20. M.A.Ahmed, N. Ocasha, and A. I. Ali, J. Applied Physics D, 2007 (under publication).

21. JCPDS- ICDD (c) (1991).

22. N. Rezlescu, E. Rezlescu, C. Pasnicu, M. L. Craus, J. Mag. Mag. Mater., 136 (1994)319.

23. A. A. Satar, A. M. Samy, R. S. El-Ezza, A. E. Eatah, Phys. Stat. Solidi, A. Appl. Res., 193 (1) (2002)86.

24. Hua Yang, Zichen Wang, Lizhu Aong, Muyu Zhao, Jianping Wang, Heli Luo., J. Phys. D. Appl. Phys., 29 (1996)2574.

25. Hua Yang, Lijun Zhao, Xuwei Yang, Lianchun Shen, Lianxiang Yu, Wei Sun, Yu Yan, Wenquan Wang, Shouhua Feng, J. Mag. Mag. Mater., 271 (2004)230.

26. 26 S.T. Mahmuda, A.K.M. Hossaina, A.K.M. Abdul Hakimb, M. Sekic,T. Kawaic, H. Tabata, Journal of Magnetism and Magnetic Materials 305 269–274 (2006).

27. S.A. Morrison, C.L. Cahill, R. Swaminathan, M.E. McHenry, V.G. Harris, J. Appl. Phys. 95 (11) 6392 (2004).

28. T. Nakamura, M. Naoe, Y. Yamad, Journal of Magnetism and Magnetic Materials (305)120–126 (2006).

29. S. F. Mansour, Egypt. J. Solids, Vol. (28), No. (2), (2005).

30. D. Ravinder, A.V. Ramana Reddy, Materials Letters (38) 265–269 (1999).

A FIRST-PRINCIPLES CALCULATION OF THE MAGNETIC MOMENT AND ELECTRONIC STRUCTURE FOR SELECTED HALF-HEUSLER ALLOYS

SAMY H. ALY, RIHAM SHAPARA

Physics department, Faculty of Science, Faculty of Science at Damietta, Mansoura University, Damietta, Egypt
samy.ha.aly@gmail.com

SHERIF YEHIA

Faculty of Science, Physics department Helwan University, Cairo, Egypt
sherif542002@yahoo.com

Half-Heusler alloys are half-metallic magnets which may be defined as a new state of matter between insulating and metallic materials. These materials have many important applications in spintronics or magnetoelectronics devices. We have done first-principles (ab-initio) calculation of the magnetic moment and electronic structure for selected half-Heusler compounds. All the calculations were performed using the DFT-based electronic structure packages FPLO and WIEN2K. Half-metallicity (e.g. NiMnSb and RhVSb), semiconducting (e.g. FeVSb and NiVAl) and fully-metallic (e.g. NiVTe) behavior have been found in the compounds studied.

Keywords: Heusler alloys; Electronic strucutre; (*ab initio*) calculation.

1. Introduction

In the early 1980s Rob de Groot et al [1] discovered a new type of magnetic materials and he was the first to name the phenomenon of half-metallicity. Before that hints of half-metallicity were given by G. H. Jonker and J. H. Vans Santen in 1950. In a half-metallic magnet [1, 2], a material must have a collinear magnetic arrangement with the following qualitatively different types of up and down band structures : one spin direction has partially occupied bands where the other direction has a filled set of bands that are separated from unoccupied bands by a band gap. These materials have many important applications in spintronics or magnetoelectronics devices [3-5].

The known Half-metallic materials are [6-12]: Heusler and semi-Heusler alloys, oxides Fe_3O_4 and CrO_2 (rutile structure), manganites $La_{0.7}Sr_{0.3}MnO_6$, the double provskite compound Sr_2FeMoO_6 , zinc-blende compounds like CrAs and CrSb, (In, Mn)AS, (Ge, Mn)N, Co substituted TiO_2 and ZnO, and diluted magnetic semiconductors. Half-metallicity has been studied by several experimental techniques including positron annihilation, optical spectroscopy and normal state transport [13].

2. The Half–Heusler Alloys

Half-Heusler alloys XYZ crystallize in the face-centered cubic structure $C1_b$ with one formula unit per unit cell and their space group is F4/3m (#216). The positions of atoms are:

$$X = (0\ 0\ 0) \qquad Y = (\frac{1}{4}\ \frac{1}{4}\ \frac{1}{4}) \qquad Z = (\frac{3}{4}\ \frac{3}{4}\ \frac{3}{4})$$

3. Magnetic Moment of XMnSb Compounds

These compounds are known experimentally to be ferro(ferri)magnets with high Curie temperatures ranging from 500 to 700 K for compounds with X = Co, Ni, Pd and Pt. For Rh compounds, the Curie temperature is around room temperature [14].

The DOS of these compounds is characterized by a large exchange splitting of the Mn d-states (Fig.1 and 2), which is around 3 eV in all cases. The large exchange splitting of the Mn d-states leads to a large localized spin moments at the Mn site, the existence of the localized moments has been verified also experimentally. The localization comes from the fact that although d electrons of Mn are itinerant, the spin-down electrons are almost excluded from the Mn site. The lattice constants used in the present work are the experimental values 5.925 Å and 5.9 Å for X = Ni and Co respectively [15]. For FeMnSb we used the theoretical lattice constant 5.88 Å. For all these compounds, the Sb atom carriers a moment antiparallel to that of Mn. Experimental values for Mn spin moment have been deduced from experiments done by Kimera et al [16], by applying the sum rules to their x-ray magnetic circular dichroisme spectra. The band structure of NiMnSb using FPLO is shown in Fig.3. One can conclude from this figure that this compound carries a magnetic moment and has a clear gap in the spin-down DOS indicating its half-metallic character. The spin-down band is close to E_f at the point Γ but with no intersection with E_f.

44

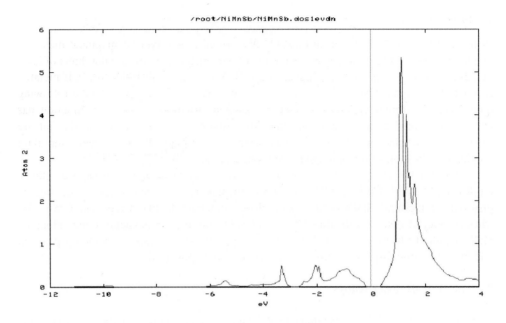

/root/NiMnSb/NiMnSb.dos1evdn

Fig. 1: The spin-down DOS of d-state of Mn in NiMnSb (using WIEN2K).

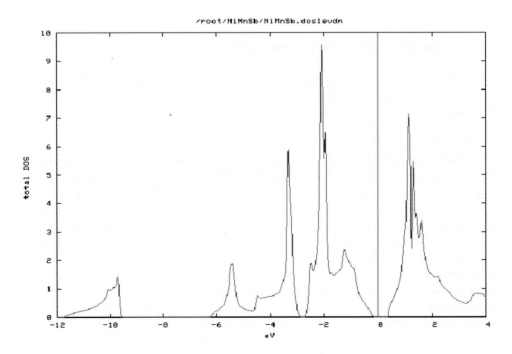

/root/NiMnSb/NiMnSb.dos1evdn

Fig. 2: The total spin-down DOS of NiMnSb (using WIEN2K).

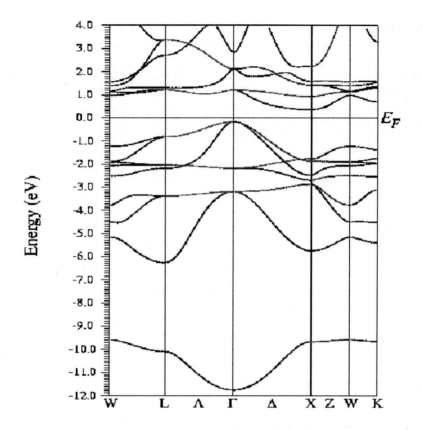

Fig. 3: The spin-down band structure of NiMnSb in WIEN2K.

4. NiMnX Compounds

We turn our attention to NiMnX compounds, where X is an sp- atom. These compounds crystallize in the C1$_b$ structure. In our calculation we used the lattice constant of the NiMnSb compound for the rest of the systems shown in table (1). The magnetic moments of these compounds are largely due to the spin magnetic moment of Mn. The spin magnetic moment of compounds with X ≡ Ge and Sn is antiparallel to Mn moment. If X atom is replaced by Cu or Si, the net magnetic moment reduces to nearly zero. For X≡ Sb, Bi and As atoms, the compounds are half- metallic with E$_f$ located in the middle of the gap.

The corresponding band structures using both of FPLO and WIEN2K (spin-down only) are shown in Fig.4 for NiMnBi. High DOS peaks correspond to flat bands. One can conclude from these figures that this compound carries a magnetic moment and has a clear gap in its spin-down band structure indicating its half-metallic character.

Table (1): The magnetic moments for NiMnX compounds. M ≡ metallic, H.M ≡ half-metallic and H.M* ≡ it may behave as a half-metal at different lattice constants or different temperatures.

Compound	Ni (μ_B)	Mn (μ_B)	X (μ_B)	Tot (μ_B)	State
NiMnAl	0.13	3.65	-0.23	3.55	M
NiMnBi	0.26	3.84	-0.98	4.00	H.M
NiMnCl	0.44	4.20	0.06	4.70	M
NiMnCu	0.02	-0.01	0.00	0.01	H.M*
NiMnGe	-0.06	3.62	-0.28	3.28	M
NiMnPb	0.00	3.44	-0.22	3.22	H.M*
NiMnSb	0.23	3.88	-0.11	4.00	H.M
NiMnSn	-0.01	3.39	-0.24	3.15	H.M*
NiMnS	0.05	0.00	0.00	0.05	M
NiMnSi	0.25	-0.27	0.00	0.00	M
NiMnTe	0.50	4.21	0.13	4.83	H.M*
NiMnTi	0.23	3.54	-0.77	3.09	M
NiMnAs	0.18	3.97	-0.15	4.00	H.M
NiMnP	0.00	0.00	0.00	0.01	M

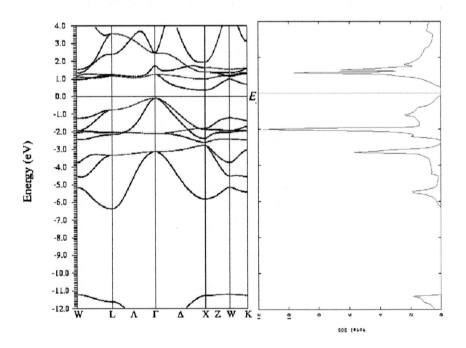

Fig. 4: The spin-down band structure and spin-down DOS of NiMnBi (using WIEN2K).

47

5. Magnetic Moment of XVSb Compounds

Both CoVSb and FeVSb compounds have been studied by X-ray diffraction and magnetic measurements [17]. The compounds XVSb with X ≡ Fe, Ni, Co, Rh, Pt, Pd and Ir crystallize in the face centered cubic structure with one formula unit per unit cell and space group F4/3m (#216). The X, V and Sb atoms are located at $(0, 0, 0)$, $\left(\frac{1}{4},\frac{1}{4},\frac{1}{4}\right)$ and $\left(\frac{3}{4},\frac{3}{4},\frac{3}{4}\right)$ positions respectively. The lattice constants used here are theoretical lattice constants obtained by minimization of energy (Fig.5 and 6). The theoretical lattice constants for NiVSb, CoVSb, FeVSb, PdVSb, PtVSb, RhVSb and IrVSb are 5.78, 5.7, 5.68, 6.1, 6.2, 6.1 and 6.05Å respectively. The experimental lattice constants for NiVSb, CoVSb and FeVSb are 5.78, 5.8 and 5.83 Å respectively [18]. The large exchange splitting of the V d-states leads to large localized spin moments at the V sites. The net magnetic moment of these compounds is smaller by nearly $2\mu_B$ than the moments of XMnSb compounds. The gap in NiVSb is shifted away from E_f. This behavior is far from half- metallicity. The largest gaps (~1.2 eV) are for Co and Pt containing compounds. FeVSb behaves as nonmagnetic semiconductor with a gap of 0.65 eV. Fermi energy is located close to the middle of the gap for compounds with X ≡ Rh, Pt and Pd (Figs.7).

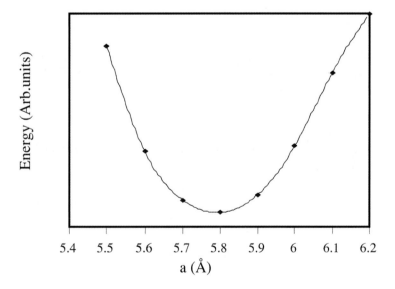

Fig. 5: The energy as function of the lattice constant for NiVSb.

48

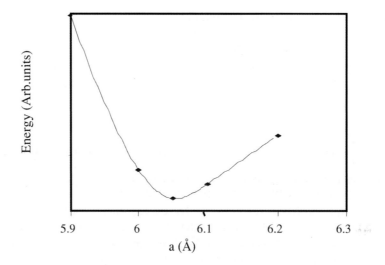

Fig. 6: The energy as function of the lattice constant for IrVSb.

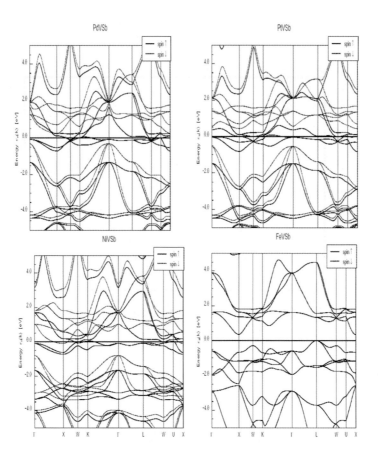

Fig. 7: The band structure of XVSb compounds.

49

References

1. R. A. de Groot, F. M. Mueller, P. G. Van. Engen and K. H. J. Buschow, Phys. Rev. Lett. 50, 2024 (1983).

2. V. Yu. Irkhin, M. I. Katsnelson, Physics-Uspekhi 37, 659(1994).

3. S. Datta and B. Das, Appl. Phys. Lett. 56, 665 (1990).

4. K. A. Kilian and R. H.Victora, J. Appl. Phys. 87, 7064 (2000).

5. C. T. Tanaka, J. Nowak and J. S. Moodera, J. Appl. Phys. 86, 6239 (1999).

6. H. Akai, Phys. Rev. Lett. 81, 3002 (1998).

7. H. Ohno, Science 281, 951 (1998).

8. J. M. De Teresa, A. Barthelemy, A. Fert, J. P. Contour, R. Lyonnet, F. Montaigne, P. Seneor and A. Vaures, Phys. Rev. Lett. 82, 4288 (1999).

9. K. L. Kobayashi, T. Kimura, H. Saweda, K. Terakura and Y. Tokura, Nature. 395, 677 (1998).

10. H. Akinaga, T. Manago and M. Shirai, J. Appl. Phys. 39, 1118 (2000).

11. H. Munekata, S. Von Molnar, H. Ohno and L. Chang, Phys. Rev. Lett. 63, 1819 (1989).

12. Y. Matsumoto, M. Murakami, T. Shono, T. Hasegawa, T. Fukumura, M. Kawasaki, P. Ahmet, T. Chikyow, S. Yakoshihara and H. Koinuma, Science. 291, 854 (2001).

13. G. Bacon, "*Neutron Diffraction*", (Oxford University Press, 1962).

14. P. J. Webster and K. R. A. Ziebeck ," *Alloys and Compounds of d-Elements with Main Group Elements*", (Springer-Verlag,Berlin Edited by H. R. J. Wijn, Landolt-Börnstein, New Series, group ш, 1988).

15. B. R. K. Nada and I. Dasgupta, J. Phys.: Condens. Matter 15, 7307 (2003).

16. A. Kimera, S. Suga, T. Shishidou, S. Imada, T. Muro, S. y. Park, T. Miyahara, T. Kaneko, and T. Kanomata, Phys. Rev. B 56, 6021 (1997).

17. K. Kaczmarska, J. Pierre, J. Beille, J. Tobola, R. V. Skolozdra and G. A. Melnik, J. Magn. Magn. Mater. 187, 210 (1998).

18. J. Tobola, J. Pierre, J. Alloys. Comp. 296, 243 (2000).

I-2 INVITED LECTURE PAPERS

A FIRST-PRINCIPLES CALCULATION OF THE ELECTRONIC STRUCUTRE, MAGNETIC MOMENT AND SPIN-DENSITY FOR SELECTED FULL-HEUSLER ALLOYS

SHERIF YEHIA[*], M. M. AHMED, M. HAMMAM and MONA A. AHMED

Physics Department, Helwan University, Faculty of science
Cairo, Helwan, Egypt
sherif542002@yahoo.com[*]

SAMY H. ALY

Physics Department, Damietta, Mansoura University, Faculty of science
Damietta, Egypt[†]
samy.ha.aly@gmail.com

Full-Heusler alloys are half-metallic materials which may be defined as a new state of matter between the insulating and metallic states. These materials have many important applications in spintronics or magnetoelectronic devices. We have done first-principles (ab-initio) calculation of the electronic structure, magnetic moment and spin density-maps for selected full-Heusler compounds. All the calculations were performed using the DFT-based electronic structure package WIEN2K. Half-metallicity (e.g. Co_2Vga), nearly half metallic (e.g. Co_2TiGa) and fully-metallic (e.g. Co_2MnGa) behavior have been found in the compounds studied.

Keywords: Heusler alloys; spin density maps; electronic structure.

1. Introduction

Heusler alloys [1–5] are generic name for a family of intermetallic compounds of composition X_2YZ or XYZ where X, Y, Z are ordered in $L2_1$ structure and Fm3m space group. Generally *X* is any element which belongs to the ends of the 3d, 4d, or 5d series, Y *a* 3d, 4d and 5d element, while Z is an sp element. For most Heusler alloys, the atoms at Y sites carry a large magnetic moment, in contrast to those at X sites even for transition metals. However, in case of Co_2YZ, the Co moments are in the range of approximately 0.1 to 1.0 μ_B as compared to 4.0 μ_B for the X_2MnZ [6]. It is considered that the magnetic moment on the Co atom depends strongly on the local environment [7]. We therefore have undertaken this study to investigate the role of Y site in the Co_2YGa alloys. We present in this paper a DFT-based study on the electronic structure, magnetic moment and spin-density maps in Co_2YGa alloys using FLAPW [8] method and Brillouin-zone integration methods as implemented in the Wien2k package [9].

CP 998, Modern Trends in Physics Research
Third International Conference MTPR-08
edited by L. El Nadi

54

2. Total density of states of Co₂YGa compounds

To examine the effect of changing Y atom on density of states(DOS), We have calculated the total density of states of Co₂YGa,(Y=Ti, V, Cr, Mn). The DOS of these compounds for the majority and minority spins are shown in figure 1 and 2, respectively.

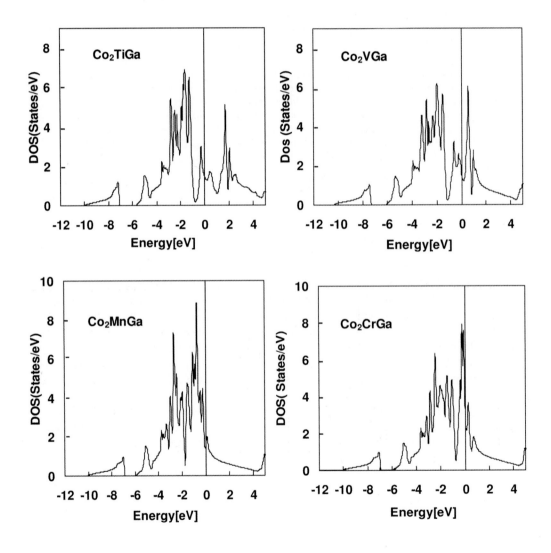

Figure 1. The calculated total DOS for the majority spins in Co₂YGa. The vertical line indicates the position of the Fermi level.

For majority and minority states, the region between -11 and -8 eV in the valence band shows the contribution from the Ga 4s states with small contributions from Co and Y 4s states, while the rest of the valence band are mainly derived from the 3d electrons of Co and Y. The same behavior was observed in Co_2FeGa [10].

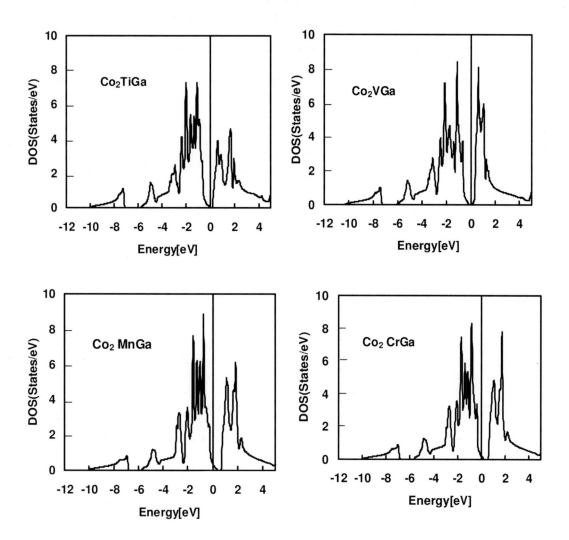

Figure 2. The calculated total DOS for the minority spins in Co_2YGa. The vertical line indicates the position of the Fermi level.

It may be remarked that there is no enough DOS above E_F in the majority spin states for Y=Mn, however more pronounced DOSs are observed for Y= Ti, V and Cr. Whereas for the minority spin (Fig.2) in the conduction band (about 1 eV above the Fermi level) Co-d state hybridize with the Y-d states. We calculated the atom and symmetry projected densities of states (PDOS) of Co and Y atoms to clarify the character of the bands. For both spin up and down states, at the Co site, the Co t_{2g} states are more dominating than the Co e_g states below E_F. Unlike, the conduction band of the minority density of states, the e_g states are more dominating than t_{2g} states. Details will be published elsewhere [11].

3. The local and total magnetic moment of Co_2YGa compounds

We have calculated the total and local magnetic moments of Co_2YGa,(Y=Ti, V, Cr, Mn).

Table (1): The calculated total and partial magnetic moments for Co_2YGa compounds

Alloy	Co(μ_B)	Y(μ_B)	Ga(μ_B)	$Tot_{cal}(\mu_B)$	$Tot_{exp}(\mu_B)$	state
Co_2TiGa	0.66	-0.21	-0.009	0.93	0.75[a]	Nearly half-metal
Co_2VGa	0.99	0.02	-0.02	2.00	1.95[b]	Half-metal
Co_2CrGa	0.75	1.67	-0.04	3.09	2.36[c]	Metal
Co_2MnGa	0.78	2.73	-0.05	4.16	4.05[a]	Metal

[a] From Ref [4] [b] From Ref [12] [c] From Ref [6]

From table (1) we notice that the calculated local magnetic moment on **cobalt** varies by varying the Y-site. In addition, both of the magnetic moment of Y-atom and the total magnetic moment increase by increasing the atomic number in Y =Ti, V, Cr, Mn . When Y = Ti and V, the obtained total magnetic moment is mainly due to Co atom. The contribution of Co atom in the total magnetic moment when Y = V is greater than when Y= Ti. The formation of magnetic properties is more complicated when **a** Heusler alloy has more than one magnetic element. On the other hand, Co_2CrGa and Co_2MnGa have the largest total magnetic moment due to contributions from Co, Cr and Mn. The Mn, and Cr moments couple ferromagnetically to the Co moment and therefore are responsible for the large magnetic moment observed in these alloys. The integer value of the magnetic moment and the presence of a gap, around E_F, in the minority spin band structure are good indications of Half-metallicity. Based on our electronic structure calculation **of** the minority spin of Co_2YGa we found that the alloys are half-metal for Y = V, nearly half-metal for Y= Ti and metallic for Y= Mn and Cr.

57

4. Spin density-maps of Co₂YGa compounds

We constructed the valence-electron density and the spin density maps of Co_2YGa, (Y =Ti and V) in the (110) plane. The minimum and maximum contours used in the plots are 0.0 and 2.0, respectively, with an interval of 0.01.

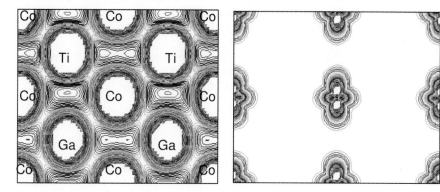

Figure (3): the electron (left) and spin density contours(right) for Co_2TiGa

Figure (3) shows the electron and spin density contours for Co_2TiGa. It is evident from this figure that Ti and Ga atoms show no contribution in the spin density maps.

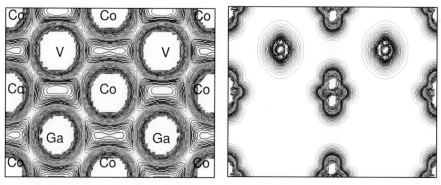

Figure (4): the electron (left) and spin density contours(right) forCo_2VGa

Figure (4) shows the electron and spin density contours of Co_2VGa. The V atoms are present in the spin density contours and this demonstrates that V, in contrast to Ti carries a magnetic moment in this system.

References

1. F. Heusler et al. Phys. Ges. 5 (1930) 219.
2. P.J. Webster, et al., Contemp. Phys. 10 (1969) 559.
3. J. Tobola, et al., J. Alloys Compd. 296 (2000) 243.
4. A. Jezierski, Phys. Stat. Sol. B 196 (1996) 357.
5. K. Kobayashi, R. Kainuma, K. Ishida and K. Fukamichi, J. Alloys Compds. 403 (2005) 161.

58

6. K.R.A. Ziebeck, et al., J. Phys. Chem. Solid 35 (1974) 1.

7. A. Jezierski, Phys. Stat. Sol. B 196 (1996) 357.

8. P. Blaha, K. Schwarz, P. Sorantin, and S.B. Tricky, Comput. Phys. Commun. 59, 399 (1990).

9. P. Blaha, K. Schwarz, G.K.H. Madsen, D. Kvasnicka, J. Luitz, WIEN2K, An Agumented Plane Wave + Local Orbitals for Calculating Crystal Properties, (K. Schwarz, Techn. Universitat Wien Austria, 2001).

10. A. Deb, M.Itou, Y.Sakurai, N.Hiraoka and N.Sakai, Phys. Rev. B 63, 064409-1 (2001).

11. Mona A. Ahmed, "Ab initio calculation of electronic band structure of Co-based Heusler alloy", M.Sc. Thesis, Helwan University, 2008.

12. R.A. Dunlap (Magnetic properties of Co-based Heusler alloys) private communications.

MAGNETIC SUPER-EXCHANGE INTERACTION AND STRUCTURE OF COPPER(II) 1, 4 BUTYLENEDIAMINE TETRACHLORIDE [NH$_3$(CH$_2$)$_4$H$_3$N]CuCl$_4$ SINGLE CRYSTAL

M.A. AHMED, I.S. AHMED FARAG* and NABILAH M. HELMY

Materials Science Lab.(1), Physics Dept., Faculty of Science, Cairo University, Giza, Egypt
* Physics Dept., National Research Centre, Dokki, Eygpt

Abstract:

Butylene diamine copper tetrachloride was prepared in a form of crystal form from aqueous solution. X-ray of single crystal and magnetic properties studies were carried out, The structure of the neutral complex [NH$_3$(CH$_2$)$_4$H$_3$N]CuCl$_4$ contains cationic [NH$_3$(CH$_2$)$_4$H$_3$N]$^{2+}$ and anionic (CuCl$_4$)$^{2-}$. Results of X-ray revealed that the Cu atoms are 4-fold coordinated by chlorine atoms to form distorted square planer. The organic group is bonded with CuCl$_4$ hydrogen bonds and Van der Waal contact. The temperature dependence of the magnetic susceptibility was measured in the temperature range from 70 to 500K at different magnetic field intensities. The results indicate that the complexes exhibit weak antiferromagnetic coupling between two metal ions

Introduction

The magnetic susceptibility of the compounds (CH$_2$)$_n$(NH$_3$)$_2$MCl$_{4-x}$Br$_x$ (n=8, 9 and x= 0, 1, 2) are measured in the temperature range from 80K up to a temperature near the melting points of each sample. The obtained values of the Curie- Weiss constant showed that the antiferromagnetic properties increase by increasing either the Br$^-$ ions or the number of carbon atoms in the complex[1].

The magnetic susceptibility of the layered compounds (CH$_2$)$_3$(NH$_3$)$_2$FeCl$_2$Br$_2$ and (CH$_2$)$_6$(NH$_3$)$_2$FeCl$_2$Br$_2$ has been measured in the temperature range 80<T<300K. The results follow the Curie-Wiess in the temperature range of 120 to 300K. The results are interpreted in terms of two-dimensional canted antiferromagnetic interaction. A comparison with the corresponding pure chloride compound is given[2].

The magnetic susceptibility of the complex (CH$_2$)$_{10}$(NH$_3$)$_2$FeCl$_4$ is measured in the temperature range from liquid nitrogen up to room temperature. The data shows that the compound is antiferromagnetic with a Neel temperature of 93K but when taking into consideration the monoclinic structure of this compound, it appears that the antiferromagnetic intra-layer exchange compound, it appears that the interaction coexist with a weak ferromagnetic interlayer interaction[3].

One of our goals in this research work is to study the structure as single crystal due to their wide range of applications.

Experimental techniques:

Preparation of single crystal:

Butylene diamine cupric tetrachloride complex was prepared by mixing 1 M aqueous solutions of CuCl$_2$ and 1 M of aqueous solutions of (CH$_2$)$_4$(NH$_3$)$_2$Cl$_2$. The mixture was heated using water bath at 80°C for several hours until it was reduced to one third of its initial volume. The sample was left several days in dark room to obtain single crystal.

Data collection

X-ray crystallographic data were collected on Enraf- Nonius, 590 Koppa Single crystal diffractometer graphatic monochromatic using

Corresponding author: M. A. Ahmed
Moala47@hotmail.com

MoKa (λ = 0.71073Å). The intensities were collected at room temperature using φ, ω scan mode, the crystal to detector distance was 40mm, further details are given in Table(1). The cell refinement and data reduction were carried using Denzo and Scalepak programs[4], the crystal structure was solved by the direct method used to solve structure: DIRDIF (Buerskens et al., (1992)[5] which revealed the position of all non- hydrogen atoms and refined by the full matrix least square refinement based on F^{2+} using maXus software package[6]. The temperature factors of

Table 1 Crystal data and details of structure determination.

Empirical formula	$C_4H_{14}Cl_4CuN_3$
M_r (gmol^{-1})	295.522
Shape/ Color	Purple/
Temperature	298K
Wavelength	0.71073(Å)
Crystal system	Monoclinic
Space groupe	$P2_1/c$
Unit cell dimensions	
a	9.2743 (4)Å
b	7.6002 (3)Å
c	7.5906 (3)Å
α	90.00°
β	103.142 (3)°
γ	90.00°
V	521.02 (4)Å^3
Z	2
Index ranges	
	h = -12 →11
	k = -9 →9
	l = -9 →9
	h = 0 →12
	k = 0 →9
	l = -9 →9
θ range for data collection	2.910—27.485°
Measured reflections	2309
Independent reflections	1240
Observed reflections	886
Refinement	fullmatrix least squares refinement on F^2
$\Delta\rho_{max}$ and $\Delta\rho_{max}$	$\Delta\rho max \rightarrow$ -2.17eÅ^3
Cell parameters from	
R_{int} Criterion:	1175
[I > 3.0 sigma(I)]	0.031

Table 2 Fractional atomic coordinates and equivalent isotropic thermal parameters ($Å^2$) for the complex $CH_2)_{4(}(H_3N)_2CuCl_4$.

	x	y	z	U_{eq}
Cu1	0.0000 (5)	0.5000 (8)	0.5000 (8)	0.0236 (3)
Cl2	-0.03378 (11)	0.29587 (14)	0.26897 (15)	0.0296 (5)
Cl3	0.24978 (12)	0.45661 (16)	0.55310 (18)	0.0332 (5)
N4	0.2250 (4)	0.4643 (5)	0.0686 (6)	0.0332 (19)
C5	0.4624 (5)	0.5833 (6)	0.0178 (6)	0.030 (2)
C6	0.3622 (5)	0.5621 (7)	0.1488 (6)	0.036 (3)
H6A	0.4146 (5)	0.4978 (8)	0.2526 (8)	0.040630
H6B	0.3354 (5)	0.6759 (8)	0.1860 (8)	0.040630
H5A	0.4039 (5)	0.6229 (8)	-0.0964 (8)	0.034753
H5B	0.5382 (5)	0.6684 (8)	0.0637 (8)	0.034753
H4A	0.1657 (5)	0.4550 (8)	0.1567 (8)	0.037990
H4B	0.2500 (5)	0.3486 (8)	0.0343 (8)	0.037990
H4C	0.170257	0.525347	-0.036132	0.037990

all non-hydrogen atoms were refined anisotropically, then hydrogen atoms were introduced as a riding model with C-H = 0.96Å and refined isotropically. The molecular graphics were prepared using ORTEP program[7]. The crystallographic Data of the structure described in this paper were deposited to the Cambridge crystallographic Data Center as supplementary publication No. CCDC 690000. The magnetic susceptibility was measured as a powder using the conventional Gouy's method. The temperature of the sample varied from liquid nitrogen up to near the melting.

Results and discussion:

Structure analysis:

The molecular graphics ORTEP of complex $(NH_3(CH_2)_2H_3N)(CuCl_4)$ is shown in Fig. (1). The Crystal data and the factional atomic coordinates and estimated standard deviations of the molecule in a symmetric unit cell are listed in table (1,2). Some selected bond lengths and bond angles are listed in table (3a). The hydrogen bonding are reported in tables in (3b).

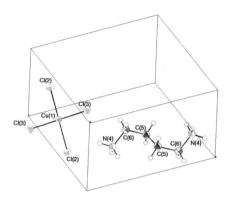

Fig. (1) The molecular structure of Butylene diamin Copper tetrachloride

It is shown from the crystal data, that the investigated complex crystallizes in monoclinic system with space group P2$_1$ /C with number of molecules in a symmetric unit cell equal two molecules This mean that the molecule itself has center of symmetry as shown in Fig. (2). It is obvious from Figs. (1, 2), the structure of the investigated complex consist of two separate parts one of them is the organic cation, $[NH_3(CH_2)_4H_3N]^{2+}$ and the other is the inorganic anion $(CuCl_4)^{-2}$. The inorganic ion is situated on the crystallographic center of space group P2$_1$ /C and surrounded by 2 sorts of Cl atoms Cl$_2$ and Cl$_3$ to form distorted square polygon with planer character. The bond length and the angles of this square polygon is Cu$_1$-Cl$_2$ bond length (2.309 Å) is greater than of Cu$_1$-Cl$_3$ bond length (2.283 Å) and Cl$_2$-Cu$_1$-Cl$_{3i}$ angle 90.34°. is grater than Cl$_2$-Cu$_1$-Cl$_3$ angle (89.66°), which reveal the planarity and small distortion of this square polygon. On the other hand, the organic cation $[NH_3(CH_2)_4H_3N]^{2+}$ lies on the center of symmetry, that deviding it into two equivalent parts with bond lengths C—C and C—N ranges between 1.516 Å– 1.48 Å respectively. The zigzag angle of the aliphatic chain of the organic cation is ranging

between 114.8°, 112.43° which in consistent with that reported in literature [8]. The inorganic and organic ions of the investigated complex are stacked in layers by succession mode, or in other words, the organic cations constitute a layer which are sandwiched between two layers of inorganic anions $(CuCl_4)^{-2}$. as shown in Fig. (3). Both layers of the structure are fixed through out the lattice by Vander-Waal forces and strengthed hydrogen bonds between Cl ions of the polygon CuCl$_4$ anions and the hydrogen of the amine group of the organic cations as shown in Fig. (4) and table (3: a,b)

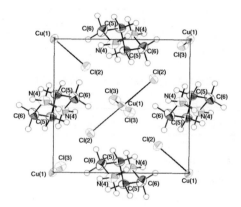

Fig. (2) The center of symmetry of molecule

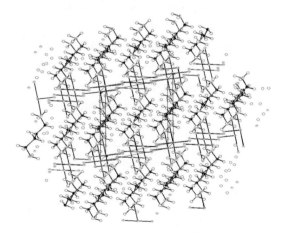

Fig. (3) The whole structure for the$(CH_2)_4(H_3N)_2CuCl_4$

62

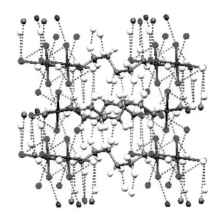

Fig. (4) The hydrogen bonding net work

Table (3.a) selected geometric parameters of Complex $(CH_2)_{4t}(H_3N)_2CuCl_4$

Cu_1-Cl_2	2.309 Å	Cl_2-Cu_1-Cl_{3i}	90.34°
Cu_1-Cl_3	2.283 Å	Cl_2-Cu_1-Cl_3	89.66°
C—C	1.516, Å		
C—N	1.48 Å		

Table (3.b) selected of hydrogen bonding of complex

Cl2—H4Ciii	2.3569 (7)
Cl3—H4Biv	2.3241 (7)

The magnetic properties of diamine complexes:

Figure (5) shows the relation between the molar magnetic susceptibility and absolute temperature as a function of magnetic field intensity for the investigated sample. The data show that χ_M decreases with increasing the temperature. This behavior is due to crystalline phase transition. Fig. (6) correlates the ($\mu_{eff.}$) and absolute temperature (80K up to 500K) as a function of the magnetic field intensity for the prepared complex. The data indicate that the general behavior of the sample is the moderate intermolecular antiferromagnetic exchange interaction, The interaction takes two paths. The first one between the layers one through Cl as a bridge and the other path through Cu–Cl–Cu within the layer. In general, one can conclude that the magnetic exchange interaction within the layer is weaker than the magnetic exchange interaction between the layers.

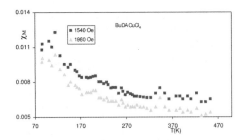

Fig. (5) Relation between the magnetic susceptibility, (χ_M) and the absolute temperature as a function of magnetic field intensity for the complexes, a:BuDACuCl₄

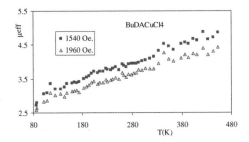

Fig. (6) Relation between the effective magnetic moment $\mu_{eff.}$, and the absolute temperature as a function of magnetic field intensity for the complex: BuDACuCl₄

Conclusion:

The structure of BuDACuCl₄ was studied by single crystal X-ray diffraction. This complex have monoclinic structure. The structure of the metallic layer (CuCl₄) is a distorted square planer and the structure of $(CH_2)_4(NH_3)_2$ in zigzag form. The metallic layers (CuCl₄) are connected with organic group by more than one type of bonds. The magnetic behavior for this complex corresponding to antiferromagnetic exchange interaction between the Cu ions pathway through Cl within the layer and between the layers through organic group

References:

1-F.A.Radwan, M.A.Ahmed, H.Mikhail, Solid State Communications, 84 (1992) 1047.
2-M.F.Mostafa, M.A.Semary, M.Abdel-Kader, Phys.Lett., 82 (1981) 350.
3-M.A.Ahmed, F.A.Radwan, M.M.El-Desoky, J.Mag.Mag.Mater., 67 (1987) 349
References:

1-F.A.Radwan, M.A.Ahmed, H.Mikhail, Solid State Communications, 84 (1992) 1047.
2-M.F.Mostafa, M.A.Semary, M.Abdel-Kader, Phys.Lett., 82 (1981) 350.
3-M.A.Ahmed, F.A.Radwan, M.M.El-Desoky, J.Mag.Mag.Mater., 67 (1987) 349

4-Otwinowski, Z. and Minor, W, (1997). In Methods in Enzymology, 276, edited by C. W. Carter, Jr. & R. M. Sweet pp. 307--326, New York: Academic Press.

5-Beurskens, P. T., Beurskens, G., Bosman, W. P., de Gelder, R. S. García--Granda, S., Gould, R. O. & Smits, J. M. M. (1996). The *DIRDIF*96 program system, Technical Report of the Crystallography Laboratory, University of Nijmegen, The Netherlands.

6-Mackay, S., Gilmore, C. J., Edwards, C., Stewart, N. & Shankland, K. (1999). maXus Computer Program for the Solution and Refinement of Crystal Structures. Bruker Nonius, The Netherlands, MacScience, Japan & The University of Glasgow.

7-Johnson, C. K. (1976). *ORTEP*--II. A Fortran Thermal—Ellipsoid Plot Program. Report ORNL5138. Oak Ridge National Laboratory, Oak Ridge, Tennessee, USA.

8-K. Jayaraman, A. Choudhury, C.N.R. Rao Solid, State Sciences 4 (2002) 413–422 Program for the Solution and Refinement of Crystal Structures. Bruker Nonius, The Netherlands, MacScience, Japan & The University of Glasgow.

RESONANT TRANSFER EXCITATION CROSS SECTIONS FOR PHOSPHORUS IONS WITH K-SHELL EXCITATION

H. HANAFY, G. OMAR* and F. SHAHIN

Phys. Dept., Faculty of Science, Beni-Suef University, Egypt
*Phys. Dept., Faculty of Science, Ain Shams University, Egypt

Abstract:

Resonant charge transfer and excitation (RTE) is an interesting process in ion-atom (I/A) collisions, which proceed through the formation of doubly-excited (d) states. If d-states stabilize by emission of x-rays, the sub-process is known as RTEX. This RTEX process is responsible for self cooling and ionization balance in thermal astrophysical and laboratory plasma. In fact, some intensive and extensive theoretical and experimental works have been done. However, many works still needed to understand the intricacies and the trends of this process for various isoelectronic and isonuclear sequences. The RTEX process in I/A collisions is identical to the dielectronic recombination (DR) in electron-ion (e/I) collisions. The DR and RTEX cross sections have been proved, Brandt (1983), that they are mathematically related.

In the present work, The RTEX cross sections are calculated for the collision of $P^{5+, 8+, 11+}$ ions with H_2/He targets, in case of K-shell excitation. It is found that, the peak values of σ^{RTEX} are 2.58×10^{-21} cm^2, 2.82×10^{-21} cm^2 and 3.02×10^{-21} cm^2 at projectile energies 119.10 MeV, 118.13 MeV and 108.74 MeV for $P^{5+}+ H_2$, $P^{8+}+ H_2$ and $P^{11+} + H_2$ collisions, respectively.

Key words: Isonuclear trends, Phosphorus Ions, RTEX cross sections

INTRODUCTION

In electron-ions collisions, a free electron may be captured to the matched ion excited states. Consequently, the bound electrons are resonantly excited due to the captured electrons causing a vacancy in the initial ground states. The doubly-excited (d) states, will be formed and then decay to fill the produced holes via radiative and/or Auger-electron emissions. Calculations of the distribution of atoms over their ionization stages and quantum states are one of the key steps in the modeling of astrophysical and laboratory plasmas. Spectral simulations of these plasmas must rely on a fast database of basic atomic and molecular cross sections and transition rates. One of the most important of these processes, and of the least accurate in the existing astrophysical database, is dielectronic recombination processes (DR). DR processes are widely studied in both theoretical [1-12] and experimental [13-14] sides. The bound-state wavefunctions and energies required in DR cross sections are generated using the single configuration Hartree-Fock (SCHF) program (Fischer's code) and the matrix program (Hahn's code).

The RTEX cross sections are carried out from DR cross sections by the method of folding [4]. The aim of this work is the study of isonuclear trends of RTEX cross sections (σ^{RTEX}) for P-ions at different charges in case of K- shell excitations.

THEORY

In electron-ion collisions, DR process may be represented as follows:

$$e + A^{q+} \xleftrightarrow{V_a} (A^{(q-1)+})^{**}$$

$$(i) \quad A_a \quad \Big|(d)$$

$$\xrightarrow{A_r} (A^{(q-1)+})^* + x - rays \qquad (1)$$

$$(f)$$

where: A^{q+} denotes the ionic target with degree of ionization (q+), and e is the incident continuum electron. V_a is the radiationless capture probability of an electron from the continuum state by the positive ion. A_a stands for the Auger decay probability for a d-state to emit an electron and stabilize back to the initial state (i). A_r is the probability of stabilization of a d-state to final state (f) by emission of X-rays.

The DR cross sections for K- shell excitation are calculated for each ion in the isolated resonance approximation (IRA) with the adapted angular momentum average (AMA) scheme [7]. The DR cross section is simplified to the form:

$$\sigma^{DR}(i \to d) = C_0 V_a(i \to d) w(d) \frac{1}{e_c \Delta e_c} \tag{2}$$

where $w(d)$ is the fluorescence yield, and $C_o = 4\pi^2\tau_o(a_o)^2 = 2.68 \times 10^{-32}$ cm^2 .sec.

The value of $V_a(i \to d)$ can be obtained from Auger decay rate to the ground (i)state, $A_a(d \to i)$ using the formula:

$$V_a(i \to d) = \frac{g_d}{2g_i} A_a(d \to i) \tag{3}$$

where g_d and g_i are the statistical weights of the d-state and i-state, respectively.
The fluorescence yield $w(d)$ is defined as:

$$w(d) = \frac{\Gamma_r(d)}{\Gamma_r(d) + \Gamma_a(d)} \tag{4}$$

Where, $\Gamma_r(d) = \sum_f A_r(d \to f)$ and $\Gamma_a(d) = \sum_{i,j} A_a(d \to i,j)$ refer to the radiative and Auger decay widths of a d-state.

In fact, there are many d-states which are allowed in the DR process. In IRA approximation, we can simply group them together in small energy bins, Δe_c, as:

$$\sigma^{DR}(i \to d, average) = \frac{1}{\Delta e_c} \int_a^b \sigma^{DR}(i \to d) de_c \tag{5}$$

where: $a = e_c - \Delta e_c/2$ and $b = e_c + \Delta e_c/2$

In addition, the RTEX process may be represented as:

$$A^{q+} + B \xleftarrow{V_a} (A^{(q-1)+})^{**} + B^+$$
$$(i) \quad A_a \Big| (d) \tag{6}$$
$$\xrightarrow{A_r} (A^{(q-1)+})^* + B^+ + x - rays$$
$$(f)$$

where: B and B$^+$ refer to the atomic or molecular target before and after the collisions

From schemes (1) and (6), it is clear that DR and RTEX cross sections are identical processes. Hence, the RTEX cross sections, σ^{RTEX} for the P$^{(5+,8+,11+)}$ ions can be obtained from the corresponding DR cross sections, σ^{DR}, using Brandt theory (1983). In such case, the DR cross sections are folded over the Compton profile of the momentum distribution of the electrons in molecular H$_2$ or the atomic He targets, Lee (1977). In impulse approximation (IMA), the total RTEX cross section, σ^{RTEX} (i;tot), may be written in terms of energy-averaged DR cross section, σ^{DR} (i;d) as follows:

$$\sigma^{RTEX}(i;tot) = \sum_d J(Q).\sigma^{DR}(i;d).(\Delta e_c / 2I_{ion}).(MI_{ion}/E)^{1/2} \qquad (7)$$

where $J(Q)$ is the Compton profile of the target with Q given by:

$$Q = 1/(2I_O)[e_c + E_t - (Em/M)][MI_O/E]^{1/2} \qquad (8)$$

where E is the projectile energy in the laboratory frame, e_c is the energy of the resonant-captured electron, E_t is the binding energy of the target, both in the rest frame of the projectile. M and m are the ionic projectile and the electron masses, respectively. I_O denotes the ionization potential energy of the hydrogen atom.

The contribution of the high Rydberg states (HRS) are estimated using the semi-empirical formula:

$$\sum_{n=n_c}^{\infty} \sigma^{DR}(n) = \frac{1}{2}n_c \left[1 + \frac{1}{n_c} + \frac{1}{2n_c^2}\right] \cdot \left[\frac{n_c - 1}{n_c}\right]^3 \sigma^{DR}(n_c - 1) \qquad (9)$$

Where, the calculations are usually carried out in details for d-states until A_a's and A_r's start to scale as $1/n^3$. At this point, we consider $n=n_c$ in equation (9) and estimate the contribution of the HRS [7].

RESULTS AND DISCUSSION

The DR cross sections for P-ions (P^{5+}, P^{8+}, P^{11+}) with K-shell excitations are computed for $\Delta n \neq 0$ transitions with energy bin of size $\Delta e_c = 1$Ry. To simplify the discussion of the DR cross sections for ions, we deal with each ion individually:

P^{5+}- ions:

In P^{5+}-ion the ground state is written as $1s^2 2s^2 2p^6$, while the intermediate state are divided into three groups $d_1 = 1s^1 2s^2 2p^6 3sn_l\ell_1$, $d_2 = 1s^1 2s^2 2p^6 3pn_l\ell_1$ and $d_3 = 1s^1 2s^2 2p^6 3dn_l\ell_1$ (n_l is the principle quantum number; $n_1 = 3,4,5....$ and ℓ_1 is the orbital quantum number; $\ell_1 = 0,1,2$ and 3 of the most outer electron). The second group (d_2) of the d-states has the dominant contribution to the total DR cross sections. The reason of this dominancy is the strong radiative rates for $2p \rightarrow 1s$ and $3p \rightarrow 1s$ transitions in the dipole approximation. The d-states with $2p^6 4\ell\, n_l\ell_1$ are found to have small contribution to the total σ^{DR} and σ^{RTEX}. Therefore, their cross sections may be neglected in future calculations in other complicated coupling schemes (e.g. LS-, JJ- coupling). However, the d_1, d_2, d_3 - states with $n_1 \geq 7$ are calculated from equation (9). The DR cross sections for P^{5+} ion are shown in figure (1a). The DR cross sections are represented by one sharp peak and some small peaks around 160 Ry. These small peaks are known as DR-satellite lines.

P^{8+}- ions:

In P^{8+}-ion the ground state takes the form $1s^2 2s^2 2p^3$, and the d- states are divided into three groups $d_1 = 1s^1 2s^2 2p^5$, $d_2 = 1s^1 2s^2 2p^4 n_l\ell_1$ and $d_3 = 1s^1 2s^2 2p^3 n_1\ell_1 n_2\ell_2$ (n_1 & $n_2 = 3,4,5....$ and ℓ_1 & $\ell_2 = 0,1,2$ and 3). The first and second groups (d_1&d_2) of the d-states have the dominant contribution to the total DR cross sections. This may be attributed to the strong radiative rates for $2p^5 \rightarrow 1s$ and $2p^4 \rightarrow 1s$ transitions and less subjected to cascade effect. It has to be noted that, the d-states of the form $d_1 = 1s^1 2s^2 2p^5$ contributed by 36% of the total DR cross section. Hence, σ^{DR} for this state must be calculated carefully in other coupling schemes. The two electrons in $2p^5$ are allowed to decay by Auger-electron emission. They are two equivalent electrons with the same n, . However, their radiative decay rate is very large. Thus, this d_1- group has large fluorescence yield, $w(d)$, and high σ^{DR}. The d_3-group of states has small contribution because they are subjected to strong many Auger decays and cascade decays. The DR cross sections for P^{8+} ion are shown in figure (1b). It has, beside small peaks, two pronounced peaks; one around 130 Ry and the other around 140 Ry. The two peaks are completely separated. HRS- states are included in the calculations.

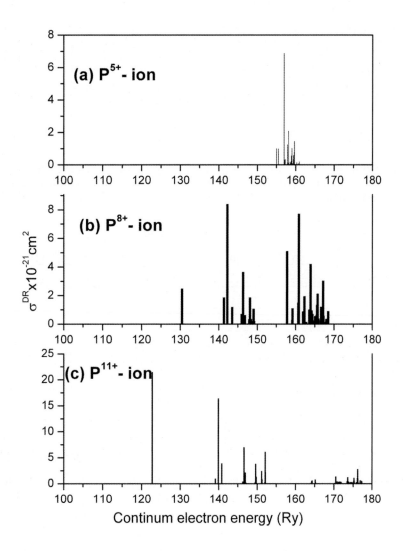

Figure (1): DR cross sections as a function of the continuum electron energy for P^{5+}, P^{8+} and P^{11+} ions with K- shell excitation.

P^{11+}- ions:

This ion is Be-like P^{11+}, with ground state $1s^22s^2$. The allowed d-states can be divided into three groups $d_1=1s^12s^22p^2$, $d_2= 1s^12s^22p^1n_1\ell_1$ and $d_3=1s^12s^2n_1\ell_1n_2\ell_2$ (n_1 & n_2 =3,4,5.... and ℓ_1 & ℓ_2= 0,1,2 and 3). Again, the groups (d_1&d_2) of the d-states have the dominant contribution to the total DR cross sections. In addition, the d_2 group has high contribution to total DR cross sections, σ^{DR}. The contribution of d_2 is $\sigma^{DR}= 32.69\times10^{-21}$

cm^2, which represent 55% from the total σ^{DR} for this ion. The group $\mathbf{d_2} = \mathbf{1s^1 2s^2 2p^1 np}$ alone has $\sigma^{DR} = 35.5 \times 10^{-21} cm^2$ (41% of the total σ^{DR}). Figure (1c) shows the DR cross sections for P^{11+} ion. In this figure, DR cross sections have two pronounced peaks at continuum-electron energies 120 Ry and 140 Ry. In addition, there are two considerable peaks at 145 Ry and 150 Ry.

In general for the $P^{(5+, 8+, 11+)}$ ions, we summarized the data for σ^{DR} of the three phosphorus ions in Table (1). This table shows the dominant groups of d-states for each ion. In addition, table (1) shows the variation of total σ^{DR} with the degree of ionization for each group. The results in table (1) clarify the trends of σ^{DR} with the effective charge Z_{eff} ($Z_{eff} = (q+z)/2$; (where: q is the degree of ionization of the P-ion, z represents the nuclear charge of ion). It is clear that, the total σ^{DR} increases with the increase in the effective charge, Z_{eff}.

Table (1): Summary of σ^{DR} (in cm^2) for $P^{(5+,8+,11+)}$ ions with K-shell excitation. The three groups (d_1, d_2 and d_3) of d-states were previously defined for each ion individually.

Ions Group	P^{5+} $\sigma^{DR} \times 10^{-21} cm^2$	P^{8+} $\sigma^{DR} \times 10^{-21} cm^2$	P^{11+} $\sigma^{DR} \times 10^{-21} cm^2$
Group one (d_1)	0.34	2.48	21.42
Group two (d_2)	1.55	20.37	47.89
Group three (d_3)	0.08	32.69	18.11
Total σ^{DR}	1.97	55.54	87.41

Isonuclear trends for P- ions:

The RTEX cross sections, σ^{RTEX} for P^{5+}, P^{8+} and P^{11+} ions with K-shell excitation are obtained by folding of the DR cross sections over the momentum distribution of the electrons in (H_2/He) targets[8]. The calculated σ^{RTEX} are carried out via the collision of P-ions with He and H_2 targets because these targets are more available experimentally. Figure (2) shows the RTEX cross sections, which are plotted as a function of the ionic projectile energy. These spectra, in figure 2, are individually integrated for each ion. Some atomic parameters such as: full width at half-maximum (FWHM), effective charge (Z_{eff}), peak centre, maximum values of the RTEX cross section σ_m^{RTEX}, and the integrated area are obtained. These data are presented in Table (2) and Table (3). In other words, Table (2) shows the integrated spectra of the σ^{RTEX}, when $P^{(5+, 8+, 11+)}$ ions collide with H_2 molecules. It is found that, full width at half-maximum (FWHM) increases as the degree of ionization increases. This indicates that the resonance lines increases. In addition, the FWAH increased monotonically as the effective charge increases. FWHM increases as $(Z_{eff})^{1.35}$ in case of P^{5+} and P^{8+}, while in case P^{11+}, it increases as $(Z_{eff})^{1.45}$. The peak energy (peak centre) decreases to small energy values as the degree of ionization increases. As the effective charge (Z_{eff}) increase the energy range are found more broader. In other words, the energy range of the spectra for P^{11+} is the widest comparing with the energy range of the spectra for P^{5+} and P^{8+}. The peak values of the RTEX cross section, σ_m^{RTEX}, is also increase as the effective charge increases. σ_m^{RTEX} in case of P^{8+} increase from P^{5+} by a factor of 1.52, while P^{11+} increase from P^{8+} by a factor of 2.14.

It is clear that, as the FWHM and the peak values of the RTEX cross section increase the total integrated area increase.

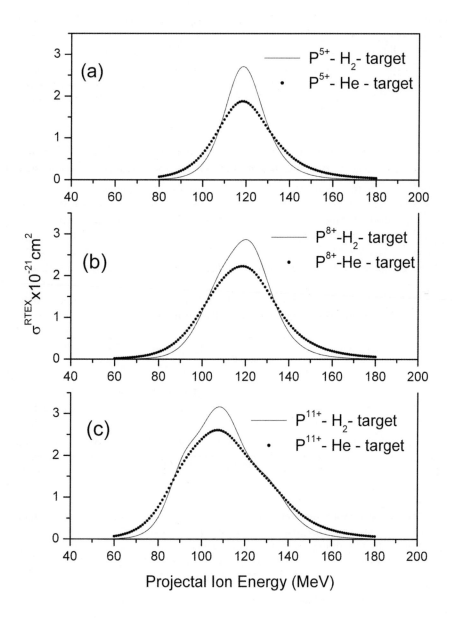

Figure (2): **Isonuclear trends of the RTEX cross sections for the collision of P- ions with H₂ (solid line) and He (dot point) targets. σ^{RTEX} (in 10^{-21} cm²) versus projectile energy (MeV). K-shell excitation is considered.**

Table (2): Summary of the integrated spectra of σ^{RTEX} for $P^{(5+,\ 8+,\ 11+)}$ ion, which are collided with H_2- target with K-shell excitation.

Atomic parameters P-ions	Z_{eff}	peak center	Intg. area of σ^{RTEX} $x10^{-21}$ cm^2	Max. value of σ^{RTEX} $x10^{-21}$ cm^2 (σ_m^{RTEX})	FWHM
P^{5+}	10	119.10	60.57	2.58	18.74
P^{8+}	11.5	118.13	92.07	2.82	26.04
P^{11+}	13	108.74	129.63	3.02	34.28

Table (3) shows the summary of RTEX cross sections for the collision of P^{q+} ions with atomic He target. The maximum values of σ^{RTEX} (σ_m^{RTEX}) for He target behave most likely as in the previous H_2 target. These σ^{RTEX} are also found to increase as the degree of ionization increases. However, the FWHM is wider for He target. This reflects the feature of the Compton profile for the momentum distribution of He- electrons. The integrated area in case of He target is approximately equal that for H_2 target. This means that the integrated area of σ^{RTEX} (total σ^{RTEX}) depends on the ionic projectile and independent of the kind of the target. The target works only as a source of electrons to be captured by the positive ions.

Table (3): Summary of the integrated spectra of σ^{RTEX} for $P^{(5+,\ 8+,\ 11+)}$ ion, which are collided with He- target with K-shell excitation.

Atomic parameters P-ions	Z_{eff}	peak center	Intg. area of σ^{RTEX} $x10^{-21}$ cm^2	Max. value of σ^{RTEX} $x10^{-21}$ cm^2 (σ_m^{RTEX})	FWHM
P^{5+}	10	119.15	54.92	1.73	25.38
P^{8+}	11.5	117.94	84.82	2.14	31.62
P^{11+}	13	109.11	119.69	2.49	38.40

In general, the three phosphorous ions have the same nucleus (z; core charge) and have different ionization charge (degree of ionization) q=5, 8 and 11. It is obvious that, as the degree of ionization (q) of the ions in the same isonuclear sequence increase; the RTEX cross section peak position is shifted towards lower values of the projectile energy range. In addition, the maximum value σ_m^{RTEX} for K-shell excitation increase with the increase in the degree of ionization. This behavior of σ^{RTEX} for K-shell excitation is different from that of σ^{RTEX} for L – shell excitation[10].

CONCLUSIONS

In this work, it is concluded that: RTEX cross sections (σ_m^{RTEX}) for K-shell excitation increase as the degree of ionization increases. It is also founded that, σ^{RTEX} is more brooder for all P^{q+} +He than that with H_2 target. This reflects the nature of the momentum distribution for electrons (Compton profile) in He and H_2 targets. The peak centre values of σ^{RTEX} are slightly shifted towards small projectile energies. σ^{RTEX} results are noted to peak at the same projectile energy for different targets (H_2/He) in the same Isonuclear sequence. This indicates that σ^{RTEX} depends only on the ionic projectile (P^{q+}) itself and independent of the type of the target.

REFERENCES

1- Badnell N. R., *Phys. Rev.* **A 42**, 209 (1990).
2- Hahn Y., *Phys. Rev.* **A 40**, 2950 (1989).
3- Omar G., and Hahn Y., *Phys. Rev.* **E 62**, 4096 (2000).
4- Brandt D., *Phys. Rev.* **A 27**, 1314 (1983).
5- Burgess A., *Astrophysics J.*, **139**, 776 (1964).
6- Gau J. and Hahn Y., *JQSRT*, **23**,121 (1980).
7- Hahn Y., *Adv. Atom. Molec. Phys.*, **21**,123, (1985).
8- Lee J. S., *J. Chem. Phys.*, **66**, 4906 (1977).
9- Omar G., et al., *Egypt. J. Phys.* **32**, 39 (2001).
10- Hanafy H., et al, *Amer. Inst. of Phys. (AIP)*, **748**, 118 (2005).
11- Hahn Y., and LaGattuta K., *Phys. Reports*, **166**, 195-268 (1988).
12- Tanis J.A, *Nucl. Instr. And Meth.*, **A262**, 52 (1987).
13- Bernstein E. M., et al, *Phys. Rev.* **A 40**, 4085 (1989).
14- Zong W., et al, *Phys. Rev.* **A 65**, 386, (1997).

I-3 CONTRIBUTING PAPERS

I-3 CONTRIBUTING FACTORS

INVESTIGATION OF THE DEVELOPED PRECIPITATES IN AlMgSiCu ALLOYS WITH AND WITHOUT EXCESS Si

E. F. ABO ZEID

Department of Physics, Assiut University, Assiut, 71516, Egypt

The effect of temperature on the sequence of hardening precipitates in Al-1.15 Mg2Si-0.34Cu (wt %) balanced containing Cu and Al-1.14Mg2Si-0.34Cu balanced containing Cu with Si in excess alloys has been investigated by hardness measurement (HV), differential Scanning calorimetry (DSC) and Transmission electron microscopy (TEM) techniques. The values of the hardness number which corresponding to the hardening precipitated particles in the alloy containing Si in excess are higher than that in the alloy without Si. The results showed that, the difference in the hardening precipitation peaks positions may match the fact that, the excess Si increases the density of $\beta``$ metastable phase and also, reduces the Mg/Si ratios in the early stage of GP zones and co-clusters formation. After the complete formation of the metastable needle shaped precipitates $\beta``$ is taken place the strengthening of the alloy would take place as a result of the formation of the semi-coherent rod shaped precipitates $\beta`$ and/or $Q`$ phases.

1. Introduction

Using aluminum as a metal of choice in the automotive industry will help society in its demands for increasing commitment to the environment. Al-alloys are increasingly attractive as a candidate for material substitution because of its strength and stiffness to weight ratio compared with traditional steels, rolled aluminum products for doors, closures or wings can achieve a weight reduction of up to 50%. This improves the vehicles fuel efficiency and it is also used to adjust the weight balance of the automobiles [1]. After the thermal treatments applied in the aluminum production factory, further processing takes place at the car factories, first, the Al-alloy should be easily deformable. After it has been painted, it undergoes the final heat treatment which is the bake hardening process at a temperature of 180 ºC. After this thermal treatment, the material should be hard and stiff for passenger protection [2]. In order to ensure that the material is easily deformable during stamping, hardly deformable after bake hardening and that the material properties do not degenerate during storage, the composition of the alloy and the sequence of thermal treatments need to be fine-tuned very carefully. A detailed knowledge of the precipitation sequences during all industrial processing steps is of great importance. Among several aluminum alloys that met these requirements Al-Mg-Si alloys with and without Cu stand out owing to a remarkable strengthening potential during paint bake cycle [3-8]. Aluminum sheet for body panels is thus predominantly made of the age hardenable Al-Mg-Si alloys. They are shipped and formed in T4 temper while still formable and are subsequently given cycle which increases strength by age hardening while curing the paint [3]. The precipitation sequence in Al-Mg-Si-Cu alloys has been reported and the crystal structures of quaternary Q and Q` phases have been identified to determine the effect of the addition of Cu in these alloys [9-11]. Addition of copper to Al-Mg-Si alloys enhances hardness and refines microstructure [12]. Numerous papers report precipitates of Q-, Q`- and B`-phases in Cu containing 6000 series aluminum alloys, and many discuss which precipitate contributes to mechanical properties of these alloys. The crystal structure of the Q`-phase in Al-Mg-Si-Cu alloys has been discussed in detail [13-14]. Matsuda et al. [15] reports the existence of the Q`-phase in Cu-containing Al-Mg-Si alloys and they conclude that the Q`-phase is a quaternary metastable phase distinct from the ternary Type-C precipitate found in the Al-Mg-Si alloys with excess Si. Also, they found that, the two phases have identical crystal lattice [15].

However, alloys with an excess of silicon (Mg: Si <1.73 wt %) contain, the $\beta``$ phase at peak hardness instead of a metastable form of the Q phase, as would be predicted by the equilibrium phase diagram [16]. Hence, peak hardness was attributed to the presence of $\beta`$ platelets lying on {001}$\alpha$ planes and $\beta``$ laths oriented along <100> α directions. Those alloys with an excess of silicon were found to exhibit the highest levels of

76

strengthening and absolute hardness [17]. To identify the effect of Si on the formation of these hardening precipitates, we compared between two Al alloys one balanced and the other is balanced with Si in excess by DSC, HV microhardness and TEM.

2. Experimental work

The chemical composition of the studied alloys is given in table (1). The DSC was used to follow the precipitation sequence that takes place in the studied alloys during the continuous heating at constant heating rates of the quenched specimens from solid solution state. Disc shaped DSC samples of 5mm diameter and 0.5mm thickness of average mass of ~26.5mg were machined from the alloy ingot. The specimens were solution heat treated for one hour at 803K in a standard convection furnace and then quenched into iced water (~273K). An annealed pure aluminum disc of similar shape and mass was used as a reference. Non–isothermal scanning techniques for the as-quenched specimens using a DSC thermal analyzer (DSC-DO 8T-12TG01, Shemadzu, Japan), at heating rates of 5, 7.5, 10, 15, 20, 30, 40 and 50 Kmin-1 were performed. DSC scans were started at room temperature and completed at 773K under a purified nitrogen gas atmosphere flowing at a rate of 30 ml min-1. The output was in mw and the net heat flow to the reference material (pure Al) relative to the sample was recorded as a function of temperature. The peak temperatures of the reaction processes were determined by using the microprocessor of the thermal analyzer.

Disc shaped specimens of about 15 mm diameter and 2 mm thickness were used for microhardness measurements, HV. Prior HV measurements the specimens were subjected to polishing process. The microhardness measurements were performed using Vickers method. Each HV value is obtained from the average of at least ten readings distributed over the whole surface of the specimens without paying attention to the positions. The error in HV arises from the measurements of the diagonal of impression, which leads to an error in HV of about ± 3%.

In order to confirm the results obtained by DSC and HV measurements, transmission electron microscopy (TEM) examinations were performed at Toyama University, Japan. For microstructural examinations, a 200 kV-TEM (TOPCON-002B) was used. It was operated at 120 kV

to avoid sample damage by the electron beam. Thin foils were prepared from the quenched and aged specimens. Thin discs of 3mm diameter were punched from the foils and then electropolished by the twin-jet technique using a solution of 25% nitric acid and 75% methanol cooled to 243 K. The electropolishing was performed using a DC current source operating at ~20V.

Table 1. Chemical composition (weight %) of the studied alloys.

Sample Type	Balanced	Exc. Si
Si	0.45	0.76
Fe	0.2	0.20
Cu	0.34	0.34
Mg	0.73	0.72
Mn	0.01	0.01
Zn	0.01	0.01
Cr	0.21	0.21
Mg2Si	1.15	1.14
Exc. Si	0.03	0.34
Al	Bal.	Bal.

3. Results and Discussion

Figure (1), presents the DSC thermograms obtained at a heating rate 20 K min^{-1} for the studied alloys immediately after solutionizing and quenching in ice and water. The DSC curve for the alloy with excess Si shows five exothermic peaks. The interpretation of the observed reaction peaks can be as follows:

Exothermic reaction peak (I): This reaction may be attributed to the formation of quenched-in vacancies and Si-Mg-vacancy clusters. The abundant concentration of the quenched-in vacancies at this stage of heating and the high binding energy of Mg-vacancy (17.34 kJ mol^{-1}) and Si-vacancy (27 kJ mol^{-1}) [18], results in the nucleation of Mg-Vacancy and Si-vacancy clusters which act as a nucleation sites for GP zones.

Exothermic reaction peak (II): This reaction which centered near 475 K may be ascribed to the formation of GP zones. The high resolution TEM and atomic probe field ion microscopy (APFIM) analysis of Dutta and Allen [19] and Edwards et al. [11], show that the clustering of Mg, Si and Mg-Si atoms and GP zones occur at the early stage of precipitation in Al-Mg-Si alloys. Therefore, this reaction peak II is probably related to atomic clustering and GP zones formation.

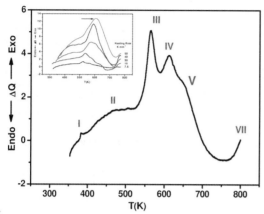

1.

Figure 1. Typical DSC scan at a heating rate of 20 K min^{-1} for Al-1.14wt% Mg$_2$Si-0.34wt%Cu with excess Si. The insert to the figure shows a change in a peak temperature with heating rate.

Exothermic reaction peak (III): is interpreted by the precipitation of β`` phase it is obvious that this reaction is relatively rapid. Therefore, this peak has a tendency to overlap with the next reaction peak as the heating rate increases. An increase in the density of β`` meta-stable precipitates obtained as a result of the excess Si, which would be the reason for the observed increasing height of this peak. Exothermic reaction peak (IV): After the precipitation of needle-shaped β``-phase took place, a transformation to the rod-shaped semi coherent β`and /or Q` precipitates have been developed.

A confirmation of this phase obtained from the TEM micrograph given in figure (2-a), For a specimen heated at the same rate 20 K min^{-1} to the corresponding peak temperature 610 K, kept there for 20 min and then cooled to room temperature. Careful inspection of the TEM micrograph show rod shaped precipitates characterizing the evolution of β` and /or Q` precipitates. Figure (2-b) presented the selected area diffraction pattern (SADP). The β` and/or Q` rod-shaped precipitates are aligned along the <100> direction of the Al lattice. The HRTEM investigation of the cross-sections of the developed rod-shaped precipitates revealed that the crystal structure of the β` and Q` is hexagonal with a=0.705 nm and c=0.405 nm [15].

Exothermic reaction peak (V): This shoulder refers to the precipitation of the non-coherent precipitates β (Mg$_2$Si) +Q lath shaped phase

2. (a)   (b)

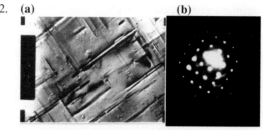

Figure 2. (a) TEM micrograph for a sample of alloy with composition Al-1.14wt% Mg$_2$Si-0.34wt%Cu with excess of Si heated at a rate of 20 K min^{-1} up to 610 K and kept there for 20 min. (b) selected area from the diffraction pattern.

The observed reaction processes for the balanced alloy can be determined as follows.

The Exothermic peak (II) can be due to the formation of Si-Mg-vacancy clusters followed by the nucleation of GP zones. The exothermic reaction peak (III) is interpreted by the precipitation of β`` phase [20]. It was observed that the exothermic peak I centered around 385 K in the DSC curve of the excess Si alloy is missing in the balanced alloy curve figure (3). The missing of the peak I results may be from that, the excess Si in the sample increases the density of Si-Mg-Vacancy clusters, and also, reduces the Mg/Si ratios in the early stage of GP zones and co-clusters formation [21]. Gupta et al. [22], suggested that the excess Si promotes the formation of large amount of Mg-Si clusters and zones and increases the β`` to β` ratio.

The exothermic reaction peak IV can be ascribed to the precipitation of rod shaped β` and /or Q` precipitates. According to figure (3), there is no endothermic reaction processes between peaks III and IV which indicates that the β`` precipitates were not dissolved, but more likely transformed to β` and /or Q`. It should be noted that both β` and Q` phases have hexagonal crystal structure and such transformation may not be predictable [23]. However, it can be explained by the fact that the habit plane of Q` has been determined to be {150} of the aluminum matrix [24]. The repeat distance along the <150> directions of the aluminum matrix is 1.03 nm which is about the same as the lattice parameter of the Q` phase [25]. Therefore, the precipitates tend to form as a lath so as to minimize the misfit in its surface and hence its energy. This may also, explain the lower height of thermal reaction for peak IV in the balanced alloy in comparison with the excess Si alloy. While, in the alloy with excess Si this reaction peak is due to the formation of both β` and /or Q` precipitates, but in the case of the balanced alloy is only due to the transformation of β` to Q`. After the transition from β` to Q` took place the precipitation of β (Mg₂Si) phase take place with the increase of temperature which indicated by the shoulder V in Figure (3).

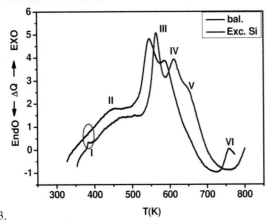

3.

Figure 3. Typical DSC scans at a heating rate of 20 K min⁻¹ for Al-1.14wt% Mg₂Si-0.34wt%Cu with excess Si alloy and Al-1.15wt% Mg₂Si-0.34wt%Cu balanced alloy.

3.1. Microhardness Results

3.1.1 Behavior of Vickers hardness during isochronal annealing of the studied alloys

An isochronal annealing regime of previously solution heat treated specimens has been suggested to follow the decomposition behaviour of the supersaturated alloys as a function of temperature. Normally the specimens were aged for 30 min at selected temperatures above that of the room temperature separated by 20 K intervals up to 803 K and subsequently quenched into chilled water at ~273 K. After each quench the microhardness measurements have been performed at room temperature. Figure (4-a, b) Shows the variation of HV as a function of the aging temperature of the balanced and Excess Si alloys. It is clear that the general behaviour of HV is characterized by five reaction peaks labeled I through V.

The first precipitation peak I, may be ascribed to, the nucleation of quenched-in vacancies and solute (Si and Mg) atoms as was concluded from the DSC thermograms. This hardening precipitation peak start to appear only in the excess Si at 350 K, figure (4-a) and disappeared in the balanced one figure (4- b). This

difference in the behaviour of the two studied alloys at this early stage of aging may be attributed to that, the excess Si modifying the Mg/Si ratio in the clusters/zones [21]. The second hardening precipitation peak II, which appeared almostly at the same temperature around ~ 390 K can be ascribed to the formation of Si-Mg-vacancy clusters and nucleation of GP zones. The third hardening precipitation peak III, takes place at 495 K for the excess Si alloy figure (4-a) and at 475 K for the balanced alloy figure (4-b), which may be ascribed to the precipitation of coherent β`` precipitates. As a result of the coherency of these precipitates, strengthening of the alloy should take place. The behviour consistency of both results, HV and DSC of the studied alloy is evident. Where as, the inconsistency of the precipitates temperatures in some cases can be explained by the lack of thermal equilibrium during the DSC scans. It was observed from figure (4-a, b) that, the values of the hardness number which corresponding to the hardening precipitated particles in the excess Si alloy are higher than that in the balanced alloy. Also, the difference in the hardening precipitation peaks positions may match the fact that, the excess Si increases the density of the β`` metastable phase and also, reduces the Mg/Si ratio in the early stage of GP zones and co-clusters formation [25]. Gupta et al. [21], suggested that the excess Si promotes the formation of large amount of Mg-Si clusters and zones and increases the β`` to β` ratio. After the complete formation of the metastable needle shaped precipitates β``, the strengthening of the alloy would take place at the forth peak IV, as a result of the formation of the semi-coherent rod shaped precipitates β` and/or Q` phases. This result consists with that of the DSC scan of these samples. Moreover, this result matches the suggestion of Miao and Laughlin [26] that, lath shaped precipitates would appear if Cu is added to the Al-Mg-Si alloys. Both coherent and semi-coherent precipitates have a positive contribution to the strengthening of the alloy [2]. Where, β`` has a monoclinic structure with composition Mg₅Si₆ that was found by means of quantitative electron diffraction [27]. After the growth and completion of β` and /or Q` phases at certain temperatures the equilibrium phases β (Mg₂Si) and Si particles for excess Si alloy and β (Mg₂Si) +Q phases for the balanced alloy have taken place. A steep decrease in the microhardnees was detected up to 700 K for the two studied alloys. This decrease ascribed to the dissociation of the all precipitated particles. Another increase in the microhardness was found with increasing the temperature up to 790 K. the abundant concentration of quenched-in vacancies in this range of temperature and the formation of vacancy clusters explains this feature. These vacancy clusters have a limited contribution to the microhardness. This result implies that the material should return to the quenched state. As a comparative result it was found from figure (4-a, b) that the microhardness values of the alloy with excess of Si is higher than that of balanced alloy in this range of temperature. This result may be attributed to that the addition effect of Cu in these two studied alloys was different, where, Cu is believed to contribute to form the Q` phase in addition to β` in the matrix. As aging proceeds, Cu atoms are attracted to the region of lattice mismatch at the Q`/α-Al interface [12]. Cu segregation probably limits the diffusional growth of Q`-phase and hence produces finer microstructure than that found in an alloy without Cu [12]. Since, in our studied alloys the Cu concentration is identical, then the net effect in the sample microhardness results comes from the addition of excess Si. The excess Si enhances the precipitation kinetics and improves the strength of the material. This result is in agreement with that proposed by Matsuda et al. [12]. Also, the increase of the absolute values of the peak hardness of these alloys which is observed in figure (4) may be attributed to the existence of Cr and Fe atoms. It can be suggested that the addition of Cu is useful to improve the hardness of these alloys [20]. It is also observed that the density of Si precipitates decreased as a result of the addition of Cr and Fe due to the formation of AlCrSi or AlFeCrSi dispersoids. The formation of the dispersoids reduces the amount of solute Si in the matrix, and the interface between dispersoids and the matrix act as sinks for super-saturated quenched-in vacancies [20]. It is observed that the precipitation peaks had been shifted towards the lower temperatures in the alloy with Si in excess. This means that, the excess Si enhances the precipitation processes.

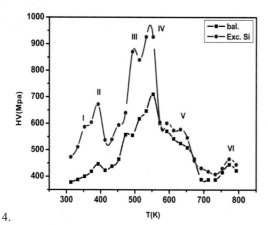

4.

Figure 4. HV vs. aging temperature for; (a) Al-1.14wt% Mg₂Si-0.34wt%Cu with excess Si alloy; (b) Al-1.15wt% Mg₂Si-0.34wt%Cu balanced alloy.

4. Conclusions

From the obtained results of the DSC, TEM and HV measurements of the studied alloys, the following conclusions can be drawn out:

1. Strengthening of the alloy depends on the precipitation of coherent and semi-coherent precipitates. After the complete formation of the metastable needle shaped precipitates β``, the strengthening of the alloy would take place.

2. The values of the mocrohardness which is corresponding to the hardening precipitated particles in the excess Si alloy are higher than that in the balanced alloy.

3. The excess Si enhances the precipitation processes enhances and also promotes the formation of large amount of Mg-Si clusters and zones and increases the β`` to β` ratio.

Acknowledgments

The author would like to thank Prof. Dr. K. Matsuda, Department of materials science, Toyama University, Japan for his kind donation of the samples. Prof. Dr. A. Gaber, Physics Dept. Faculty of Science, Assiut University is also acknowledged for his encouragement throughout this work.

References

1. K. Fukui, M. Takeda and T. Endo: Mater. Lett., **59**, 1444 (2005).

2. M. A. V. Huis, J. H. Chen, M. H. F. Sluiter and H. W. Zandbergen: Acta Mater., **55**, 2183 (2007).

3. Y. Birol: Mater. Sci. Eng., **391A**, 175 (2005).

4. T. Moons, P. Ratchev, P. D. Smet, B. Verlinden and P.V. Houtte: Scr. Mater., **35**, 939 (1996).

5. G. B. Burger, A. K. Gupta, P. W. Jeffrey and D. J. Lloyd: Mater. Charac., **35**, 23 (1995).

6. A. Perovic, D. D. Perovic, G. C. Weathlery and D. J. Lloyd: Scr. Mater., **41**, 703(1999).

7. K. Raviprasad, C. R. Hutchinson, T. Sakurai and S. P. Ringer: Acta Mater., **51**, 5037(2003).

8. S. C. Wang, M. J. Starink and N. Gao: Scr. Mater., **54**, 287 (2006).

9. A. Gaber, M. A. Gaffar, M. S. Mostafa and E. F. Abo Zeid: Mater. Sci. Tech., **22**, 1483(2006).

10. D. J. Chakrabarti, B. Cheong and D. E. Laughlin: Automotive alloys II. Warrendale, PA: TMS, 27 (1998).

11. G. A. Edwards, K. Stiller, G. L. Dunlop and M. J. Couper: Acta Mater., **46**, 3893 (1998).

12. K. Matsuda, D. Teguri, Y. Utani, T. Sato and S. Ikeno: Scr. Mater., **47**, 833 (2002).
13. C. Cayron and P. A. Buffat: Acta Mater., **48**, 2639 (2000).
14. C. Wolverton: Acta Mater., **49**, 3129(2001).
15. K. Matsuda, Y. Uetani, T. Sato and S. Ikeno: Metall. Mater. Trans., **A 32**, 1293 (2001).
16. D. Kent, G. B. Schatter and J. Drennan: Mater. Sci. Eng., **405A**, 65(2005).
17. D. G. Eskin: J. Mater. Sci., **38**, 279(2003).
18. S. Esmaeili and D. J. Lloyd: Mater. Charac. **55**, 307(2004).
19. I. Dutta and S. M. Allen: J. Mater. Sci. Lett., **10**, 323(1991).
20. A. Gaber, M. A. Gaffar, M. S. Mostafa and E. F. Abo Zeid: J. Alloys and Compounds, **429**, 167(2007).
21. A. K. Gupta, D. J. Lloyd and S. A. Court: Mater. Sci. Engin., **A316**, 11(2001).
22. S. D. Dumolt, D. E. Laughlin and J. C. Williams: Scr. Metall., **18**, 1347(1984).
23. D. J. Chakrabarti and D. E. Laughlin: Progress in Materials Science, **49**, 389(2004).
24. K. Matsuda, T. Noi, K. Fujii, T. Sato, A. Kamio and S. Ikeno: Mater. Sci. Eng., **A262,** 232(1999).
25. C. D. Mariora, S. J. Andersen, J. Jansen and H. W. Zandbergen: Acta Meter., **49**, 321(2001).
26. W. F. Miao and D. E. Laughlin: Metall. Mater. Trans., **31A,** 361(2000).
27. K. Matsuda, S. Taniguchi, K. Kido, Y. Uetani and S. Ikeno: Mater. Trans., **43,** 2789(2002).

EVIDENCE OF JAHN-TELLER DISTORTION and PHASE SEPARATION IN Ca DOPED LaMnO$_3$

M. A. AHMED and S. I. EL-DEK

Materials Science Lab. (1), Physics Dept., Faculty of Science, Cairo University, Giza, Egypt

Abstract:

A series of Ca doped LaMnO$_3$ (La$_{1-x}$Ca$_x$MnO$_3$; 0.10≤x≤0.50) was prepared using conventional solid-state reaction. IR spectroscopic analysis was carried out for all samples. All investigated samples were ferromagnetic with the Curie temperature increasing with Ca content. The experimentally calculated values of the effective magnetic moment agree well with those computed theoretically. The largest value of the magnetic susceptibility as well as magnetic moment was achieved at x=0.30 pointing to a typical ferromagnetic character. The hysteresis which appeared in the magnetization during heating and cooling runs for the sample with x=0.30 enhances the choice of the sample x=0.30 as the optimum Ca content.

Keywords: Perovskites- La manganite – IR- magnetic susceptibility.

Introduction:

Perovskite rare earth manganites doped with bivalent or monovalent cations R$_{1-x}$A$_x$MnO$_{3+\delta}$ (where R is rare earth cation and A is doping cation) were intensively studied over the last decade as materials possessing colossal magnetoresistance. Many attempts were done to find the doping ratio providing the highest sensitivity of the electrical resistivity to the magnetic field at room temperature which is the most important challenge for the applications of manganites as magnetic field sensors or movement sensors. Unfortunately, this sensitivity drops rapidly with the increase in the Curie temperature of the manganite materials doped with divalent cations like Ca^{2+}, Sr^{2+} or Pb^{2+}. But for A=Na$^+$ the higher sensitivity was found at room temperature [1–6].

The cubic colossal magnetoresistive manganites present a very rich phase diagram, with a variety of antiferromagnetic, canted and ferromagnetic spin structures, as well as insulating charge ordered and metallic states [7]. A characteristic feature is the coexistence of different phases, caused by the tendency of charge carriers to cluster in nanodomains of the ferromagnetic metallic type, due to the presence of many competing interactions (phonon couplings, effective bandwidths, magnetic energies and Coulomb repulsions) of comparable strength [8, 9]. NMR [10-12], muon experiments and neutron scattering [13-15] provide complementary views, which can identify these inhomogeneous states.

The end member, LaMnO$_3$, is an orthorhombic (Pbnm) layer antiferromagnet with a tiny canting, originated by a Dzyaloshinski-Moriya antisymmetric interaction. The layer arrangement of the Mn^{3+} Jahn-Teller ions, each carrying 4 μ_B, consists of a stacking of ferromagnetic planes, antiferromagnetically coupled along the c-axis. The spin lies almost along the b-axis, with a minute component along c. The saturation field for the weak ferromagnetic component is very small: a few 100 Oe already provides a single weak ferromagnetic domain. Ca substitution La$_{1-x}$Ca$_x$MnO$_3$ introduces xMn^{4+} ions per formula unit, each carrying 3μ_B. Reminding that at high doping metallic ferromagnetism sets in, both x=0.08 and 0.15 samples are insulating, although the latter already displays an almost full ferromagnetic moment.

82

In the present study, we have focused our attention to the unsolved questions on the magnetic characterization and Jahn-Teller distortion of the manganites $La_{1-x}Ca_xMnO_3$. The answers of these questions will help in improving the physical properties those related to their applications.

Experimental Techniques:

The samples of the general formula $La_{1-x}Ca_xMnO_3$ $0.10 \leq x \leq 0.50$; were prepared by the double sintering ceramic technique from oxides of analar grade form (BDH), La_2O_3, CaO and MnO_2. Stoichiometric ratios were good mixed, grounded using agate mortar for three hours and transferred to electric ball mill for another three hours. The samples were pressed into pellets form using uniaxial press of pressure 1.9×10^8 N/m^2. The pellets were presintered in air at $950^{\circ}C$ for 15 hours with a heating rate of $1^{\circ}C/min$. and the pellets were grounded again, pressed into pellets and finally sintered at $1250^{\circ}C$. Regrinding was carried out again for the third time, then the powder was sieved and pressed into discs of diameter 1cm and thickness of $\cong 1.5$ mm, fired at $1250^{\circ}C$ for another 4 hours in air with the same above rate using Lenton furnace type (UAF 16/5, UK). X-ray diffraction analysis was carried out using Philips Diffractometer model "PW3710" with a Cu target of wave length ($\lambda=1.54060$ $^{\circ}A$) to assure the complete solid-state reaction and single phase formation. IR absorption spectra were measured using a Fourier transform spectrometer (Jasco model 470) in the spectral range (4000-400cm^{-1}) and FTIR-FT (Nicolet U.S.A),was used for measuring the IR band in the range (400-200 cm^{-1}) using KBr. The dc magnetic susceptibility measurements were carried out using Faraday's method in which a very small amount of the powdered sample was inserted in a cylindrical glass tube at the point of maximum gradient. The measurements were performed from 78 K - 350K

at three different magnetic field intensities. The temperature was measured using a temperature controller of the type ITC 503 (Oxford) with an accuracy 0.01K from 77-200K and 0.1 for T>200K.

Results and Discussion:

a. IR Spectroscopic Analysis:

In the manganites of compositions $La_{1-x}Ca_xMnO_3$, Jahn-Teller distortion due to Mn^{3+} ions plays an important role [16-18] in discussing their physical properties. More informations of this effect can be available through light transmission experiments [19-22]. The crystal structure of the rare earth manganites is known to be of the distorted $GdFeO_3$ type structure that a central Mn atom is octahedrally surrounded by its nearest neighbor six O anions. The near ideal MnO_6 octahedra has the symmetry of the point group, which has six vibrating modes. Wang et al. [23] found that the band around 600 cm^{-1} corresponds to the stretching mode (υ_s) involves the internal motion of a change in length of the Mn-O bond. They also found that, the band around 400 cm^{-1} corresponds to the bending mode (υ_b), which is sensitive to any change in the Mn-O-Mn bond angle Fig. (1: a, b). These two bands are related to the environment surrounding the MnO_6 octahedra.

The infrared spectra for the investigated manganite $La_{1-x}Ca_xMnO_3$; $0.10 \leq x \leq 0.50$ are shown in Fig.(2:a, b). As it was reported [23], the band at 600 and 400 cm^{-1} in the compounds under study exist at nearly the same positions. Another two high frequency bands; υ_1 at ≈ 970 cm^{-1} and υ_2 at ≈ 927 cm^{-1} were appeared, table (1). The first band υ_1 may be due to the La-O stretching vibration in the <La-O> polyhedra as recently reported[24].

Table (1): The bands positions in the infrared absorption spectra of $La_{1-x}Ca_xMnO_3$ $0.10 \leq x \leq 0.50$ recorded in the range (1100-400 cm^{-1}).

x	υ_1	υ_2	s	υ_3	υ_4	υ_5
0.10	970.01	927.59	875	607.43	507.5-s	400.51
0.20	970.98	928.55	875	600.71	510.52-s	400.92
0.25	970.02	929.50	875	602.57	500.52-s	398.58
0.30	970.02	927.60	875	597.82	510.51-s	397.59
0.50	970.08	928.01	875	616.14	501.40	403.25

s : shoulder

The second band (υ_2) at about 927 cm^{-1} is accompanied by a shoulder at about 875 cm^{-1}. The band and shoulder can be ascribed to the redistribution of the La-O bond lengths due to the substitution of La^{3+} of mass number (138.91) with Ca^{2+} (40.08). The band (υ_3) appeared at about 600 cm^{-1} is ascribed to the stretching vibration of the Mn-O bond.

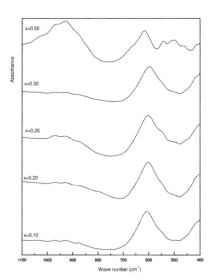

Fig. (2.a) IR absorption spectra for the samples $La_{1-x}Ca_xMnO_3$, $0.10 \leq x \leq 0.50$.

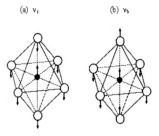

Fig. (1: a, b) Schematic infrared-active vibrations [23] of a MnO$_6$ octahedron. Open balls indicate surrounding oxygen ions and closed ball Mn ion. ν_s: Higher frequency stretching mode; ν_b: lower frequency bending mode.

The small variation in the position of this band indicates that, the predominant mode of vibration is out of plane (orthorhombic) as the result of Jahn-Teller (JT) distortion.

The band (υ_3) was shifted slightly; table (1); to lower wave number with increasing Ca content except at x=0.50, this band (υ_3) was shifted to higher wave number as a consequence of the small tilting of the MnO$_6$ octahedron. The tolerance factor is a simple characterization of the size mismatch that occurs when the A-site ions are too small to fill the space in the three dimensional network of MnO$_6$ octahedra. For a perfect size match (t=1), the Mn-O-Mn bond angle would be 180 $^\circ$. For t<1, rather than a simple contraction of bond distances, the octahedra tilt and rotate to reduce the excess space around the A site, resulting in θ< 180° [25-33]. The frequency of vibration is influenced by the change of the bond lengths and angles of Mn-O-Mn and thus the stretching mode related to the atoms vibrating periodically along the chemical bond connecting Mn and O will be decreased. The shoulder appeared at $\approx$500 cm^{-1} for the samples with x$\leq$0.3 can be discussed as the change induced by the Ca^{2+} substitution as the result of the generation of a small amount of Mn^{4+} which is necessary for the charge neutralization. This shoulder becomes a band by increasing Ca content for the samples with x>0.30 and the number of Mn^{4+} ions increase thereby changing the Mn-O-Mn bond lengths.

84

The band appeared at about 400 cm^{-1} corresponds to the bending mode of vibration which is sensitive to the change in the Mn-O-Mn bond angles. The distortion of the MnO$_6$ octahedra is commonly designated as Jahn-Teller distortion. The appearance of bending and stretching modes of vibrations is an evidence for the existence of JT distortion. There is no remarkable shift in υ_b that reflects an internal motion of the Mn^{3+} and O^{2-} ions against the other O^{2-} ions in a plane perpendicular to their direction and it indicates that no noticeable change in the Mn-O-Mn bond angle is observed. Two small bands were observed at ≈350 and at ≈325 cm^{-1} and could be assigned to the bending mode of vibration of Ca-O bond and to the lattice vibrations respectively.

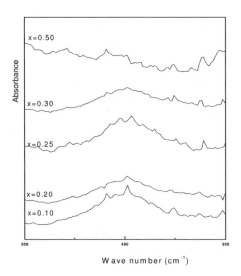

Fig. (2.b) IR absorption spectra for the samples La$_{1-x}$Ca$_x$MnO$_3$, 0.10≤x≤0.50.

b. Magnetic Susceptibility:

Figure (3:a-f) illustrates the dependence of the molar magnetic susceptibility (χ_M) on the absolute temperature; for La$_{1-x}$Ca$_x$MnO$_3$ samples; 0.10≤x≤0.50; as a function of the magnetic field intensity. The figure clarifies that for x=0.10, χ_M increases slight with decreasing

temperature down to about 200K. After that, it increases rapidly reaching a maximum value (peak) at ≈135K. The peak position shifts towards high temperature with increasing the magnetic field intensity because the saturation of the spins takes place at higher fields. Further decreasing temperature, χ_M increases again, more than seven times of its value at room temperature. Also, it was observed that the value of χ_M decreases with increasing the magnetic field intensity for all samples because the field tries to orient more spins in its direction where the magnetization reaches its saturation value. A closer look to the data clarifies that, the ferromagnetic character is the predominant behavior in samples 0.10≤x≤0.50 with a small peak appeared at x=0.10 and 0.20 which is transformed into a hump at x=0.25. The positive values of the Curie-Weiss constant (θ) reported in table (2) enhances the above mentioned statements. Papavassiliou et al. [33] observed mixed phase tendency in La$_{1-x}$Ca$_x$MnO$_3$ using Mn^{55} NMR technique. The appearance of coexisting peaks at x=0.10, 0.25 and 0.50 was clear in their study and these peaks correspond to either ferromagnetic (FM) metal, (FM) insulator or antiferromagnetic (AF) states. Phase separation in x=0.50 polycrystalline samples obtained under different thermal treatments was also reported by several authors [34-41]. At x=0.30; Fig. (3.d); the value of χ_M is large as compared with that at lower Ca content. In addition, the sample of x=0.30 behaves as a typical ferromagnet with T$_C$≅200K below which χ_M increases reaching maximum value that remains stable down to ≅77K. Increasing Ca content to x=0.35 increases T$_C$ to about 230K keeping a clear ferromagnetism and almost ideal with χ_M value lesser than that at x=0.30. Moreover, it is significant to note that the increase of χ_M below T$_C$ is gradual at x=0.35, while it is sharper at x=0.30. At x=0.50, T$_C$ shifted to higher values ≅

250K. At this concentration, the number of Mn^{3+} ions is expected to be nearly equal to that of Mn^{4+} ions. Accordingly, more than one interaction exist: i. The superexchange interaction, Mn^{3+}-O-$Mn^{3+}(J_{33})$, ii. The

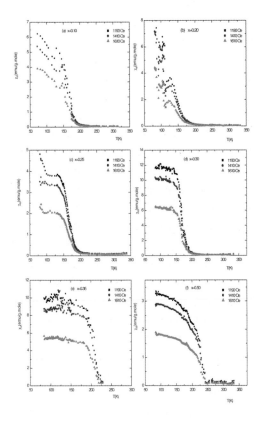

Fig. (3:a-f): The dependence of the molar magnetic susceptibility (χ_M) on the absolute temperature as a function of the magnetic field intensity for $La_{1-x}Ca_xMnO_3$ samples; $0.10 \leq x \leq 0.50$

antiferromagnetic superexchange interaction Mn^{4+}-O-Mn^{4+} (J_{44}), and iii. The ferromagnetic double exchange interaction Mn^{3+}-O-$Mn^{4+}(J_{34})$. As the interaction value J_{34} becomes strong, ferromagnetic ordered structures with an increased number of (Mn^{3+}, Mn^{4+}) neighbors is observed below the Curie temperature. From the shape of χ_M vs T of this sample, it is instantly recognized that the resultant of the two

antiferromagnetic interactions J_{33} and J_{44} was masked by the ferromagnetic J_{34} interaction. In other words, the two super exchange interactions (J_{33} and J_{44}) in cooperation with each other affect on the spin order within the layers (in the ac plane) keeping the sample ferromagnetic, i.e. a ferromagnetic coupling between the antiferromagnetic layers (along the b-axis). Consequently, the high positive value of J_{34} overcomes the sum of the small values of both J_{33} and J_{44}.

The calculated values of the effective magnetic moment table (2) shows that the Bohr magneton per Ca content is small in the 1st region up to x=0.30 (maximum magnetic moment) after that the value of μ_{eff} decreases with increasing Ca content. The theoretical magnetic moment is calculated from the assumption of charge balance to give $La_{1-x}Ca_xMn^{3+}_{1-x}Mn^{4+}_xO_3$ [42-44]. This was an obvious result, from which the Ca content of x=0.30 can be considered as the optimum concentration because it gives the highest magnetic moment and largest values of magnetic susceptibility in addition to its typical ferromagnetic character.

Table (2): Values of the magnetic constants as calculated from χ_M^{-1} vs T plots at 1610 Oe.

x	χ_M (emu/g.mole) at 100K	μ_{eff} (B.M.) Exp.	μ_{eff} (B.M.) Theor.	C (emu/g.mole)K	θ (K)	T_C (K)
0.10	3.47	4.94	4.72	3.03	163	173
0.20	2.89	4.71	4.62	2.76	171	181
0.25	2.00	4.56	4.58	2.60	175	193
0.30	6.42	5.62	4.54	3.94	210	200
0.35	5.34	5.50	4.49	3.77	225	220
0.50	1.80	5.26	4.34	3.46	248	245

The magnetic properties of the manganites [31] are governed by the exchange interactions between the spins of Mn ions. These interactions are relatively large between two Mn spins separated by an oxygen atom and are controlled by the overlap between the Mn d-orbitals and the O p-orbitals. The corresponding superexchange interactions depend on the orbital configuration that follows the rules of Goodenough [45]. Generally, for Mn^{4+}–O–Mn^{4+}, the interaction is AF, whereas for Mn^{3+}–O–Mn^{3+} it may be ferro- or AF [31, 45], such as in $LaMnO_3$ where both F and AF interactions coexist. Considering the Mn^{4+}–Mn^{4+} AF superexchange and Mn^{3+}–Mn^{4+} F DE interaction to be of same magnitude, the mean field approximation leads to $T_C \approx 2x(1 - x) - x^2$, which is maximum for $x = 1/3$. The fact that the FM phases are generally found around $x = 1/3$ in manganites is in agreement with this crude model [31, 45].

Figure (4) shows the hysteresis effect in the sample of $x=0.30$ in the temperature range from 170-220K which enhances its erromagnetic behavior with a first order phase transition. From the data reported in table (2) it is observed that, increasing the Ca content in the samples increases the Curie temperature as well as the calculated values of the Curie Weiss constant (θ). Also, it was found that θ is slightly lower than T_C for the samples with $0.10 \leq x \leq 0.25$, while for $x=0.30$ and 0.35 θ is higher than T_C as it is expected for a ferromagnetic material. The fact that $\theta < T_C$ indicates the existence of weak antiferromagnetic interactions [46] in the samples resulting from J_{33} for $0.10 \leq x \leq 0.25$ and J_{44} for $x \geq 0.50$. Consequently, one may expect a small canted ferromagnetic order in the compositions $0.10 \leq x \leq 0.25$.

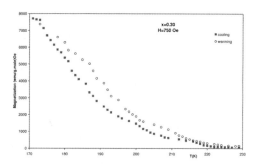

Fig (4) Effect of heating and cooling runs on the magnetization of the sample $La_{1-x}Ca_xMnO_3$; $x=0.30$

The values of the observed moment (exp.) are slightly higher than those calculated theoretically due to the contribution of the double exchange interaction (J_{34}) which enlarges the ferromagnetic region on the expense of the paramagnetic one. Since the Mn^{3+} ion ($3d^4$) [47, 48] is present in an octahedral crystal field and the pairing energy is larger than the crystal field splitting, then it favors the high spin configuration ($t_{2g}^3 e_g$) consequently, this agrees well with the small differences between both experimental and theoretical values of the magnetic moment.

References:

1. W.H. McCarroll, I.D. Fawcett, M. Greenblatt, K.V. Ramanujachary, J. Solid State Chem. **146**, 88, (1999).

2. S. Nakamura, K. Nanba, S. Iida, J. Mag.Mag. Mater. **177–181**, 884 (1998).

3. M. Sahana, R.N. Singh, C. Shivakumara, N.Y. Vasanthacharya, M.S.Hegde, S.Subramanian, V. Prasad, S.V. Subramanyam, Appl. Phys. Lett. **70**, 2909, (1997).

4. G.H. Rao, J.R. Sun, K. Baerner, N. Hamad, J. Phys.: Condens. Matter **11**, 1523 (1997).

5. S.L. Ye, W.H. Song, J.M. Dai, K.Y. Wang, S.G. Wang, J.J. Du, Y.P.Sun, J. Fang, J.L.Chen, B.J. Gao, J. Appl. Phys. **90**, 2943 (2001).

6. O.Yu. Gorbenko, O.V. Melnikov, A.R. Kaul, A.M. Balagurov, S.N. Bushmeleva, L.I. Koroleva, R.V. Demin, Materials Science and Engineering B **116**, 64–70, (2005).

7. A. Millis, Nature **392** , 147, (1998).

8. V.J. Emery, S.A. Kivelson, Physica C **209**, 597 (1993).

9. S. Yunoki, A. Moreo, E. Dagotto, Phys. Rev. Lett. **81**, 5612 (1998).

10. G. Allodi et al., Phys. Rev. Lett. **81**, 4736 (1998).

11. G. Allodi et al., Phys. Rev. B **57**, 1024, (1998).

12. M.M. Savosta et al., Phys. Rev. Lett. **79**, 4278, (1997).

13. M. Hennion et al., Phys. Rev. Lett. **81**, 1957, (1998).

14. J. De Teresa et al., Nature **386**, 256, (1997).

15. F. Moussa, Phys. Rev. B **54** , 15149, (1996).

16. A.J.Millis, B.P. Littlewood, B.I.Shraiman, Physical Review Letters **74**, 5144, (1995).

17. H.Roeder, J.Z.Hang, A.R.Bishop, Physical Review Letters **76**, 1356 (1996).

18. J.B.Goodenough, Annual Review Materials Science **28**, 1, (1998).

19. K.H.Kim, J.H.Jung, T.W.Noh, Physical Review Letters **81**, 1517 (1998).

20. A.Arulraj, C.N.Rao, J.Solid State Chemistry, **145**, 557 (1995).

21. C.Roy, R.C.Budhani, J.Applied Physics. **85**, 3124, (1999).

22. P.Calvani, G.De Marzi, P.Dore, S.Lupi, P.Maselli, F.D, Amore, S.Gagliardi, S.W.Cheong, Physical Review Letters, **81**, 4504, (1998).

23. Xin Wang, Qiliang Cui, Yuewu Pan and Guangtian Zou, J.Alloys and Compounds, **354**, 91, (2003).

24. M.A.Ahmed, R.Seoudi and S.I.El-Dek, J. of Molecular Structure, **754,** 1-3, 41-44, (2005).

25. J. P.Chapman, J.Paul Attfield, Lide M.Rodriguez Martinez, Luis Lezama and Teofilo Rojo, Dalton Transactions, **19**, 3026, (2004).

26. Z. Shengming, Z. Hong, Shilei, Z.Zongyan, Z. Guien, Z. Yuhery, J.Phys: Condensed Matter, **11**, 6877, (1999).

27. R.Mahesh and M.Itoh, Australian J.of Physics, **52**, 235, (1999).

28. T. Nakajima, H. Yoshizawa and Y.Ueda, J. Physical Society of Japan, **73**, 8, (2004).

29. T.Nahajima, H.Kageyama, Y.Ueda, J.Magnetism and Magnetic Materials, **272-276**, 405, (2004).

30. J.C.Loudon, N.D.Mathur, P.A.Midgley, J.Magnetism and Magnetic Materials **272-276**, 13, (2004).

31. A-M Haghiri Gosnet and J-P Renard, J.Physics D: Applied Physics, **36**, R127, (2003).

32. R.V.Demin, L.I.Koroleva and Ya M. Mukovskii, J.Physics:Condensed Matter, **17**, 221, (2005).

33. G.Papavassiliou, M.Faradis, M.Belesi, M.Pissas, L.Panagiotopoulos, G.Kallias, D.Niarchos, C.Dimitropoulos and J.Dolinsek Physical ReviewB, **59**, 63910, (1999).

34. P.Levy, F.Parisi, G.Polla, D.Vega, G.Levya, H.Lanza, R.S.Freitas and L.Ghivelder, Physical ReviewB, **62**, 6437, (2000).

35. G.Kallias, M.Pissas, E.Devlin, A.Simopoulos, D.Niarchos, Physical ReviewB, **59**, 1272, (1999).

36. G.Allodi, R.De Renzi, G.Guidi, Physical ReviewB, **57**, 1024, (1998).

37. G.Allodi, R.De Renzi, F.Licci, M.W.Pieper, Physical Review Letters **81**, 4736, (1998).

38. C.H.Chen and S.W.Cheong, Physical Review Letters, **76**, 4042, (1996).

39. J.Dho, I.Kim, S.Lee, K.H.Kim, H.J.Lee, J.H.Jung, T.W.Noh, Physical ReviewB, **59**, 492, (1999).

88

40. M.Paraskevopoulos, F.Mayr, J.Hemberger, A.Loidl, R.Heichele, D.Maurer V.Muller, A.A.Mukhin, A.M.Balbashov, J.Physics: Condensed Matter, **12**, 3993, (2000).

41. E.Dagotto, T.Hotta and A.Moreo, Physics Reports, **344**, 1-153, (2001).

42. G.H.Jonker and J.H.Van Santen Physica **16**, 337, (1950).

43. G.H.Jonker and J.H.Van Santen Physica **19**, 120 (1953).

44. H.J.Zeiger and G.W.Pratt, "Magnetic Interactions in Solids" Clarendon University Press. Oxford, (1973).

45. J.B.Goodenough, Physical Review **100**, 564, (1955).

46. A.Zouari, C.Boudaya and E.Dhahri, Physica Status. Solidi (a) **188**, 3, R1177, (2001).

47. L.F.Bates, "Modern Magnetism", Cambridge University Press, (1948).

48. F.Albert Cotton and G. Wilkinson, "Advanced Inorganic Chemistry" Interscience Publishers 2nd edition, (1966).

ELECTRICAL CONDUCTION MECHANISM AND OPTICAL PROPERTIES OF POLYVINYL ACETATE AND CELLULOSE ACETATE PROPIONATE BLENDS

F. H. ABD EL-KADER, A. M. SHEHAP, A. F. BASHA and N. H. EL-FEWATY

Cairo University, Faculty of Science, Physics Department-Giza-Egypt

Abstract: Films of Polyvinyl acetate (PVAc), cellulose acetate propionate (CAP) homopolymers and their blends of compositions 0.85/0.15, 0.7/0.3, 0.5/0.5, 0.3/0.7 and 0.15/0.85 (wt/wt) were prepared to investigate the type of electrical conduction mechanism. The current-voltage characteristics have been studied under different conditions. Also, ultraviolet/visible spectra of all samples have been studied according to their different composition ratios.

The conduction mechanisms at different temperatures and voltage ranges appear to be essentially a space charge limited current for the two individual polymers, while for the blend samples the predominance mechanism is Poole-Frenkel type. Ultraviolet/visible studies of the investigated samples showed that the blend sample of 0.5/0.5 (wt/wt) has the smallest absorption edge (4.58 eV) and highest band tail (0.61 eV).

The composition blend sample 0.5/0.5 (wt/wt) has the most proper conduction and optical properties which has attractive attention in the view of its application in electronic and optical devices.

Keywords: *CAP/PVAc blends; space charge; Poole-Frenkle emission; Schottky emission; absorption edge; band tail.*

1-Introduction:

The blending of two or more polymers has gradually become an important technique to improve the cast-performance ratio on commercial product [1]. The manifestations of the superior properties of polymer blends depend upon the miscibility of its component at the molecular scale. Since the polarities of CAP and PVA_c are closely matched so, we can expect some interactions between these two macromolecules which lead to form thermodynamically compatible system.

Polyvinyl acetate has been shown to be miscible with other bacteriological polymer which is used for sutures, drug delivery system and implants for bone fixation [2, 3]. The blending has been found to be miscible over complete range that modifies the properties and biodegradability. A dynamical mechanical analysis of PVAc/CAP have been studied and birefringence is found in the most blend samples resulted from the molecular orientation [4].

Electrical and optical studies of polymers have attracted much attention in view of their application in electronic and optical devices. Electrical conduction in polymers has been studied aiming to understand the nature of the charge transport prevalent in these materials while, optical properties are aimed to achieving better reflection, antireflection, interference and polarization properties. The aim of this work was to study the effect of composition ratio of CAP and PVAc on the conduction mechanism and the optical parameters such as absorption co-efficient and band tails

CP 998, Modern Trends in Physics Research
Third International Conference MTPR-08
edited by L. El Nadi

2-Experiment

Cellulose acetate propionate and polyvinyl acetate were supplied by Acros organic, USA, had a molecular weight 75000 and 170000 respectively. The polymers were dissolved separately in chloroform using a magnetic stirrer at 40°C for 8 hours. Solutions of CAP and PVAc were mixed together with different weight/weight percentage (1.0/0.0, 0.85/0.15, 0.7/0.3, 0.5/0.5, 0.3/0.7, 0.15/0.85 and 0.0/1.0). Films of appropriate thickness about 60μm were cast onto stainless steel patrie dishes and were finally kept in oven at 50°C for 24 h to evaporate residual solvent.

Current-voltage (I-V) characteristics were studied with an effective film area of about 1 cm^2 through voltage range 10-400 volt at constant temperatures 40, 50 and 60°C. The specimens were sandwiched between two copper electrodes in a sample holder and kept in a temperature – controlled cell. The current measurements were made by means of an electrometer (Keithley 617 Aurora Rood Cleveland, Ohio, USA) by allowing sufficient time to elapse after application of the voltage. The ultraviolet/visible absorption spectra of the samples were recorded on Berkin Elmer 4B spectrophotometer over the range 200-800 nm.

3-Results and Discussion

3.1. Current-voltage characteristics: To understand the actual transport mechanism operating in the present samples, the isothermal current-voltage (I-V) characteristics were studied.

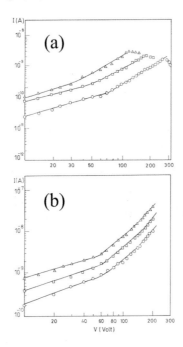

Fig. (1) Log I - Log V for (a) pure CAP and (b) pure PVA$_c$ at temperatures 40°C($\bigcirc$), 50°C($\square$) and 60°C ($\triangle$). Figures 1 and 2 show logI versus logV plots at three different temperatures 40, 50 and 60°C for homopolymers and their blends 0.85/0.15, 0.7/0.3, 0.5/0.5,0.3/0.7 and 0.15/0.85 (wt/wt) CAP/PVAc respectively.

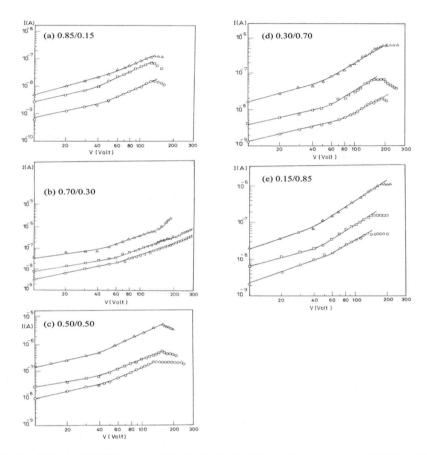

Fig. (2) Log I – Log V for the CAP/PVA$_c$ (wt/wt) blends (a), (b), (c), (d) and (e) at temperatures 40℃ (O), 50℃ (□) and 60℃ (△).

Evidently, all the plots of logI versus logV indicate three different regions follow the power law (I α V^s) where s is an exponent [5]. For electric field below about 12kV/cm (first region) the curves show usual ohmic behavior (s = 1). At higher electric fields up to about 33kV/cm (second region), it deviates from the Ohmic behavior and square law region was obtained of s≤2 (see Table 1). The value of s is equal to 2 for individual polymers while it is less than 2 for all blend samples at the various temperatures. The voltages at the beginning and at the end of the square law region have different values depending upon the temperature and the ratio of blend compositions (see Fig 1). This region shows the presence of other modes of conduction such as (a) tunneling, (b) space charge effects, (c) schottky emission, and (d) Poole- Frenkle effect. On further increase of the applied field (third region) the current becomes unstable and changes irregularly in all samples except pure PVAc in which the behavior is linear with s=2.2 to 3.2. It may be mentioned that the current increases with increasing temperature for the same applied voltage in Ohmic and square law regions.

As the thickness of the samples are quite large (about 60 µm), the conduction by tunneling is out of question. Space charge effect may be present and contributing to overall current. In the high field region the possible mechanisms that could be applied to explain the conduction are Richardson-Schottky and Poole-Frenkel mechanisms.

92

Table (1): Calculated value of the exponent s in the second region and dielectric constant at 1kHz for different samples at temperature 40°C, 50°C and 60°C.

CAP/PVAc wt/wt	40°C		50°C		60°C	
	s	$\acute{\epsilon}$	s	$\acute{\epsilon}$	s	$\acute{\epsilon}$
1.00/0.00	2.00	3.31	2.00	3.43	2.00	3.54
0.85/0.15	1.73	3.60	1.71	3.62	1.69	3.74
0.70/0.30	1.71	2.70	1.74	2.80	1.75	2.93
0.50/0.50	1.69	2.88	1.56	3.00	1.78	3.20
0.30/0.70	1.71	2.34	1.75	2.10	1.71	2.20
0.15/0.85	1.51	2.10	1.71	2.20	1.76	2.20
0.00/1.00	2.00	2.00	2.00	2.10	2.00	2.10

Figure 3 shows a plot of current as function of thickness (d) on log-log scale at temperature 40°C in the square law region at 80V for CAP and PVAc homopolymers and their blend samples. The plots are straight lines with slope -0.3 for individual polymers while equal nearly 2.8 for the blend samples. So, we get $I \alpha V^2$ and $I \alpha d^3$ for individual polymers which provide supporting evidence for existence of space charge conduction. Thus, for individual polymers this region can be well explained using standard space charge limited current (SCLC) theory [6]. In SCLC process, in the absence of traps in the polymer, the current density can be written as [7].

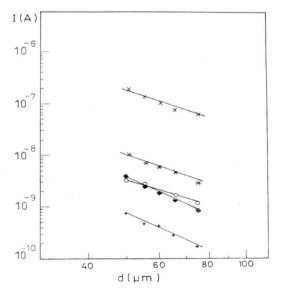

Fig. (3) The current versus the thickness of (+) pure CAP, (♦) Pure PVAc (✱) 0.7/0.3, (✕) 0.5/0.5, (O) 0.3/0.7 (wt/wt) CAP/PVAc at temperatures 40°C for applied voltage 80 volt.

$$J = \frac{9}{8} \frac{\mu_0 \varepsilon' \varepsilon_0 V_{tr}^2}{d^3} \qquad (1)$$

where $\acute{\varepsilon}$ is the dielectric constant of the film material determined from capacity measurements at $1kH_z$ (see table 1) , ε_0 is the permittivity of free space , μ_0 is the mobility of free charge carriers and V_{tr} is the voltage at which transition from the Ohmic to square law region takes place . If traps are present in the polymer, the SCLC may be decreased by several orders of magnitude. Rose argued [8] that neither the space – charge density nor the field distribution should be modified by a trap- limiting parameter θ relating the proportion of trapped charges (n_t) to free charges (n_0) and J is now written as [9]

$$J = \frac{9}{8} \frac{\mu_0 \varepsilon' \varepsilon_0 \theta V_{tr}^2}{d^3} \qquad (2)$$

where θ is given by

$$\theta = \frac{n_0}{n_0 + n_t} = \frac{I_1}{I_2} \qquad (3)$$

Thus, for the trap-free case , $n_{t=0}$, therefore $\theta=1$; with traps present θ is always less than unity . Experimentally θ is the ratio between the current density at the beginning, (I_1) of the square law region, and the end of the rise up (I_2). The free carrier mobility μ_0 can now be calculated from equation (1) using experimental value of θ. It may be mentioned that when a trap level exists, the electron mobility is reduced by $1/\theta$, and the effective electron drift mobility in an insulator with traps is therefore

$$\mu_e = \mu_0\, \theta \qquad (4)$$

This equilibrium concentration of charge carrier in the conduction band n_0 can also be obtained using relation

$$n_0 = (\acute{\varepsilon}\, \varepsilon_0\, \theta/\, e\, d^2)\, V_{tr} \qquad (5)$$

The free carrier density n_0 may be using in equation (3) to determine the value of n_t, the trapped carrier density.

The values of μ_e, μ_0, n_0 and n_t for CAP and PVAc homo polymers at three different temperatures 40, 50 and 60°C are given in Table 2. The obtained values are quite comparable with those usually reported in literature for this class of polymeric material [10-14].

Table (2): Value of μ_0 and μ_e (m^2 V^{-1} S^{-1}), n_0 and n_t(m^{-3}) for CAP and PVAc homopolymers at temperature 40°C, 50°C & 60°C.

sample	T=40°C				T=50°C				T=60°C			
	μ_o x 10^{-11}	μ_e x 10^{-12}	n_o x 10^{-17}	n_t x 10^{-18}	μ_o 10^{-11}	μ_e x 10^{-1}	n_o x 10^{-17}	n_t x 10^{-18}	μ_o x 10^{-11}	μ_o x 10^{-12}	n_o x 10^{-17}	n_t x 10^{-18}
CAP	3.19	2.18	2.10	2.85	5.28	7.00	3.50	1.96	1.15	1.72	3.27	2.00
PVAc	42.82	36.80	3.04	1.12	36.40	66.80	2.20	1.25	3.87	11.10	4.58	1.14

Chutia and Barua [11] have reported four regions in I-V characteristics curves for pure PVAc film ; the first region is Ohmic , second and fourth regions are trap and trap-free square law and the third region has a slope about 4 to 6.3 . Hence the variation in behavior of present data shown in Figs 1,2 and the values of μ_0, μ_e, n_0 and n_t in table 2 with that reported [11] may be due to the difference of PVAc molecular weight, purity and method of preparation of the samples.

The values of μ_0 and μ_e are very low for both CAP and PVAc but they increase with increasing temperatures; indicate that polaron hopping between localized sites is involved in the process of carrier transport. Besides, the values of n_0 and n_t are changed irregularly with increasing the temperatures for individual polymers.

For the blends of the different compositions, the values of s is between 1 and 2 and in the mean time the relationship between I versus d has a slope < 3, which are conflicted with space charge theory. So we may say that the SCLC model may not be the predominant conduction process in all blend samples. Therefore, it will be worthwhile to examine the fit of the present results in close relation to electronic process mainly Richardson- Schottky or Poole-Frenkle type conduction.

These two mechanisms may be fully described as [15]

$$I \; \alpha \; \exp (e\beta V^{1/2} / K_B T d^{1/2}) \tag{6}$$

where K_B is the Boltzmann constant, T the absolute temperature and β is a parameter , which determines the nature of conduction process. The theoretical value of β in case of Richardson – Schottky emission is given as,

$$\beta_{RS} = e(e/4\pi \; \acute{\epsilon} \; \epsilon_0 \;)^{1/2} \tag{7}$$

while in case of Poole-Frenkle emission , it is $\beta_{PF} = 2 \; \beta_{RS}$

95

Figure 4 shows the plots of log I versus $V^{1/2}$ for the blends at temperatures 40, 50 and 60°C. The linearity of log I-$V^{1/2}$ for plots in the field region (about 4 kV/cm up to 33 kV/cm), points to an electronic- type conduction mechanism. Here, the charge carrier is released by thermal activation over a potential barrier The physical nature of such a potential barrier can be interrupted in two ways. It can be the transition of electrons over the barrier between the cathode and the dielectric (Schottky emission). Alternatively, **charge carriers can be released from traps in to the dielectric (Poole-Frenkle effect).**

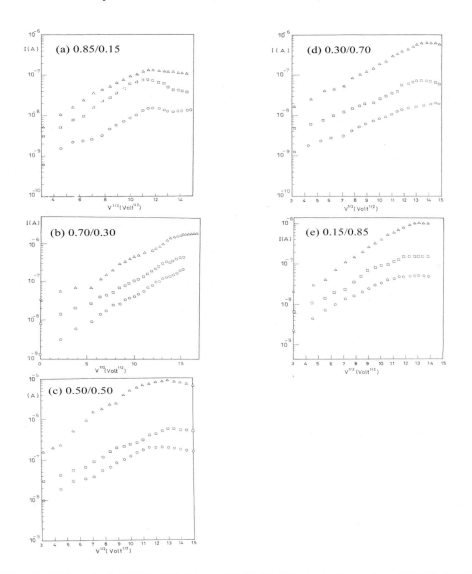

Fig. (4) Log I – V1/2 for the CAP/PVAc (wt/wt) blends (a), (b), (c), (d) and (e) at temperatures 40℃ (O), 50℃ (□) and 60℃ (△).

But, the curves of log I –$V^{1/2}$ show an appreciable deviation from linearity at high field region (≥35 kV/cm) which may be due to space charge build up, giving rise to non uniformity of field distribution between the electrodes [16,17] in addition , there is a deviation from linearity at lower fields (≤4 kV/cm) at certain temperatures which may be attributed to the accumulation of space charge at the electrode [18,19]. The experimental value of β (β_{exp}) at different film temperatures were deduced from the slope of plots of Fig. 4, while theoretical values of β were calculated using equation (7) and dielectric data in Table 1. Both experimental and theoretical values are listed in Table 3.

It can be seen that there is a close agreement between β_{exp} and theoretical value of β_{pf} leading to the conclusion that poole-Frenkel emission is the dominant charge transport mechanism operating in these films.

Table (3): The comparison between the β (eV m1/2V-1/2) constant for all investigated samples at three different temperatures.

CAP/PVA$_c$(wt/wt)	T=40°C		T=50°C		T=60°C	
	β_{exp}	β_{the}	β_{exp}	β_{the}	β_{exp}	β_{the}
0.85/0.15	5×10^{-5}	$\beta_{RS=2 \times 10^{-5}}$ $\beta_{PF=4 \times 10^{-5}}$	5.8×10^{-5}	$\beta_{RS=2 \times 10^{-5}}$ $\beta_{PF=4 \times 10^{-5}}$	4.3×10^{-5}	$\beta_{RS=2 \times 10^{-5}}$ $\beta_{PF=4 \times 10^{-5}}$
0.7/0.3	3.4×10^{-5}	$\beta_{RS=2.3 \times 10^{-5}}$ $\beta_{PF=4.6 \times 10^{-5}}$	4.1×10^{-5}	$\beta_{RS=2.25 \times 10^{-5}}$ $\beta_{PF=4.5 \times 10^{-5}}$	3.6×10^{-5}	$\beta_{RS=2.2 \times 10^{-5}}$ $\beta_{PF=4.4 \times 10^{-5}}$
0.5/0.5	4×10^{-5}	$\beta_{RS=2.2 \times 10^{-5}}$ $\beta_{PF=4.4 \times 10^{-5}}$	4×10^{-5}	$\beta_{RS=2.2 \times 10^{-5}}$ $\beta_{PF=4.4 \times 10^{-5}}$	4.1×10^{-5}	$\beta_{RS=2.1 \times 10^{-5}}$ $\beta_{PF=4.2 \times 10^{-5}}$
0.3/0.7	4×10^{-5}	$\beta_{RS=2.65 \times 10^{-5}}$ $\beta_{PF=5.3 \times 10^{-5}}$	5.6×10^{-5}	$\beta_{RS=2.6 \times 10^{-5}}$ $\beta_{PF=5.2 \times 10^{-5}}$	5.8×10^{-5}	$\beta_{RS=2.55 \times 10^{-5}}$ $\beta_{PF=5.1 \times 10^{-5}}$
0.15/0.85	4.6×10^{-5}	$\beta_{RS=2.6 \times 10^{-5}}$ $\beta_{PF=5.2 \times 10^{-5}}$	4.3×10^{-5}	$\beta_{RS=2.6 \times 10^{-5}}$ $\beta_{PF=5.2 \times 10^{-5}}$	4.8×10^{-5}	$\beta_{RS=2.55 \times 10^{-5}}$ $\beta_{PF=5.1 \times 10^{-5}}$

3.2. Optical Properties:

Figure 5 shows the absorbance spectra of both CAP and PVAc homopolymers and their blends 0.85/0.15, 0.7/0.3, 0.5/0.5, 0.3/0.7 and 0.15/0.85 (wt/wt) CAP/PVAc in the wavelength range 200-700 nm. The spectra indicate that there is no sharp absorption edge, which is a characteristic of the glassy state [20]. This is consistent with the amorphous nature of the homopolymers as noticed from x-ray diffraction pattern that reported previously [21]. As seen from these spectra all samples have the same absorbance behavior, except the change in the absorption values and position of the observed band

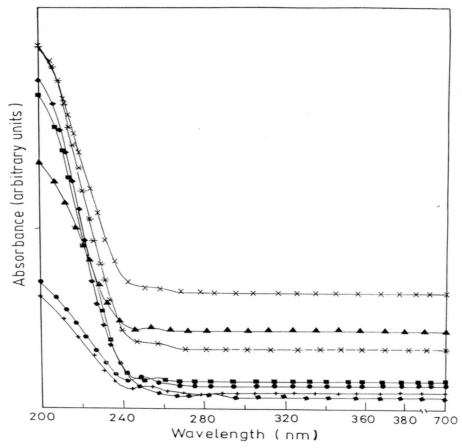

Fig. (5) The relation between absorbance versus 1 for:- (+) pure CAP, (♦) Pure PVa$_c$, (•) 0.85/0.15, (✳) 0.7/0.3, (✕) 0.5/0.5, (▲) 0.3/0.7, and (■) 0.15/0.85(wt/wt) CAP/PVA$_c$.

Spectra are composed of almost flat baseline and a steep cut off. The absorbance values of homo polymers are lower than those for blend compositions in the wavelength range 200-700 nm. Besides, the absorbance values for blend samples change irregularly with their compositions. Among these samples, the value for the blend samples of composition 0.5/0.5 (wt/wt) is the highest one. The spectra of all samples show shoulder-like band, which may be the $\pi \rightarrow \pi^*$ electronic transition (k-band) of carbonyl chromophoric groups in both CAP and PVA$_c$ polymers [22,23]. The position of this band appears at 258 and 282 nm for CAP and PVA$_c$ respectively. But, the band position of blend samples lies between those of homo polymers in an irregular way, except that 0.85/0.15 (wt/wt) CAP/PVAc blend sample which is shown to be shifted towards higher energy side compared to the band CAP.

The absorption coefficient α(ν) was determined from the spectra using the formula,

$$\alpha(\nu) = \frac{1}{d}\ln(\frac{1}{T}) \qquad (8)$$

where T is the transmittance. The calculated absorption co-efficient for the samples under investigation are relatively small (50-1800 cm^{-1}) as in most low carrier concentration amorphous materials [24].

Figure 6 show the variation of absorption co-efficient with incident photon energy for all samples under investigation. The curves show a linear dependence of α(ν) versus (hν) near the band edge. The extrapolation of the linear portion of the curves to the abscissa has been used to find the values of the absorption edg These values are listed in Table 4.

Table (4): The values of absorption edge and band tail for CAP, PVAc and their blends.

CAP/PVAc wt/wt	Absorption Edge (eV)	Band tail Ee (eV)
1.0/0.0	5.17	0.30
0.85/0.15	4.93	0.44
0.7/0.3	4.90	0.36
0.5/0.5	4.58	0.61
0.3/0.7	4.84	0.56
0.15/.085	4.98	0.27
0.0/1.0	5.05	0.20

It is clear that the values of the absorption edge for blend samples are less than those for pure polymers, therefore, one may suggest that there is certain degree of miscibility in the blend system. Also, the blend sample of composition0.5/0.5 (wt/wt) has the minimum absorption edge of 4.58eV, indicating its most proper conduction when compared to the rest of samples.

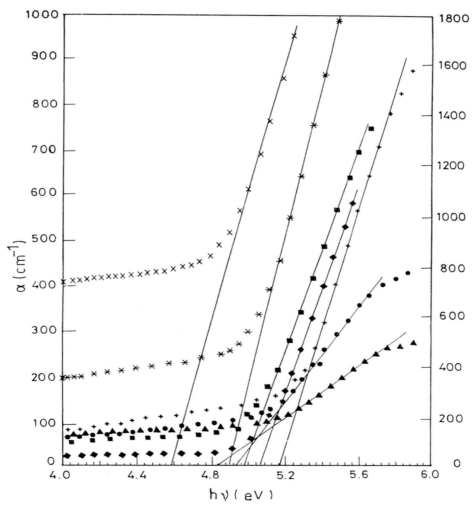

Fig. (6) The relation between absorption coefficient α versus hν for:- (+) pure CAP, (♦) Pure PVaₑ, (•) 0.85/0.15, (✳) 0.7/0.3, (✕) 0.5/0.5, (▲) 0.3/0.7, and (■) 0.15/0.85(wt/wt) CAP/PVaₑ.

According to Tauc [25] there are three distinct regions in the absorption edge spectrum of amorphous semiconductors. The first is the weak absorption tail, which originates from defects and impurities, the second is the exponential edge region, which is strongly related to the structural randomness of the system and the third is the high absorption region that determines the optical energy gap. In the exponent edge where the absorption coefficient, α, lines in the absorption region $1 < α < 10^4$ cm^{-1}, The absorption coefficient is governed by Urbach relation [26] as,

$$α = α_0 \exp(hν/E_e) \qquad (9)$$

where $α_0$ is a constant and E_e characterizes the slope of the exponential edge region and it is the width of the band tails of the localized states existing in the films.

100

Figure 7 shows the relation between lnα and hν for individual polymers and their blend samples. In the current case, the exponential dependence of absorption coefficient on photon energy suggests that these materials obey the Urbach Rule.

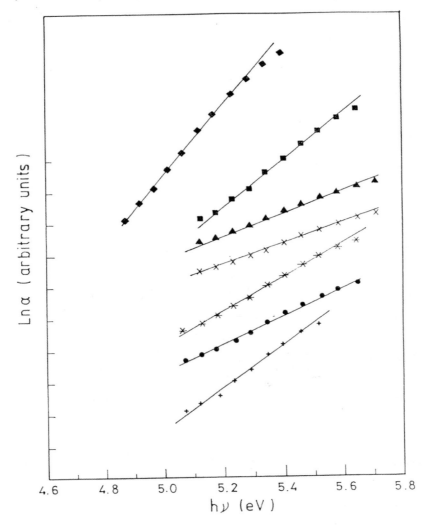

Fig. (7) The relation between ln α versus hν for:- (+) pure CAP, (♦) Pure PVa$_c$, (●) 0.85/0.15, (×) 0.7/0.3, (✗) 0.5/0.5, (▲) 0.3/0.7, and (■) 0.15/0.85(wt/wt) CAP/PVA$_c$.

The values of band tail E$_e$ were calculated from the slopes of these lines and are listed in Table 4. It is clear that the values of E$_e$ for blend samples are higher than those for both homo polymers and they vary largely with compositions, but in an irregular trend. In addition, the blend sample 0.5/0.5 (wt/wt) has the highest value of the band tail which reflects the reduction in band gap.

Conclusion:

Analysis of the I-V characteristics measurements of individual polymers and their blends show that at low fields the current is ohmic and conduction is due to hopping of charge carriers between localized states. At higher fields, it is revealed that the type of conduction mechanism is space charge limited for individual polymers and the Poole-Frenkel for blend samples.

The optical absorption spectra revealed that the blend sample of composition 0.5/0.5 (wt/wt) CAP/PVAc has a minimum absorption edge and maximum band tail. Consequently, it is expected that this blend has the most proper conduction compared to other samples.

Finally, these complementary studies indicate that by varying the blend composition the electrical and optical properties of poly blend films can be controlled to some extent. Also, it point to the fact that a composition 0.5/0.5 is the critical concentration that gives optimum values for electrical and optical properties.

References:

1-J.R. Fried, *Polymer Science and Technology* (Prentice Hall International Editions, New Jersey, 1995) A, P.263.

2-G.W. Adams, J.M.Cowie, Polymer, 40, 1993(1999).

3-N. Haiz, L. Sharma, Polymer, 41, 5749(2000).

4-M. Yamaguchi, K. Masuzawa, Eur.Polym.J, 43, 3277(2007).

5-G. Leditzky and T. Leising, J.Phys.D; Appl.Phys. 27, 2185(1994).

6-M.A. Lampert, P.Mark, *Current Injection in Solids* (Academic, New York, 1970).

7-K.C. Kao, W. Hwang, *Electrical Transport in Solids* (Pergamon, Oxford, 1981).

8-A. Rose, Phys. Rev. 97, 1938(1955).

9-M.A. Lampert, Phys. Rev. 103, 1648(1956).

10-D.A. Seanor, *Polymer Science* ed. Jenkins A.D., Vol.2. (North–Holland and Pub. Co. Amsterdam, 1972).

11-J. Chutia and K. Barua, T. Phys. D; Appl. Phys. 13, 9(1980).

12-A.F. Basha, H.A. Abd El-Samed and M. Amin, Ind. J. Phys. 59a, 213(1985).

13-F. Zanora and M.C. Gonzalez, Polymer, 38, No.2, 263(1997).

14-S. Besbes, A. Bouazizi, H. Ben Ouada, H. Maaref, A. Haj Said and F. Matoussi, Material Science and Engineering, C21, 273(2002).

15-J.G. Simmons, in: *Handbook of Thin Film Technology*, eds. L.I. Maissel and R. Glang (McGraw – Hill, New york, 1970).

16-N.G. Belsare and V.S. Deogaonkar, Indian J. of Pure and Appl. Phys. 36, 280(1998).

17-M.J.A. Sarkar and M.A.R. Sarkar, Indian J. of Pure and Appl. Phys. 38, 190(2000).

18-P.C. Machendru, N.L. Pathak, K. Jain and P. Mahendru, Phys. Stat. Sol. (a) 42, 403(1977).

19-N. Nagara, ch.V. Subba Reddy, A.K. Sharma and V.V.R. Narasimha Rao, J. Power Sources, 112, 236(2002).

20-C.A. Mogarth and A.A. Hosseini, J. Mat. Sci. 18, 2679(1983).

21-F.H. Abd El-Kader, A. Shehap, A.F. Basha and N.H. El-Fewaty (in publication Materials of Physics and Chemistry 2008).

22-F.H. Abd El-kader, S.A. Gaafar, K.H. Mahmoud, S.I. Bannan, M.F.H. Abd El-kader, Current Applied Physics, 8,78(2008).

23-F.H. Abd El-kader, S.A. Gaafar, K.H. Mahmoud, S.I. Bannan, M.F.H. Abd El-Kader, J. Appl. Polym. Sci. (accepted for publication, in press, 2008).

24-N.F. Mott and E.A. Davis, *Electronic Process in Non-crystalline Materials*, 2nd edn. (Oxford University Press, Oxford, 1977).

25-J. Tauc, *Amorphous and Liquid Semiconductors* (Plenum, New-York, 1974) Chap 4.

26-F. Urbach, Phys. Rev. 92, 1324(1953).

A NEW TECHNIQUE FOR INCREASING EFFICIENCY OF SILICON SOLAR CELLS

A. IBRAHIM

Faculty of Science, Physics department, Tanta University, Egypt
E-mail: ali_02us@yahoo.com

Abstract

A new type of photovoltaic system with higher generation power density has been studied in detail. The feature of the system is a V-shaped module (VSM) with two tilted monocrystalline solar cells. Compared to solar cells in a flat and tilted at a fixed 30^0 from horizontal orientation, the VSM enhances external quantum efficiency and leads to an increase of 30 to 35% in power conversion efficiency. Due to the VSM technique, short-circuit current density was raised from 25.94 to 34.7 mA/cm^2, but both fill factor and open-circuit voltage were approximately unchanged. For the VSM similar results (about 35% increases) were obtained for solar cells fabricated by using mono-crystalline silicon wafers with only conventional background impurities.

Keywords: Silicon; Solar cells; V-shaped structure; Light confinement

1. Introduction

Silicon is one of the most abundant elements in the earth's crust, and there appears to be plenty of potential for developing silicon solar cell techniques with mass production. The light absorption properties of silicon play a key role in the design of silicon photovoltaic (PV) devices. Every possible advantage must be taken of the light-absorbing capabilities of a solar cell in order to obtain the best possible PV performance from silicon. For silicon wafers, the material defects associated with background impurities (e.g., oxygen and carbon impurities that are hardly eliminated in silicon purifying processes) can create electron–hole pairs through absorbing infrared photons via sub-band-gap excitation processes [1–3].

Numerous PV system designs with improved optical approaches have been developed to boost the output of solar cells [4–14]. Recently, we obtained a result that a V-shaped module (VSM) technique can lead to an increase of 24% in power conversion efficiency (η) for polycrystalline silicon solar cells [15], and the idea also resulted in enhancing photocurrent for monocrystalline silicon solar cells with polished surfaces [16]. Furthermore, a scheme of fabricating cells by using both-side polished wafers has been proposed to execute back surface reflectors [16]. The V-shaped module consists of two tilted cells and was constructed to have a V-shaped structure, as shown in Fig. 1(b). In this letter, we report the η of the V-shaped structure monocrystalline silicon solar cell

CP 998, Modern Trends in Physics Research
Third International Conference MTPR-08
edited by L. El Nadi

module for the first time. In addition, the external quantum efficiency (EQE) of the module has been investigated as well.

Area of collecting light

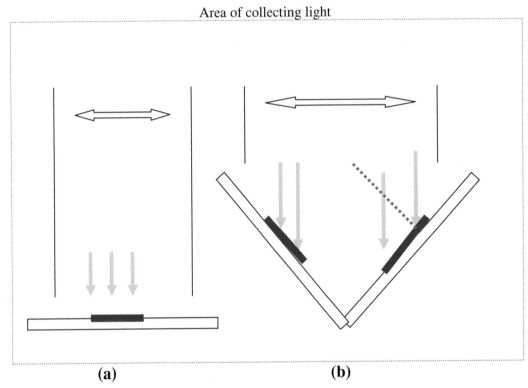

(a)   **(b)**

Fig.1. the schematic cross sections of the flat (a) and V-shaped (b) photovoltaic cells under illumination.

2. Results and discussion

Two 0.5-cm-square monocrystalline silicon solar cells n^+pp^{++} (PESC Passevated Emitter Solar Cell) having identical characteristics were chosen as experimental samples in this study. The two cells were fabricated by using polished, 1-3 Ω cm, p-type (1 0 0) Czochralski (CZ) silicon wafers with defects [16]. Boron was diffused into the front surface of the wafers. The p^{++} emitter was characterized by a sheet resistivity of 0.7 Ω /cm^2 and a penetration depth (junction depth) of 0.6 mm. The front-grid metallization consisting aluminum covered approximately 7% of the area of the cell and made contact directly to the p-type silicon emitter region. The back ohmic contact metallization was also aluminum. An antireflection (AR) coating film consisting of $Si_3 N_4$ with thickness of 1000A ° thickness was used. In general, the processing conditions for fabricating the cells were similar to the well-known conventional cell process.

To know the optimum V-shape angle a movable holder was designed and a relation between the photo-generated current and the V-shape angle was measured as in Fig. 2.

At the beginning of this work, the two cells were measured in a planar way (flat orientation) in order to make a comparison. In this way, incoming sunlight was vertical to the surfaces of the cells, as shown in Fig. 1(a). One of the cells was measured under 1sun (of light intensity of 860 W/m^2), air mass 1.5 (AM1.5) illumination and exhibited an η of 9.97%, together with an open-circuit voltage (Voc) of 549mV, a short-circuit current density (Jsc) of 25.94 mA/cm^2, and a fill factor (FF) of 0.73. Fig. 3 demonstrates the current–voltage (I–V) characteristic of the cell in a flat position. In addition, blue points curve in Fig. 5 shows the EQE. The second cell was also measured and its characteristics were approximately the same as those of the first cell. After the measurements were made as described above, the two cells were tilted and were further installed to create a V-shaped structure (see Fig. 1(b)). The opening angle is the angle between the two tilted cells and was 60^0 in this study. For the measurement of the module, incoming light was projected along the symmetrical line after the measurements were made as described above; the two cells were tilted and were further installed to create a V-shaped structure (see Fig. 1b). The opening angle is the angle between the two tilted cells and was 60^0 in this study. For the measurement of the module, incoming light was projected along the symmetrical line of the V-shaped structure, as shown in Fig. 1(b). Therefore, the sunlight struck each tilted cell with an incidence angle of 60^0. As described in Ref. [15], the comparison between the measured results of the V-shaped structure and the planar configuration should be carried out under the condition of the same collecting light area. Moreover, the area of collecting light is 0.45 cm^2 for the V-shaped structure module with an opening angle of 60^0 and 0.45 cm^2 or a 0.5-cm-square cell in the planar way.

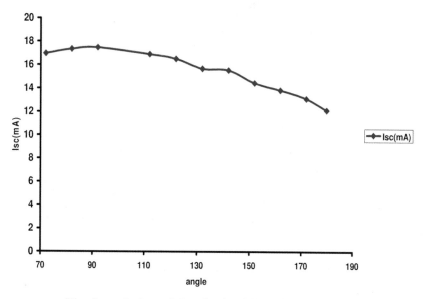

Fig. 2: variation of the obtained Isc with the V-shape angle.

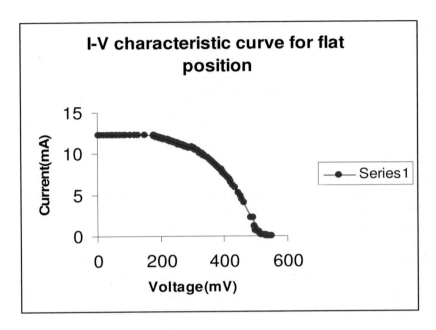

Fig. 3: The characteristic I-V curve for silicon solar cell in a flat orientation.

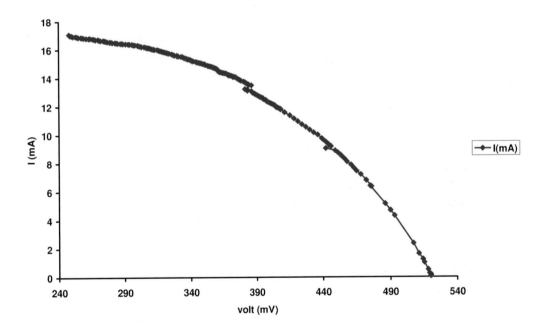

Fig. 4: The characteristic I-V curve for silicon solar cell system in V-shape position.

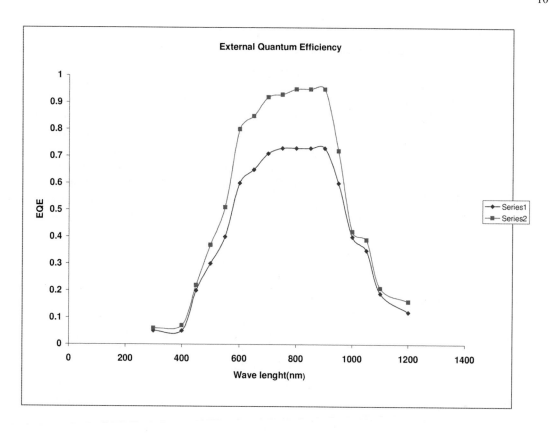

Fig. 5: The External Quantum Efficiency (EQE) of the cells in flat and V-shape positions.

The V-shaped structure module was measured under the same conditions adopted in the measurement for the cells in the planar way. One of the tilted cells exhibited a FF of 0.75, a Voc of 553 mV, and a short-circuit current per unit collecting light area of 34.7 mA/cm^2, resulting in an η of 13.05%. Fig. 4 demonstrates the I–V characteristic of the cell, and red points curve in Fig. 5 shows the EQE. Another tilted cell in the V-shaped structure module was also measured. The measurements showed that the output parameters of the two tilted cells were identical. Furthermore, the two tilted cells in the V-shaped structure module were electrically connected in parallel. In view of the luminous flux into the area of collecting light (in this case, the area of collecting light is equivalent to an aperture over the V-shaped structure), the FF, Voc, and photocurrent density of the module with parallel connection were identical to those of each tilted cell. Note that the short-circuit current per unit area of collecting light has been raised from 24.94 to 33.7 mA/cm2. Also note that the V-shaped structure technique can cause the η of the cells to increase from 9.97% to 13.05% under the condition of same collecting light area (aperture area). For the VSM, similar results (about 30% increases) were also obtained for solar cells fabricated by using mono-crystalline silicon wafers with only conventional background impurities. From these results, the following explanation can be proposed. As discussed in Ref. [15], each cell in the V-shaped structure module receives both sunlight and infrared photons emitted from the opposite cell due to a transformation mechanism in which a photon is down-converted into two or more photons with lower

108

energies. According to the effect of PV infrared response enhancement owing to intermediate levels associated with impurities and defects [1–3], an infrared photon could induce an electron–hole pair because an intermediate-level impurity (or defect) could be photo ionized via a sub-band-gap excitation by the infrared photon. Due to the transformation mechanism an incident visible photon could be down converted into two or more infrared photons in each cell of the V-shaped structure system, and some of the infrared photons could subsequently be absorbed via sub-band-gap excitation processes to create electron–hole pairs in the opposite cell. This means that more infrared-induced carriers could be created and some visible photons could indirectly create electron–hole pairs in the V-shaped structure system. Based on the transformation mechanism, an infrared photon could as well be down-converted into two or more infrared photons with lower energies. With respect to incident solar light, the V-shaped structure enhances the creation of electron–hole pairs not only for infrared light but also for visible light. Because of trapping more photons, the use of the V-shaped structure could boost η and EQE of the cells.

3. Conclusions

The V-shaped structure increased generation power density and has been proven to cause an increase of 30% to 35% in power conversion efficiency. Compared to some cell configuration designs [10], the V-shaped structure with two tilted cells has larger area of whole surface and has larger surface area for receiving photons projected from the opposite face. The V-shaped structure solar cell technique is thus believed to be more effective for trapping infrared emission of cells. Therefore, the V-shaped structure offers a new opportunity to tap the potential of solar cells for increasing η: In addition, it is very easy to manufacture the V-shaped structure. However, the V-shaped structure system requires more cells than the planar way. In the case where the opening angle is 60^0, the cost per watt for the V-shaped structure is twice as high as that for the flat orientation. Although the V-shaped structure solar cell module scheme causes an increase in cost, the scheme could offer opportunities to develop some PV technologies such as space solar cells, concentrator solar cells, etc.

References

[1]   M. Wolf, Proc. IRE 48 (1960) 1246.
[2]   G. Guttler, H.J. Queisser, Energy Convers. 10 (1970) 51.
[3]   A. Luque, A. Marti, Phys. Rev. Lett. 78 (1997) 5014.
[4]   J. Haynos, J. Allison, R. Arndt, A. Meulenberg, International Conference on Photovoltaic Power Generation, Hamburg, Germany, September, 1974, p. 487.
[5]   A.W. Blakers, M.A. Green, Appl. Phys. Lett. 48 (1986) 215.
[6]   P. Campbell, M.A. Green, Sol. Energy Mater. Sol. Cells 65 (2001) 369.
[7]   M.A. Green, Adv. Mater. 13 (2001) 1019.
[8]   M.A. Green, Prog. Photovoltaics Res. Appl. 10 (2002) 235.
[9]   J.D. Levins, G.B. Hotchkiss, M.D. Hammerbacher, Proceedings of the 22nd IEEE Photovoltaic Specialists Conference, Las Vegas, 1991, p. 1045.

[10] Toshiro Maruyama, Hiroshi Minami, Sol. Energy Mater. Sol. Cells 79 (2003) 113.

[11] M. Lipinski, P. Panek, E. Beltowska, H. Czternastek, Mater. Sci. Eng. B 101 (2003) 297.

[12] T. Uematsu, Y. Yazawa, Y. Miyamura, S. Muramatsu, H. Ohtsuka, K. Tsutsui, T. Warabisako, Sol. Energy Mater. Sol. Cells 67 (2001) 415.

[13] Toshio Matsushima, Tatsuyuki Setaka, Seiichi Muroyama, Sol. Energy Mater. Sol. Cells 75 (2003) 603.

[14] A. Cuevas, R.A. Sinton, N.E. Midkiff, R.M. Swanson, IEEE Electron Device Lett. 11 (1990) 6.

[15] Jianming Li, M. Chong, Y. Heng, J. Xu, H. Liu, L. Bian, X. Chi, Y. Zhai, Electron. Lett. 40 (2004) 1219.

[16] J. Li, M. Chong, J. Xu, X. Duan, M. Gao, F. Wang, Digest Book of Fifth International Conferences on Thin Film Physics and Applications, Shanghai, China, May, 2004, p. 135.

STUDY OF THE RELATIONSHIP BETWEEN ELECTRICAL AND MAGNETIC PROPERTIES AND JAHN–TELLER DISTORTION IN $R_{0.7}Ca_{0.3}Mn_{0.95}Fe_{0.05}O_3$ PEROVSKITES BY MÖSSBAUER EFFECT

E. K. ABDEL-KHALEK [†]

Physics Department, Faculty of Science, Al Azhar University, Nasr City, Cairo, Egypt

W. M. EL-MELIGY, E. A. MOHAMED, T. Z. AMER AND H. A. SALLAM

Mössbauer Laboratory. Physics Department, Faculty of Science, Al Azhar University, Nasr City, Cairo, Egypt

Abstract

In this work structural, magnetic and electrical properties of $R_{0.7}Ca_{0.3}Mn_{0.95}Fe_{0.05}O_3$ (R= Pr and Nd) perovskite manganites are presented. Structural characterization of these compounds shows that both have orthorhombic (Pbmn) phase. The Mössbauer spectra show clear evidence of the local structural distortion of the $Mn(Fe)O_6$ octahedron on the basis of non-zero nuclear quadrupole interactions for high-spin Fe^{3+} ions. It was found that the local structural distortion decreases significantly with replacing Pr^{3+} by Nd^{3+}. This replacing dependence of the Jahn–Teller coupling strength estimated from the Mössbauer results was found to be consistent with the electrical and magnetic properties.

Keywords: Rare-earth manganite, X-ray diffraction, Mössbauer, Electrical and magnetic properties

1. Introduction

Rare-earth manganite perovskites have been extensively studied because of their interesting electrical and magnetic properties [1]. The substitution of rare-earth site with ones of smaller radii is expected to enhance the microstructural inhomogeneity that may play an important role in the electrical and magnetic properties of the perovskite. One of the factors that are thought to be responsible for the variations is the structure tuning induced by the small ionic radius of the interpolated cation in the R site. Another one is the magnetic ionic characteristic of each rare-earth. Early studies have shown that Mn^{3+} ions are mainly replaced by Fe^{3+} ions in this Fe-doping range, and that both ions have identical ionic radii in six-fold octahedral coordination [2]. Therefore, the substitution of Fe^{3+} for Mn^{3+} does not change the structure and, consequently, the Jahn–Teller effect can be investigated from the quadrupole interaction of the Fe nucleus although the Fe^{3+} ion is not a Jahn–Teller one. Thus can be assumed that the values of the second-order crystal electric field (CEF) coefficients for FeO_6 and MnO_6 octahedra are approximately the same in these compounds. Therefore,

few per cent of Fe, which substitutes for Mn in $R_{0.7}Ca_{0.3}Mn_{0.95}Fe_{0.05}O_3$ compounds can be used as a micro-probe to detect the symmetry of the nearest-neighbor O^{2-} ions in the $Mn(Fe)O_6$ octahedron and its influence on the magnetic and electric properties.

2. Experimental

The $R_{0.7}Ca_{0.3}Mn_{0.95}Fe_{0.05}O_3$ (R= Pr and Nd) perovskites, were prepared by co-precipitation method using ammonium carbonate. The XRD patterns were collected in the range of 20° to 80° with 0.02° steps using Cu Kα radiation. The analysis of the phases was carried out by the program FULLPROF based on the method of Rietveld. The size of the crystallographic grain has been deduced by applying the Scherrer formula [3]. The Mossbauer absorption spectra were recorded at RT by using a conventional constant acceleration spectrometer with ^{57}Co (Cr) radioactive source and a metallic iron foil was used for calibration. Electrical resistance was measured as a function of temperature in the range of 77–300K using the four-probe DC method. B-H curve was obtained applying a vibrating sample magnetometer (VSM) with 0.8T applied field at 77K.

[†] Email: eid_khalaf0@yahoo.com

CP 998, Modern Trends in Physics Research
Third International Conference MTPR-08
edited by L. El Nadi

3. Results and discussion

3.1. *X-ray measurements*

The X-ray diffraction (XRD) patterns for polycrystalline samples shows that are of single phase. Structural investigations by using the standard Rietveld profile refinement showed that the synthesized samples crystallized in the distorted orthorhombic perovskite structure with space group Pbnm. From table 1 it is noticed that the lattice parameters decrease with

Table (1). Structure parameters for RE^{3+} samples.

RE^{3+}	Pr^{3+}	Nd^3
Lattice parameters	a = 5.356 b = 5.355 c = 7.530	a = 5.350 b = 5.345 c = 7.525
V(Å3)	216.01	215.26
‹Mn-O-Mn›º	130.43	132.38
‹Mn-O›Å	1.997	1.965
Cry Size L(nm)	80.05	81.02

replacement of Pr^{3+} ions by the smaller Nd^{3+} ions. This can be attributed to the decrease of the average ionic radius ‹r_A› where the octahedral tilt and rotate to reduce the excess space around the A site, leading to the decrease of the distances between two Mn ions as well as the decrease of unit cell volume [4]. The orthorhombic phase presents a ratio c/a < √2 characteristic of a cooperative Jahn–Teller deformation [5]. There is one long Mn-O bond and two short ones in the orthorhombic MnO_6 octahedral, which reveal the presence of the Jahn-Teller distortion. The increase in particle size with the replacement of Pr^{3+} ions by smaller Nd^{3+} ions can be attributed to the decrease of the average ionic radius <r_A> and unit cell volume.

3.2. *Mossbauer measurements*

Fig.(1). Shows the Mossbauer spectra for the $R_{0.7}Ca_{0.3}Mn_{0.95}Fe_{0.05}O_3$ where R is Pr^{3+} or Nd^{3+}. Hyperfine parameters are listed in Table 2. On account of the isomer shift (IS) values the charge state of iron is exclusively 3+, without coexistence of the 4+ charge state. Thus the Fe ion in the samples occupied octahedral site with 3+ valence only [6]. While the small quadrupole splitting (QS) indicates the orthorhombic distortion of the ideal cubic perovskite structure. The broadening in the line width (Table 2) for the spectra of the samples may be due to several

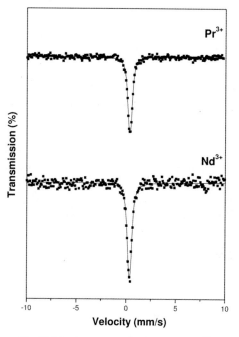

Fig. (1). Mossbauer spectra for the different RE^{3+}.

Table (2). Mossbauer Parameters as a function of the RE^{3+}content.

RE^{3+}	QS(mm/s)	IS(mm/s)	LW(mm/s)	E_{JT}(ev)
Pr^{3+}	0.21	0.34	0.41	0.73
Nd^{3+}	0.18	0.33	0.44	0.64

QS is quadrupole splitting (mm/s); IS is isomer shift relative to Fe at room temperature (mm/s); A is relative intensity (%).

reasons. The most common are superparamagnetic behaviour due to nanocrystalite size and local environment effects [7]. The decrease of the QS value of Fe^{3+} ions on replacing Pr with Nd can be attributed to the increase of the ‹Mn-O-Mn›. The decrease of the IS value of Fe^{3+} ions on replacing Pr with Nd can be attributed to the decrease in the Mn/Fe-O bond length as observed from the XRD results. The Jahn–Teller effect can be investigated directly from the obtained QS-values. The local structural distortion of MnO_6 octahedra resulting from the Jahn–Teller effect of high-spin Mn^{3+} ions removes the degeneracy of the e_g and t_{2g} orbitals so as to make some energy levels more stable. The e_g orbital group is separated into two energy levels, dz^2 and d_{x2-y2}. The energy separation of the upper-level orbitals e_g has been shown to be larger than that of the lower-level orbitals. Since the Jahn–Teller distortion strongly influences the electron-hopping process between the upper-level orbitals of Mn^{3+} and Mn^{4+} ions. It has proved that [8] the energy separation between dz^2

112

and d_{x2-y2} orbitals arises mainly from the contribution of the second-order crystal electric field (CEF) coefficient of the distorted MnO_6 octahedra, $A_{20}(MnO_6)$, and that the relationship between the Jahn–Teller coupling, E_{JT}, and $A_{20}(MnO_6)$ can be written as

$$E_{JT} = \frac{2e}{7}\sqrt{\frac{5}{\pi}} A_{20}(MnO_6)\langle r^2 \rangle_{Mn}$$

where $\langle r^2 \rangle$ is the expectation value of the square of the radial distance of the Mn^{3+} ions' 3d orbital from the nucleus. Mossbauer spectroscopy can be employed to determine the value of A_{20} by a measurement of the quadrupole splitting, as described below. Since QS at the nucleus of a Fe^{3+} ion is solely the result of contributions from the surrounding ions, the value of $A_{20}(FeO_6)$ can be obtained from

$$A_{20}(MnO_6) = \sqrt{\frac{4\pi}{5}} \frac{\Delta(Fe^{3+})}{eQ\left[1+\frac{\eta^2}{3}\right]^{1/2}}$$

where Q is the electric quadrupole moment of the ^{57}Fe nucleus and η is the asymmetry parameter. Since the Mn^{3+} and Fe^{3+} ions have identical ionic radii (0.645Å) in six-fold octahedral coordination [2], it is assumed that $A_{20}(MnO_6)\approx A_{20}(FeO_6)$. Therefore, the relationship between E_{JT} and the quadrupole splitting at the Fe^{3+} ion, $\Delta(Fe^{3+})$, can be written as

$$E_{JT} = \frac{4\Delta(Fe^{3+})}{7Q\left[1+\frac{\eta^2}{3}\right]^{1/2}}\langle r^2 \rangle_{Mn}$$

The quadrupole splitting for the Mossbauer experiment is generally in unit of mm/s. Here it should be converted to ev by a factor of $E\gamma/c$, where $E\gamma = 14.4$ kev for the I = $3/2\leftrightarrow1/2$ transition of ^{57}Fe and $c = 3\times10^{11}$ mm/s is the velocity of light. In the case of axial symmetry, $\eta = 0$. Using approximately the expectation value $\langle r^2 \rangle$ of free Mn^{3+} ions, $\langle r^2 \rangle= 0.3535$ Å^2, and Q = 0.28 $\times10^{-24}$cm^2, the Jahn– Teller coupling, E_{JT}, in each perovskite was estimated (table2). From Table 2, it is found that the energy separation between dz^2 and d_{x2-y2} decreases with replacement Pr^{3+} ions by smaller Nd^{3+} ions. This can be attributed to the decrease of local structural distortion which due to the increase of the ‹Mn-O-Mn›.

3.3. Electric measurements

The results of the resistivity versus temperature of the samples in zero magnetic fields is illustrated in Fig. (2).

All samples behave as semiconductor in the whole temperature range studied. Absence of Metal–Semiconductor transition attributes to grain boundary effect due to nanocrystallites and the small kink in the

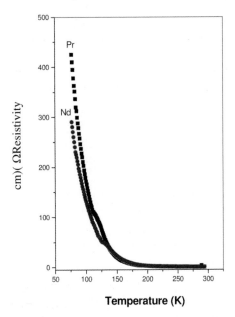

Fig. (2). Resistivity dependence on T for the different RE3+.

resistivity of the samples due to the structural transition [9]. From Fig. (2) it can be seen that the electrical resistivity decreased with replacement of Pr^{3+} ions by the smaller Nd^{3+} ions. This can be due to the increase of the Mn-O-Mn bonds and the decrease of the Mn-O distance.

In order to explain the electronic conduction, two models viz; variable range hopping model ($T < \theta_D/2$) where θ_D is Debye temperature and the small polaron hopping ($T > \theta_D/2$) are generally used. The Mott's equation for VRH mechanism [10] is given by

$$\sigma = \sigma_0 \exp(-T_0/T)^{1/4}$$

where σ_0 is the pre factor. The resistivity data is fitted to above equation by plotting ln (σ) vs. $T^{-1/4}$ and from the best fits $\theta_D/2$ values are estimated as the temperature at which deviation from linearity occurs in the temperature region (Fig. 3). Further T_0 values for each sample were calculated from slopes of ln (σ) vs. $T^{-1/4}$ plot. Finally, using the T_0 values and the equation

$$T_0 = 16\alpha^3/k_B N(E_F)$$

$N(E_F)$, the density of states at the Fermi level for each material was also obtained. Here, the value of α =2.22

nm^{-1} has been used for calculations. All the calculated parameters are given in Table 3 and are found to be in

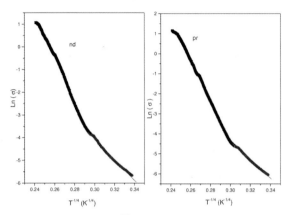

Fig. (3) Plot of ln (σ) vs. T$^{-1/4}$ for, the solid line indicates the best fit for the VRH model equation.

Table (3). The best fit parameters obtained from the experimental resistivity data in the paramagnetic insulating region

RE^{3+}	θ_D(K)	N(E$_F$)(eV^{-1}cm^{-3})	T$_0$(10^{6}K)	Ep(meV)
Pr^{3+}	230	1.80×1016	4.432	127.0
Nd^{3+}	258	1.90×1016	4.180	126.4

agreement with those reported in the literature for other manganite materials [11]. While the conduction mechanism of these materials at high temperatures (T > θ_D/2) is governed by small polarons and the polaronic models could be due to either adiabatic or nonadiabatic approximations. According to Jung et al. [12], higher values of N(E$_F$) in the present manganite system could be due to their relative high value of conductivity. These higher values of N(E$_F$) are clear signatures of the applicability of the adiabatic hopping mechanism. Based on this fact, the adiabatic small polaron hopping model can be used in the present investigation. The temperature dependence of the electrical resistivity arising out of adiabatic approximations is given by

$$\rho = \rho_\alpha T \exp (Ep/k_BT)$$

where ρ_α is the residual resistivity, and E$_P$ is the activation energy. The E$_P$ obtained from the slopes of the ln(ρ/T) vs. 1/T plots (see Fig. 4) are given in Table 3

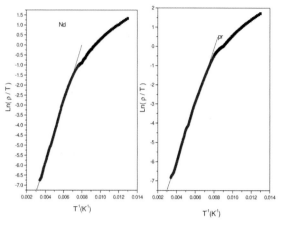

Fig. (4). Plot of ln (ρ /T) vs. T-1for, the solid line indicates the best fit for the equation ρ= ρ_αT exp (Ep/K$_B$T).

It can be seen from the table that the activation energy values are found to decrease and N(E$_F$) increases with replacement of Pr^{3+} ions by the smaller Nd^{3+} ions which is in agreement with Jahn–Teller energy separation obtained from Mössbauer results. This may be due to the fact that with increasing grain size interconnectivity between grains increases, which cooperates the conduction electron to hop the neighboring sites. In addition to the increase in Mn–O–Mn bond angle and a decrease in Mn–O bond length, then the better the overlap between the relevant orbitals, and the higher the probability of hopping occurring

3.4. Magnetic measurements

The hysteresis curves at 77K with 0.8T applied field are shown in Fig. (5). The shape of the hysteresis loop is characteristic of weak ferromagnetism. The saturation Bs, coercive field H$_C$ and remanent Br are listed in

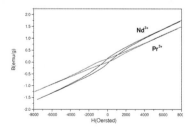

Fig. (5). The magnetic hysteresis (B-H) measurements for the different RE^{3+} at 77K.

114

Table (4). Bs, Br and Hc for the different RE^{3+} content.

RE^{3+}	Bs(emu/g)	Br(emu/g)	Hc(Oe)
Pr^{3+}	1.367	0.074	510.1
Nd^{3+}	1.695	0.104	277.4

Table 4. It can be seen from the table that the saturation Bs increase and a decrease of coercive field are observed with replacement of Pr^{3+} ions by the smaller Nd^{3+} ions. This may be due to that the ferromagnetic (FM) contripution for the Nd sample is much higher than that of the Pr sample. The phenomenon is possibly due to the existence of field induced Nd moment, which interacts with the Mn moment and destabilizes the AFM structure, and leads to the development of FM double exchange interaction [13] which is consistent with the decrease in Jahn–Teller energy separation obtained from Mössbauer results.

4. Conclusions

Mossbauer results show clear evidence of the local structural distortion of the $Mn(Fe)O_6$ octahedron can be effectively employed to estimate the Jahn–Teller coupling of these perovskites. The change of rare-earth site (Pr or Nd) dependence of the Jahn–Teller coupling strength estimated from the Mossbauer results was found to be consistent with the electric and magnetic properties. The variable range hopping and small polaron models results show that the changes in the behavior of the density of states at the Fermi level and activation energy with change of rare-earth are in agreement with the change in the Jahn–Teller coupling energy. Consequently, the connection between the local structural distortion and the electrical and magnetic behavior of this material can be well explained.

Acknowledgments

The authors would like to thank Prof. Dr. Said Farag Mostafa, Powder Technology Dept, Central Metall. R & D Institute, El-Tebbin, Helwan, Cairo, Egypt, for providing the facility for the magnetic measurements.

References

1. S. Jin, T.H. Tiefel, M. Mecormack, R. A. Fastnacht, R. Ramesh and L. H. Chen, Science 264, 413 (1994).
2. Z. Kou, X. Ma, N. Di, Q. Li, and Z. Cheng, Phys. Stat. Sol. (b) 242, 2930, (2005).
3. A. Banerjee, S. Pal, S. Bhattacharya, B.K. Chaudhuri, J. Appl. Phys. 91 (2002) 5125.
4. B. Hannoyer, G. Marest, J. M. Greneche, Ravi Bathe, S. I. Patil and S. B. Ogale, Phys. Rev. B 61, 9613 (2000).
5. R. Dhahri, and F. Halouni, Journal of Alloys and Compounds 385, 48,(2004)
6. M. Kopcewicz, V. A. Khomchenko, I. O. Troyanchuk and H. Szymczak, J. Phys.: Condens. Matter 16, 4335 (2004).
7. A. G. Mostafa, E. K. Abdel-Khalek, W.M. Daoush and M. Y. Hassaan. Accepted for ICAME 2007
8. H. Cheng, Z. H. Wang, N. L. Di, Z. Q. Kou, G. J. Wang, R. W. Li, Y. Lu, Q. A. Li, B. G. Shen, and R. A. Dunlap, Appl. Phys. Lett. 83, 1587 (2003).
9. A. Urushibara, Y. Moritomo, T. Arima, A. Asamitsu, G. Kido, and Y. Tokura, Phys. Rev. B, 51, 14103, (1995).
10. N.F. Mott, Metal–Insulator Transitions, Taylor & Francis, London, (1990).
11. G. Venkataiah, P. Venugopal Reddy, Journal of Magnetism and Magnetic Materials 285, 343, (2005).
12. W. H. Jung, J. Mater. Sci. Lett. 17 (1998) 1317.
13. D. Zhu, P. Cao, W. Liu, X. Ma, A. Maignan and B. Raveau, Materials Letters 61 (2007) 617–620

ELECTRONIC BAND STRUCTURE AND MAGNETIC PROPERTIES OF YCo₅

SAMY H. ALY,   FATEMA ALZAHRAA HASSAN

Physics department, Faculty of Science, Faculty of Science at Damietta, Mansoura University, Damietta, Egypt
samy.ha.aly@gmail.com

SHERIF YEHIA

Faculty of Science, Physics department Helwan University, Cairo, Egypt
sherif542002@yahoo.com

We present a first-principles study on the magnetic properties and electronic structure of YCo₅ using the two well-known electronic structure packages FPLO and WIEN2k. Our results have been compared with the results of experiments and other ab initio calculations. The comparison shows a fair agreement between the present work and other published investigations.

Keywords: Electronic structure; YCo₅; *ab initio* calculations

CP 998, Modern Trends in Physics Research
Third International Conference MTPR-08
edited by L. El Nadi
Copyright @ 2011 by World Scientific Publishing Co. 978-981-4317-50-4 / 981-4317-50-0

116

1. Introduction

Over the past few years a new class of magnetic materials has come into prominence. Intermetallic compounds of the general formula RCo_5, where R is a rare-earth metal or yttrium proved to have a great technical importance in the fabrication of permanent magnets, and have been studied extensively both experimentally[1,2] and theoretically[3-11]. YCo_5 is the simplest to study because it has no 4f electrons. Yttrium has many properties in common with the rare-earths. Its outer configuration is $4d^1 5s^2$ compared to the usual $5d^1 6s^2$ of the rare-earths, and its size is comparable[12]. Since 4f levels are tightly bound core levels, they do not form energy bands; details of their electronic structure are due to their multiplet structure. Its existence as a typical RCo_5 material with large moment implies that 4f electrons are not the only source of the magnetic moment. YCo_5 has a hexagonal $CaCu_5$-type structure with six atoms in a unit cell, its space group is P6/mmm. This structure is quite close packed and the packing fraction is 67 % if the space is filled up by touching rigid spheres. The construction of the touching spheres gives an ideal c/a ratio of 0.816 and an ideal ratio of 0.764 between the radii of Cu and Ca atoms[12]. Co atoms in YCo_5 occupy two crystallographically different sites: 3g and 2c sites.

2. Method

FPLO package is a Full-Potential Local-Orbital minimum basis code to solve the Kohn-Sham equations on a regular lattice using the local spin density approximation (LSDA). This method is not the only possible one but it is distinguished by the combination of three advantages: accuracy, efficiency, and straightforward interpretation of its outcomes in chemical terms. Much effort has been spent to achieve a level of numerical accuracy which is comparable to advanced full-potential LAPW implementations though the basis set is one order of magnitude smaller. This makes highly accurate full potential calculations for elementary cells of up to 100 atom feasible on single-CPU machines and is a good starting point for approaches beyond the LSDA.

3. Calculations and Discussion

The YCo_5 Compounds atoms in YCo_5 occupy two crystallographically deferent sites: 3g and 2c sites. The magnetic moments of Co on 3g and 2c sites were measured by neutron diffraction to be 1.72 and 1.77 μ_B / atom, respectively[13]. These local moments have also been calculated by the spin-polarized band method[3,4,14]. The magnetic properties and the electronic structure of YCo_5 were calculated by the self-consistent linear muffin-tin orbital (LMTO) method within the atomic sphere approximation (ASA)[15].

Fig.1 shows the calculated results of the total energy as a function of the volume of the unit cell V. The calculated minimum energy is achieved at V = 75 $Å^3$ for the nonmagnetic phase and 77 $Å^3$ for the magnetic phase (compared to 75 $Å^3$ calculated by Yamada[15]). Both values are smaller than the observed volume of 84 $Å^3$[15].

The lattice parameters used in the present calculation are taken from the observed volume with the ideal c=a ratio. Fig.2 shows the dependence of the local moments of Y and the 3g and 2c Co sites in YCo$_5$ on the unit cell volume. As shown the Co moment decreases rapidly at V = 72.3 Å^3, a transition from a high-moment to a low-moment state occurs at this critical volume. The change in Co moment on the 3g site is found to be larger than that on 2c site. For volumes larger than 72.3 Å^3 the moment of Co atom on the 3g site is larger than the moment of Co atom on the 2c site. While at volume smaller than 72.3 Å^3 the moment of Co atom on the 2c site is larger than the moment on the 3g site.

Fig.3 shows the dependence of total magnetic moment per formula unit on volume. A transition from a high-moment state to a low-moment one occurs at a critical V= 72.3 Å^3. Calculated local magnetic moments of Y, Co(2c) and Co(3g) of YCo$_5$ compound are listed in Table 1. Other theoretical and experimental results are also presented for comparison.

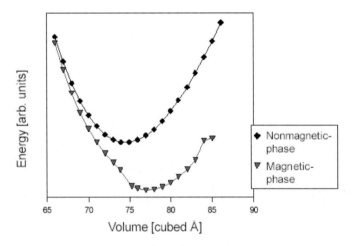

Fig.1 Dependence of the total energy on volume for magnetic and nonmagnetic hcp YCo$_5$.

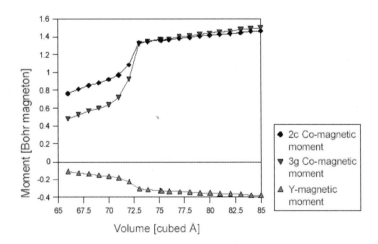

Fig.2 Dependence of the local magnetic moments on the volume for YCo$_5$.

118

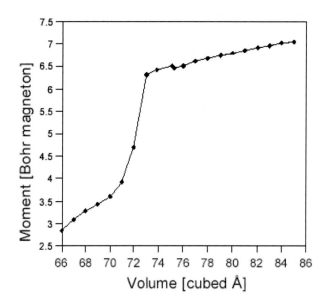

Fig.3 Dependence of the total magnetic moments on the volume for magnetic hcp YCo₅.

Table 1: Calculated local magnetic moments of Y, Co(2c) and Co(3g) of YCo₅ compound are listed , other theoretical and experimental results are also presented for comparison.

Sites	Present (μ_B)	WIEN2K	Ref [13]	Ref [14]	Ref [15]	Ref [16]
Y	-0.3	-0.14	-------	-0.27	-0.3	-0.61
Co (2c)	1.47	1.61	1.77	1.44	1.52	1.68
Co (3g)	1.49	157	1.72	1.37	1.47	2.04
Total	7.01	7.16	8.70	6.79	7.15	8.87

We have studied the energy dependence on the magnetic moment for hcp YCo₅ using the experimental value of the lattice constants. Our result is displayed in Fig.4 which shows that the energy minimum occurs at a magnetic moment of 7.0 μ_B. This value is smaller than the observed value 8.70 μ_B[13].

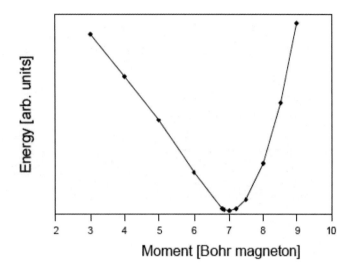

Fig.4 The total energy at fixed spin moment for hcp YCo$_5$ for a volume of 84 Å^3.

Fig.5a show the total DOS [15] for magnetic YCo$_5$. Calculated value of the total DOS for YCo$_5$ is 11.8 (states /eV. f.u.) at E$_F$. The band structure of YCo$_5$ along the specified path ΓMKΓALHA in the hexagonal BZ is shown in Fig.5b for both spin-directions. The difference in energy between bands for spin-up and spin-down direction is evident. Most of the relatively large DOS peaks are located closed to or below E$_F$ in Fig.5a which is reflected in Fig.5b by the relatively flat peaks in this region.

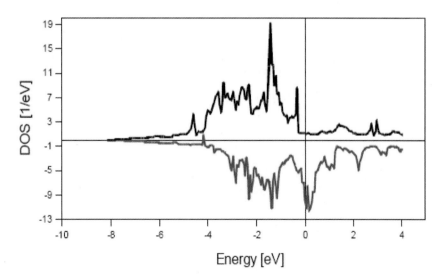

Fig.5a The total DOS for magnetic hcpYCo$_5$ for a volume of 84 Å^3.

120

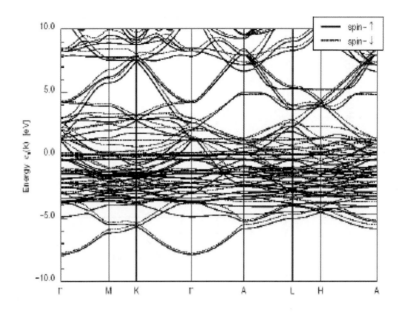

Fig.5b Energy bands for magnetic hcp YCo_5 for a volume of 84 $\mathring{A}^3$.

References

1.   H. R. Kirchmayr and C. A. Poldy. J. Magn. Magn. Mater. 8, 1 (1978).
2.   R. J. Radwanski and J .J. M. Franse. Int. J. Mod. Phys. 7, 782 (1993).
3.   S. K. Malik, F. J. Arlinghaus and W. E. Wallace. Phys. Rev. B 16, 1242 (1977).
4.   L. Nordstrom, M. S. S. Brooks and B. Johansson. J. Phys.: Condens. Matter 4, 3261 (1992).
5.   J. Trygg, L. Nordstrom and B. Johansson. .Physics of Transition Metals., Eds, P.M. Oppeneer and J. Kubler, (World Scientific, Singapore, 1993).
6.   M. Yamaguchi and S. Asano. J. Appl. Phys. 79, 5952 (1996).
7.   G.H.O. Daalderop, P.J. Kelly and M.F.H. Schuurmans. Phys. Rev. B. 3, 14415 (1996).
8.   M. Yamaguchi and S. Asano. J. Magn. Magn. Mater. 168, 161 (1997).
9.   L. Steinbeck, M. Richter and H. Eschrig. J. Magn. Magn. Mater. 226, 1011 (2001).
10.   L. Steinbeck, M. Richter and H. Eschrig. Phys. Rev. B 63, 184431 (2001).
11.   J. J. M. Franse and R. J. Radwanski, .Handbook of Magnetic Materials. 7th ed, K H. J. Buschow, (Amsterdam, Elsevier, 1993).
12.   I. Kitagawa, K. Terao, H. Yamada and M. Aoki , Crys. Res. Technol, 31, 555 (1996).
13.   J. Schweizer and F. Tasset, J. Phys. F 10, 2799 (1980).
14.   L. Nordstrom, O. Eriksson, M. S. S. Brooks and B. Johansson, Phys. Rev. B 41, 9111 (1990).
15.   I. Kitagawa, K. Terao, M. Aoki and H. Yamada, J. Phys.: Condens. Matter 9, 231 (1997).
16.   G. W. Zhang, Y. P. Feng and C. K. Ong. J. Magn. Magn. Mater. 184, 215 (1998).

INFLUENCE OF ZINC SUBSTITUTION ON SOME PHYSICAL PROPERTIES OF Co-La FERRITE

M. A. AHMED, N. OKASHA*

Materials Science Lab. (1), Physics Department, Faculty of Science,
Cairo University, Giza, Egypt
* Physics Department, Faculty of Girls, Ain Shams University, Cairo, Egypt

Abstract

Co- Zn substituted ferrites with the formula $Co_{1-x}Zn_xLa_{0.025}Fe_{1.975}O_4$ with $0.1 \leq x \leq 0.9$ was prepared by conventional solid state reaction. X- ray diffraction (XRD), scanning electron microscope (SEM) and magnetic susceptibility (χ) are utilized in order to study the effect of variation in Zn content and its impact on crystal structure, magnetic properties such as (χ), Curie temperature (T_C), and exchange interaction constant (J) between the different cations. The results of XRD reveal that, assured the single phase cubic spinel structure for all investigated samples with appearance of small peaks represented a secondary phase due to the presence of rare earth (La^{3+}) ions. The particle size (t) varied with composition and heat treatment. The value of both T_C and J decreases with increasing Zn content up to the critical concentration (0.5) then increases. The scanning electron micrographs indicate the distribution of grains in the sample with uniform size and agree well with the results of X- ray analyses. The influence of rare earth ions substitution on the structure and magnetic properties was examined.

Keywords: Co- Zn ferrites; Microstructure; magnetic susceptibility; Curie temperature; Exchange interaction constant.

1- Introduction

Spinel ferrites are technologically an important class of magnetic oxides because of their magnetic properties, high electrical resistivity, and low eddy current and dielectric loss. Ferrites are extensively used in microwave devices, computers memory chips, magnetic recording media, etc. Knowledge of cation distribution and spin alignment is essential to understand the magnetic properties of spinel ferrites. The interesting physical and chemical properties of ferro- spinels arise from their ability to distribute the cation among the available tetrahedral (A) and octahedral (B) sites [1].

Ferrites have low conductivity which influences their dielectric and magnetic behavior [2,3]. The activation energy, resistivity and Seebeck coefficient were measured [4] for semi conducting $CoFe_2O_4$. It was found that, the charge carriers are not free to move through the crystal lattice but jump from ion to ion. It is known that Zn^{2+} ions prefer the tetrahedral sites in the spinel lattice and in the mixed

Co- Zn ferrite while 80% of Co^{2+} occupies octahedral sites and 20% may go to tetrahedral sites [5]. This means that cobalt and iron have two valences; therefore, they partially occupy both the tetrahedral and octahedral sites. Cobalt ferrite is an important material not only for its magnetic properties but also for its catalytic properties which depends on the textural and morphological characteristics. Overall, Co ferrite, crystallizes in partially inverse spinel structure represented as $(Co^{2+}_x \ Fe^{3+}_{1-x}) \ [Co^{2+}_{1-x} \ Fe^{3+}_{1+x}] \ O^{2-}_4$ where (x) depends on the thermal history. Cobalt ferrite is a well- known magnetic material that has been studied in details due to its high coercivity and moderate saturation magnetization as well as its remarkable chemical stability and mechanical hardness [6].

The physical properties of ferrites depend on the preparation conditions such as sintering temperature and time as well as the heating and cooling rates. These properties depend also on the amount and type of dopant. The electron exchange

122

interaction $Fe^{2+} \leftrightarrow Fe^{3+} + e^-$ results in a local displacement of electrons during the sintering process. [7-11] and gives the origin of the conduction mechanism in ferrite material.

The magnetic behavior of the ferrite samples is of considerable interest both from a scientific and practical point of view [12, 13] as a result of both size and surface effects. The major size effects involve the reduction of domain boundaries, which leads to single- domain particles, and a thermal randomization of the total spin system, commonly denoted as super magnetism. The phenomena originating from surface effects include canted spin structure, and magnetic "dead layers" at surfaces.

Some physists found that the rare- earth (RE) oxides are good electrical insulators and have resistivity at room temperature greater than 10^6 Ω cm [14]. Little information exists in the literature about the influence of the rare- earth oxides on the physical parameters of the ferromagnetic oxide compounds. The obtained results reveal that, by introducing a relatively small amount of R_2O_3 instead of Fe_2O_3 a modification of both structure and transport properties are obtained.

In general, the RE ions are too large in radius to occupy the tetrahedral and octahedral sites [15]; during the sintering process. These will form secondary phases on the grain boundaries of the compound. However, it should be emphasized that the process of dissolution of R_2O_3 in the spinel lattice may also occur. Thus, some authors [16, 17] reported that the possibility of formation of compounds with inverse spinel structure takes place when R^{3+} substitutes Fe^{3+} on the octahedral sites.

The aim of the present work is to improve the physical and magnetic properties of a stoichiometric

La^{3+} substituted Zn ferrite by partial substitution of Co^{2+} ions through the study of its structure and magnetic characterizations.

2- Experimental procedure

The investigated samples $Co_{1-x}Zn_xLa_{0.025}Fe_{1.975}O_4$; $(0.1 \leq x \leq 0.9)$ were prepared using the solid state reaction technique where analar grade form (BDH) were mixed in a stoichiometric ratios of Fe_2O_3, ZnO, CoO and La_2O_3 in good grinded, presintered at 850°C for 30h and final sintered at 1100° C for 90h with heating and cooling rates 4°C/min. The details of the samples preparation are reported elsewhere [18]. The crystalline phases and microstructure of the prepared samples were identified using X- ray diffractometer equipped with Cu- K_α radiation source (Proker D8) λ =1.5418 Å. Scanning electron microscope (SEM) (model JEM- 100S) was used to study the morphology of the calcined samples. The measurement of magnetic susceptibility of the investigated samples was carried out using Faraday's method in which a small amount of the sample was inserted at the point of maximum force. The measurements of magnetic susceptibility were carried out at different temperatures (300K-700K) at fixed magnetic field intensity (H=2160 Oe). The temperature of the samples was measured using K- type thermocouple with junction in contact with the sample to obtain the exact temperature with the accuracy better than ± 1°C.

3- Results and Discussion
3- a. Structural properties

X-ray diffraction patterns for the investigated samples are shown in Fig. (1).

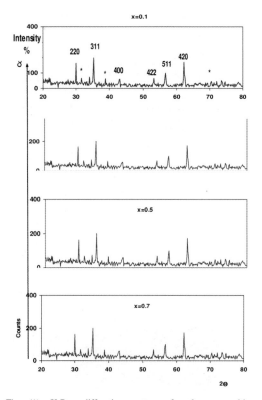

Fig. (1): X-Ray diffraction patterns for the composition $Co_{1-x}Zn_xLa_{0.025}Fe_{1.975}O_4$; $0.1 \leq x \leq 0.9$.

(0.58 Å) and Fe^{3+} (0.64 Å) ions by larger Zn^{2+} ions (0.74 Å) which increases the interionic distances such as A-O and consequently A-A.

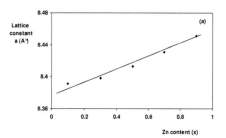

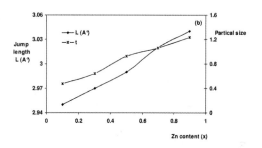

Fig. (2:a,b): The dependence of the lattice constant (a) on the Zn content (x). (b) Effect of Zn^{2+} ion substitution on the particle size (t) and the jump length (L).

The data reveal that, besides the single phase spinel structure with cubic symmetry, there are a small peaks of secondary phase of La_2O_3 which cannot enter the octahedral site [14] but form small aggregates on the grain boundary. The values of lattice constant (a) as function of Zinc content (x) as calculated from X- ray diffraction pattern are clarified in Fig. (2). It is clear that, the value of the lattice constant increases gradually with increasing Zn^{2+} content which obeys Vegard's law [19]. Usually in a solid solution in spinels within miscibility range, a linear variation in the lattice constant with the concentration of the components is observed. Similar linear dependence has also been observed in Li- Co [20], Li- Cd [21] and Li- Ferrites [22]. In other words, the linear increase in the lattice constant is due to replacement of smaller Co^{2+} ions

The variation of bulk density (D), X- ray density (D_x) and porosity (P) as a function of Zn^{2+} ion concentration (x) is reported in Fig. (2: a, b) and Table (1). The porosity of samples is calculated using the relation [23], P= 1-D/ Dx. The X- ray density is calculated according to the relation [23,] Dx= ZM/ Na^3, where Z is the number of molecules per unit cell (Z=8), M the molecular weight, N the Avogadro's number and (a) is the lattice constant. It is clear from the Table that, decreases with increasing Zn content; i. e. the bulk density nearly reflects the same general behavior of X- ray density.

The increase in porosity with increasing Zn content (x) related to rapid densification of ferrite samples and also to the difference in specific gravity

124

of the ferrite components since Co (6.79 g/cm^3) is heavier than ZnO (5.60 g/cm^3) [24].

The particle size (t) was calculated using Deby-Scherrer formula [20] and plotted versus Co content (x) in Fig. (2b). It is clear that, the particle size (t) of the investigated samples increases with increasing Zn^{2+} ions due to the larger ionic radius of Zn^{2+} ions (0.74 Å) on the expense of Fe^{3+} ions (0.64 Å) on tetrahedral sites. Also in the same figure, the jump length (L) on tetrahedral sites increases with the introduction of larger ions (Zn^{2+}), the values of (L) for octahedral sites and tetrahedral sites are given by L= ¼ (a √2), and L= ¼ (a √3) [25] respectively. This can be explained on the basis of the larger ionic radius of Zn^{2+} (0.74°A) as mentioned before which leads to an increase in hopping length. The substitution of Zn^{2+} ions causes the transfer of Co^{2+} ions to octahedral sites. Since the Co^{3+} ion of radius (0.545°A) is less than that of Fe^{3+} therefore, a decrease in the hopping length on octahedral sites is expected. The tendency of Zn cation to occupy its preference site (A- site) agrees well with Blasse [26] since their 4s and 4p or 5s and 5p electrons can form covalent bond with the 2p electron of oxygen ion.

Besides, this behavior of Zn^{2+} cation is due to its size and valency which is considered as an important factor to fulfill this tendency to occupy the tetrahedral site. This is favored by the polarization effects of the oxygen ions intermediated between A and B sites [27]. This means that the tetrahedral sites are expanded by an equal displacement of the 4 oxygen ions outwards along the body diagonal of the cube. Simultaneously, the oxygen ions connected with the octahedral site moved by the same amount as the tetrahedral sites expands. This behavior is reflected in the bond length of the tetrahedral and octahedral sites as given in Table (1).

Table (1): The calculated values of bond length of Co$_{1-x}$Zn$_x$La$_{0.025}$Fe$_{1.975}$O$_4$; $0.1 \leq x \leq 0.9$.

	Tet- Tet	Oct- Oct	Tet- Oct	Tet- bond	Oct- bond
x	A- A	B- B	A- B	A- O	B- O
0.1	3.6369	2.9695	3.4821	1.8185	2.0998
0.3	3.6364	2.9696	3.4819	1.8182	2.0996
0.5	3.6428	2.9748	3.4879	1.8214	2.1032
0.7	3.6472	2.9784	3.4922	1.8234	2.1058
0.9	3.6541	2.9841	3.4988	1.8271	2.10977

The shape and morphology of the particles were examined by direct observation using high resolution scanning electron microscopy. The scanning electron micrographs (SEM) of the prepared samples are given in Fig. (3).

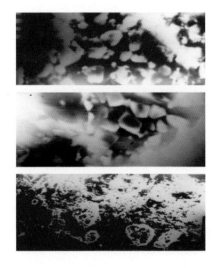

Fig. (3): SEM of Co$_{1-x}$Zn$_x$La$_{0.025}$Fe$_{1.975}$O$_4$; $0.1 \leq x \leq 0.9$ system.

The typical SEM micrographs indicate the distribution of the grains with uniform size which shows that the grain size increases with increasing Zn content and fine particles with higher Zinc concentration are agglomerated although the presence of La^{3+} ions on the grain boundaries [21]. Similar behavior was reported in case of La substituted Cu ferrite [28] and Er substituted Mn- Zn ferrite [29].

3- b. Magnetic properties

The magnetic properties of Co-Zn ferrites are strongly dependent on Zn concentration. The change in magnetic properties is due to the influence of the cationic stoichiometry and their occupancy in the specific sites as explained before [30]. Figure (4: a-d) is a typical curve showing the relation between the normalized dc magnetic susceptibility χ_T/χ_{RT} (RT = room temperature) and the absolute temperature at a selected magnetic field 2160 Oe below Curie temperature (T_C).

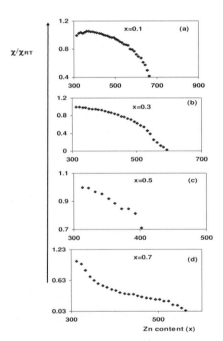

χ/χ_{RT}

Fig. (4:a-d): Variation of χ^T/χ_{RT} as a function of temperature at selected magnetic field intensity 2160 Oe.

The data in the figure gives the ferromagnetic behavior up to x = 0.5 then change to paramagnetic trend. The data show that the normalized molar magnetic susceptibility decreases with increasing temperature then dropped to a very small value. This means that, the value of magnetization decreases with increasing temperature because of the replacement of some Zn^{2+} with Fe^{3+} ions on

tetrahedral sites. Accordingly, the magnetic moments of the few remaining Fe^{3+} ions on A sites are no longer able to align antiparallel with all the moments of the B sites. But for x >0.5, the existence of magnetic ordering can not be explained on the basis of ferromagnetic to paramagnetic transition. Therefore these can be explained on the basis of the temperature evolution of the spin- spin correlation within and between groups of spins or clusters [31]. The Zn^{2+} ions have a preference for the tetrahedral sites. Therefore, at low concentration of Zn, the magnetization equal to the difference between the magnetic ions of B and A, respectively. Since Zn^{2+} ions have very small effect on the magnetization process, so, the net magnetization = M_B with negligible value of M_A where Zn^{2+} ion in A- site. After Zn content of x = 0.5, the critical region is reached to which some of Zn^{2+} ions replaces Fe^{2+} ions on tetrahedral sites which change to Fe^{3+}. Accordingly, the magnetization of the tetrahedral (A) sites increases suddenly and the net magnetization is sharply decreased [32]. This means that, the magnetic moments of the few remaining Fe^{3+} ions on A sites are no longer able to align all the moments of the B ions antiparallel to themselves, since this is opposed by the negative B- B exchange interaction which remains unaffected.

Figure (5a) shows the relation between the Curie temperature T_C and Zn content (x). According to Neel's model, the A-B interaction is the most dominant in ferrites, therefore, Curie temperature of the ferrites is determined from the overall strength of A- B interaction [33] which is a function of the number of Fe^{3+}_A- O_2- Fe^{3+}_B linkage [34] which depends upon the number of Fe^{3+} ions in the formula unit and their distribution among tetrahedral (A) and octahedral (B) sites.

126

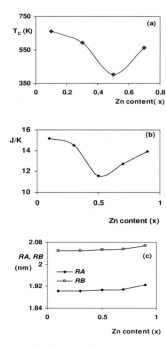

Fig. (5:a-c): a - Variation of Curie temperature (T_C) with Zn contents. b - Variation of exchange interaction (J/K) with Zn content (x). c - Variation of bond length (R_A, R_B) with Zn content (x).

The decrease in T_C can be explained on the basis of the paramagnetic region increases on the expense of the ferrimagnetic one. This was expected, because Zn^{2+} occupies half of the available A sites, leaving the other half for Fe^{3+} ions. The remaining Fe^{3+} ions reside on the octahedral sites. The Fe^{3+} moments on the tetrahedral sites orient all the octahedral sites moments antiparallel to them so that the Fe^{3+} moments on the tetrahedral sites neutralize only one- third of the octahedral Fe^{3+} ions, leaving a large percentage (the order two-thirds) oriented but uncompensated, giving a net magnetic moment. As the concentration of non magnetic Zn^{2+} ions on A sites increases the migration of some of Fe^{3+} from A to B sites takes place. The decrease in the number of magnetic ions at A sites decreases the A- B interaction thereby decreasing Curie temperature. Also, in case of rare

earth (La^{3+}) substituted ferrites, the orbital contribution is predominant and the sample shows high magneto crystalline anisotropy. Figure (5b) shows the relation between the values of exchange interaction (J/K) versus zinc content (x). The data reveals a decrease in J with the increasing Zn content. This reduction is due to the weakening of the A- B interaction. It is therefore concluded that the substitution of Zn in Co- ferrite dilute the magnetic properties of these samples. Also, the increase of J above the critical concentration (x $\geq$ 0.5), confirm the presence of large ionic radius La^{3+} on the grain boundaries and this agree well with the other magnetic parameters. Figure (5c) shows the compositional variation of the bond length R_A and R_B with zinc content. The values of R_A and R_B were calculated using the relation [35]:

$$R_A = a \sqrt{3} \, (\delta + 1/8)$$
$$R_B = a \, (3 \, \delta^2 - \delta /2 + 1/16)^{1/2}$$

Where δ is the deviation from u parameter, (δ= u- 0.375), R_A is the shortest distance between A- site cation and oxygen ion and, R_B is the shortest distance between B- site and oxygen ion. It is clear from the figure that, both R_A and R_B increase linearly with increasing Zn content. The increase of average bond length R_A with Zn content can be associated with the increase of lattice constant (a) which may be explained on the basis of the ionic volume difference between the ions as mentioned before. It is well known that [36], from bond susceptibility in spinels, that there exists an inverse relationship between the covalent character of spinels and bond length. So, as the content of Zn increase, the bond length R_B also increase, indicating the decrease of long- covalent character with the increase of Zn content. Similar results have been reported by Sawant et al [37] in Cu- Zn and S. S. Bellad et al [38] in Li- Cd ferrite systems.

4- Conclusions

Zn^{2+} ions substitution decrease sample's density and increase both porosity and lattice constant. A higher sintering temperature (1100°C) is necessary for densification with Zn and R_2O_3. Because of the preference of Zn^{2+} ions to occupy tetrahedral sites (A) and also due to its larger ionic radius (0.74°A), it causes a decrease in the hopping length on octahedral sites. Magnetic properties were studied and confirmed a decrease in T_C up to the critical concentration of Zn content (0.5).

6- References

[1] A. Tawfik, J. Solid State Communications, 94 (1995) 987.

[2] E. C. Snelling, A. D. Giles, *Ferrites for Inductors and Transformers*, Second Edn., NY, (1986).

[3] V. Blasko, V. Petkov, V. Rusanov, Li. M. Martinez, B. Martinez, J. S. Munoz, M. Mikhove, J. Mag. Mag. Mater., 162 (1996) 331.

[4] M. A. Ahmed, Phys. Stat. Sol. (a) 111 (1989)567.

[5] S. A. Mazen, A. E. Abd El- Rahim, J. Mater. Sci., 23(1988)2917.

[6] O. S. Josyulu and J. Sobhanadri, J. Phys. Stat. Sol. (a)59 (1980)323.

[7] S. Phanjoubam, D. Kothari, J. S. Boijal, Phys. Sol. (a) 111 (1989)131.

[8] M. El- Saadawy, M. M. Barakat, J. Mag. Mag. Mater., 213(2000)309.

[9] D. R. Mane, U. N. Devatwal, K. M. Jadhav, J. Mater. Lette., 44 (2000)91.

[10] H. J. Richter, J. Phys. D. Appl. Phys. 32 (1999) R 147.

[11] Musbah Ul Islam, Mazhar Uddin Rana, Tahir Abbas, J. Mater. Chem. And Phys., 57 (1998) 190.

[12] Musbah Ul Islam, K. A. Hashmi, Mazhar Uddin Rana, Tahir Abbas, J. Sol. St. Comm., 121 (2002) 51.

[13] K. Pyo Chae, Jae- G. Lee, W. Ki Kim, Y. B. Lee, J. Mag. Mag. Mater., 248 (2002)236.

[14] E. Rezlescu, N. Rezlescu, P. D. Popa, L. Rezlescu and, C. Pasxicu, J. Phys. Stat. Sol. (a) 162 (1997) 673.

[15] J. Smit and H. P. J. Wijn, Les Ferites, Danod, Paris (1961).

[16] G. I. Ciufarov and Yu. P. Vorobiev, in: Fizicheskaya Khimiya Okislov Metallov, Izd, Moscow (p. 135) (1981).

[17] H. Gleiter, J. Weissmuller, O. Wollersheim, R. Wurschum, Acta Mater., 49 (2001)737.

[18] M. A. Ahmed, N. Okasha, M. Gabal, J. Mater. Chem. Phys., 83 (2004)107.

[19] N. Rezlescu, E. Rezlescu, C. Pasnicu and, M. L. Craus, J. Phys. Condens. Matter 6 (1994)5707.

[20] B. D. Cullity, *Elements of X-ray Diffraction* (Addison-Wesley Publishing, Reading, MA, 1956).

[21] J. M. Song and J. G. Koh, J. Mag. Mag. Mater., 152 (1996)383.

[22] D. Ravinder, J. Appl. Phys., 75 (1994)6161.

[23] A. M. El- Sayed, Ceram. Int., 28 (2002)363.

[24] R. David, *Handbook of Chemistry and Physics* (CRC Press, New York, 1995) p. 985.

[25] Viswanathan and V. R. K. Murthy, "Ferrite Material Science and Technology" Narosa Publishing House (1990).

[26] G. Blasse, Philips Research Rept. Supplement, 3 (1964)13.

[27] S. A. Patil, S. M. Otari, V. C. Mahajan, M. G. Patil, A. B. Patil, M. K. Soudagar, B. L. Batil and S. R. Sawant, Solid State Commen., 78 (1991)39.

[28] S. Unnikrishnan, D. K. Chakrabarty, Magnetic properties, Phys. Stat. Sol., (a) 121 (1990) 265.

[29] B. F. Levine, Phys. Rev., B7 (1973) 2591.

[30] C. Rath, S. Anand, R. P. Das, K. K. Sahu, S. D. Kulkarni, S. D. Date, N. C. Mishra, J. Appl. Phys., 91 (4) (2002)2211.

[31] A. N. Patil, R. P. Mahajan, K. K. Patankar, A. K. Ghatage, V. L. Mathe, S. A. Patil, Phys. Ind. J. Pure Appl., 38 (2000) 651.

[32] M. A. Ahmed, N. Okasha, A. Ebrahem, J. Ceramics International, 31 (2005) 361.

[33] J. M. Hastings, L. M. Corlis, Phys. Rev., 104 (1956) 328.

[34] M. S. Ramana, J. Mater. Sci. Lett., 3 (1984) 1049.

[35] G. Winkler *Magnetic Properties of Materials* Inter. University Electronics Series, Vol. 13, edited by J. Smit (McGraw Hill Book Co., New York, 22, 1971).

[36] V. M. Talanov, Izv. Ruzov. Khimiya ikhim. Takhnol., 12 (1978) 1395.

[37] S. R. Sawant, S. S. Suryavanshi, Curr. Sci., 57 (1988) 12.

[38] S. S. Bellad, R. B. Pujar, B. K. Chougule, Mater. Chem. And Phys., 52 (1998) 166.

ROLE OF Cu²⁺ CONCENTRATION ON THE STRUCTURAL, SPECTROSCOPIC AND MAGNETIC PARAMETERS OF Y³⁺ SUBSTITUTED Ni-Zn FERRITE

M. A. AHMED*, M. M. EL-SAYED† AND S. I. EL-DEK

Materials Science Lab. (1), Physics Dept., Faculty of Science, Cairo University, Giza, Egypt
†Physics Dept., Faculty of Education, Ain Shams University, Roxy, Cairo, Egypt

Abstract

The samples under investigation of the formula $Ni_{0.7}Zn_{0.3}Cu_ZY_{0.01}Fe_{1.99-z}O_4$; $0 \leq z \leq 0.1$ were prepared using standard ceramic technique from pure analar oxides (BDH). IR spectroscopic analysis has been carried out for all samples. The magnetic susceptibility for the samples was performed using Faraday's method as a function of temperature and magnetic field intensities. The data were interpreted in view of the exchange interaction constant that takes place between the different cations. Variation of T_C with Cu content in the samples was observed.

Keywords: NiZnYCu ferrite- FTIR- magnetic susceptibility-Jahn-Teller distortion

Introduction

Spinel ferrites are commercially important materials because they possess excellent magnetic and electrical properties [1]. Recently, there has been a growing interest in low- temperature sintered Ni-Zn-Cu ferrites used as multilayer chip inductors because of their good electro-magnetic properties at high frequencies and low densification temperature [2].

Microstructure and magnetic properties of Ni-Zn ferrites are highly sensitive to composition, sintering conditions, grain size, type and amount of additives, impurities and the preparation methodology [3-5]. The addition of copper to Ni-Zn ferrites improve the magnetic
properties (e.g. high initial permeability and very low coercivity). The magnetic properties of Cu substituted Ni-Zn ferrite depend on the site distribution of Cu^{2+} ions and the strength of the exchange interaction among magnetic ions. It is reported that [6] the value of saturation magnetization and Curie temperature for copper substituted Ni-Zn ferrite remain practically unchanged due to partial migration of Cu^{2+} into A-site position.

J. J. Shrotri et al. [7] reported that Cu substituted $Ni_{0.8-x}Zn_{0.2}Cu_xFe_2O_4$ series showed a single phase ferrite for $0 \leq x \leq 0.2$ composition while for higher values of $x \geq 0.3$; Cu forms a secondary phase of $CuO/CuFeO_4$. Also, it contributes to the larger lattice cell parameter where Cu^{2+} has larger ionic radius (0.87 Å) than Ni^{2+} (0.83 Å) on the octahedral sites. Another effect of increasing Cu^{2+} ions is the increase of the grain size with increasing annealing temperature [1, 7-10] Caltun and Spinu [11] noted that increasing the copper substitution, decreased the average grain size and the microstructure became more uniform with fewer pores.

The aim of the present work is to improve the physical and magnetic properties of a stoichometric Y^{3+} substituted Ni-Zn ferrite by partial substitution of Cu^{2+} ions through the study of its structural, spectroscopic and magnetic characterizations.

---

*Corresponding author: Email: moala47@hotmail.com

CP 998, Modern Trends in Physics Research
Third International Conference MTPR-08
edited by L. El Nadi

Experimental

The samples were prepared using the standard ceramic technique[12-14], where analar grade form oxides (BDH) were mixed in a stoichiometric ratio according to the formula $Ni_{0.7}Zn_{0.3}Cu_zY_{0.01}Fe_{1.99-z}O_4$, where $0 \leq z \leq 0.1$. Grinding using agate mortar for 4 h was carried out for each sample. After that, the powder was compressed into pellets form using a uniaxial hydraulic press using pressure of $1.9 \times 10^8 \ N/m^2$. The pellets were calcinated (pre-sintered) at $900°C$ for 6 h with heating / cooling rate of $2°C/min$ in open atmosphere using Lenton furnace UAF 16/5(UK). In the final sintering process, the material was held at $1200°C$ for 8 h and slowly cooled to room temperature with the same rate of heating in open atmosphere. X-Ray diffractometer equipped with CuK_α radiation source ($\lambda = 15418$ Å) was used to confirm the crystal structure, where no extra lines were found, indicating the absence of any secondary phases. Infrared spectral analysis, using KBr was carried out by means of FTIR 300E Fourier transform infrared spectrometer, Jasco (Japan). Small amount of the samples were crushed for magnetic susceptibility measurements using Faraday's method at different temperatures (300-800 K) as a function of magnetic field intensity (1280- 2558 Oe).

Results and Discussion

X-ray diffraction patterns for the samples $Ni_{0.7}Zn_{0.3}Cu_zY_{0.01}Fe_{1.99-z}O_4$; $0 \leq z \leq 0.1$ shown in Fig. (1) reveals single phase spinel structure with cubic symmetry as compared with the corresponding ICDD card. It is well

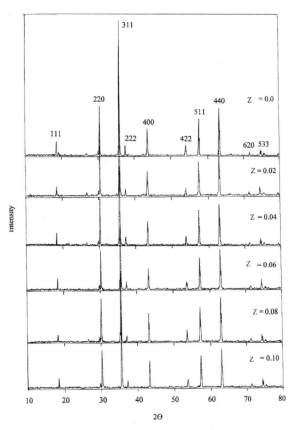

Fig.(1) X-ray diffraction patterns for the ferrite samples of the formula
$Ni_{0.7}Zn_{0.3}Cu_zY_{0.01}Fe_{1.99-z}O_4$; $0 \leq z \leq 0.1$.

130

known [15, 16] that Ni ferrite crystallizes in an inverse spinel structure. By replacing some of the Ni^{2+} by Zn^{2+} ions, the Zn^{2+} cations will prefer the tetrahedral sites while the iron ions redistribute themselves among A and B sites. Cu^{2+} ions prefer B site but up to certain content, after which some of iron ions migrate from B to A-site. Y^{3+} ions are expected to diffuse to the spinel lattice by surface diffusion since it did not giving any extra line in the XRD due to its small concentration (0.01).

The calculated lattice parameter was plotted versus Cu content, Fig. (2.a). The data clarify that at Cu^{2+} content of 0.02, the lattice parameter decreases. This can be attributed to the special preference of Cu^{2+} ions (0.87^oA) to the B sites where it replaces some of the Fe^{2+} ions (0.92^oA) [17] that were formed during the sintering process. Further increasing of Cu^{2+} content up to 0.08 in the spinel matrix, increases slightly the value of the

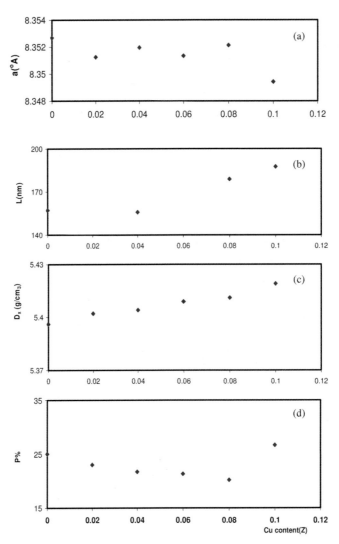

Fig.(2:a-d) Dependence of the lattice parameter, grain size, X-ray density and porosity on cu content (z).

lattice parameter. The divalent iron ions in the spinel lattice prefer the tetrahedral site due to their relatively larger size. Therefore, one can ascribe the increase in the lattice parameter with increasing Cu^{2+} content from z = 0.02 to 0.08 to the following: i) The small number of Fe^{2+} ions on the B site together with the preference of Cu^{2+} to the B site.ii) Further increase in Cu^{2+} ions , they replace some of Fe^{3+} ions (0.69 Å) with smaller radius. At z = 0.1 some of the Ni^{2+} ions (0.83 Å) transformed to Ni^{3+} ions (0.70 Å) to preserve charge neutralization, therefore decreasing the unit cell dimension.

The particle size was calculated using Sherrer equation [18] and plotted versus, Cu content (z), Fig. (2.b). The data shows that the grain size increases monotonically with z. The theoretical density was calculated using D_X =ZM/NV, [19] where Z is the number of nearest neighbors (Z=8), M is the molecular weight, N is Avogadro's number and V is the unit cell volume, the data are represented in Fig. (2.c). This means that Cu^{2+} substitution enhances the densification due to the difference between the atomic masses (Cu=63.54 and Fe=55.84) with the result of increasing D_X with z and decreasing the porosity values as in Fig. (2.d).

Figure (3.a) illustrates the infrared spectra for the investigated samples $Ni_{0.7}Zn_{0.3}Cu_zY_{0.01}Fe_{1.99-z}O_4$; $0\leq z\leq 0.1$ in the ranges (800-400 cm^{-1}). The four fundamental bands observed, indicating the spinel structure of the ferrite samples [20-22]. The assignment of the bands is reported in Table (1).

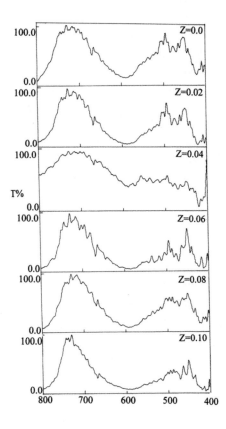

Fig.(3.a) IR transmission spectra of $Ni_{0.7}Zn_{0.3}Cu_zY_{0.01}Fe_{1.99-z}O_4$; $0\leq z\leq 0.1$.

132

Table (1): IR Transmission band for $Ni_{0.7}Zn_{0.3}Cu_zY_{0.01}Fe_{1.99-z}O_4$, where $0 \leq z \leq 0.1$.

Z	υ_1 (cm^{-1})	υ_2 (cm^{-1})	υ_3 (cm^{-1})			υ_4 (cm^{-1})
0.00	588.9	469.0	376.7	-	345.8	328.6
0.02	588.7	469.0	376.4	-	345.1	327.0
0.04	588.5, 572.7 s*	470.6	376.4	362.3	345.6	320.8
0.06	587.5	471.8	375.8	362.7	346.8	-
0.08	594.9	472.6	376.4	362.9	346.3	329.3
0.10	589.0	471.9	377.2	362.2	346.3	327.5

s* : splitting

The data in the table show also that the two high frequency bands in the ranges 588 and 469 cm^{-1} was ascribed to the stretching vibration of the tetrahedral υ_1 and octahedral υ_2 sites respectively. The ratio of the line position of υ_1 and υ_2 is given by the relation [23]

$$\upsilon_1/\upsilon_2 = (K_t/K_o)\ \sqrt{2}$$

where K_t and K_o designate the force constants associated with the unit displacement of a cation-oxygen in the tetrahedral (A) and octahedral (B) site, respectively. This ratio is nearly equal to 0.88 for all Cu content. This ratio points to the expansion of in the octahedral site is not equally compensated by the same amount of the shrinkage of the tetrahedral sites. Consequently, this leads to a slight increase in the lattice volume. This result agrees well with the increase in the lattice parameter with Cu content as obtained from XRD, Fig. (2.a).

The band υ_1 didn't show any appreciable shift by varying z between 0.0 and 0.06 which enhances our expectation for the replacement of Cu^{2+}ions on the B-sites up to this content and the stable ion content on the A-site. At z =0.04, a clear splitting and sharpening of υ_1 is observed. This can be interpreted in view of the Jahn-Teller distortion of the B-site. It is well known [24] that each oxygen anion in the spinel sub lattice is surrounded by 3 metal ions of the B-site and one metal ion of the A-site, therefore any distortion of the octahedron will be reflected as a change in the A-site size. This is in good agreement with the appearance of a splitting of υ_3 at 362 cm^{-1} at z $\geq$ 0.04 which will be discussed later. After that, some Fe^{3+} are changed to Fe^{2+} ions through A-B hopping which can result in the shift of υ_1 towards the right direction at z =0.08 predicting the lowering of the force constant as well as the electrostatic energy. This in turns leads to a slight increase in the bond lengths A-O and B-O. The result of this process is the increase in the value of the unit cell. This matches well with the results of XRD.

The band υ_3 at $\approx$ 345 cm^{-1} is ascribed to the divalent M-O stretching vibration on the octahedral site. The splitting of this band at z = 0.0 is the real consequence of the existence of Ni^{2+} with Fe^{2+} ions and the band splitted to two bands at 362 and 376 cm^{-1} at z $\geq$ 0.04. The splitting at 376 cm^{-1} does not vary in position or intensity with Cu content and may associated to the existence of a constant content of Ni^{2+} ions on the B-sites. The splitting at 362 cm^{-1} which begins to appear at z $\geq$0.04 may be associated to the Jahn-Teller distortion of the B-site due to existence of Cu^{2+} ions.

It is well known that, Cu^{2+}ion, with its $3d^9$ configuration, provides one of the best opportunities for the observation of Jahn-Teller effect. Cu^{2+} ions are found in distorted octahedra. Two modes are related directly to Jahn-Teller distortion of the (Cu^{2+}- O) octahedral with the in plane bonds differentiating in a long and short one namely (Q_1) and the tetrahedral distortion with the in plane bonds lengths shortening and the out of plane bonds extending namely (Q_2) [25-27]. The splitting at 362 cm^{-1} starts to appear at z $\geq$ 0.04 due to the pronounced

(Q₁)   (Q₂)

Fig.(3.b)

influence of Jahn-Teller effect of Cu^{2+} ion instead of iron ions on B-sites. The existence of υ_3 explicitly is a reliable indication of the formation of the inverse spinel structure. The band υ_4 at ≈ 328 cm^{-1} is ascribed to the lattice vibrations.

Figure (4:a-f) illustrates the dependence of the molar magnetic susceptibility on the absolute temperature as a function of magnetic field intensity for the samples $Ni_{0.7}Zn_{0.3}Cu_zY_{0.01}Fe_{1.99-z}O_4$; $0 \leq z \leq 0.1$. The data in the figure shows a monotonic decrease in χ_M with increasing temperature for all Cu content until reaching the Curie temperature which depends on the Cu content. The data show that the values of χ_M were decreased with increasing the magnetic field intensity due to the saturation of the moments. It is well known that [28] the mechanism of magnetization in soft ferrites results from domain wall motion which can be affected by the grain size and sintering density and spin domain rotation. But, if the grain size is less than 5μm and the cut off frequency is less than 40 MHz in weak applied fields, domain wall motion will contribute principally to the magnetic

From a closer look to Fig. (4:a-f) and (5, a) one can observe the general increase of χ_M with increasing Cu content at T= 350 K and at two different magnetic field intensities. Maximum χ_M was observed at z =0.04, this behavior gives an idea about the critical content which will be discussed later. Since the grain growth is enhanced by the densification, therefore a higher number of atoms contributing to the magnetic moment per unit volume are obtained in high density samples. This was the main reason for the increase of χ_M with increasing Cu content (z).

The increase of the Curie temperature, as illustrated in Fig.(5.b) with the increasing Cu content up to z = 0.08 is mainly attributed to the change in the superexchange interaction between the magnetic ions in the spinel lattice. It has been established experimentally [26] that the predominant exchange energies between the magnetic ions in the ferrites are negative. These ions include the ferric ions as well as the divalent metal ions substituted in the spinel ferrite samples. The magnitude of the negative exchange energies between two magnetic ions depends upon the distances from these ions M and M$^{'}$ to the oxygen anion, via which the superexchange takes place and on the angle M-O-M$^{'}$. Therefore, one could expect that by increasing Cu^{2+} ion content, the Jahn-Teller distortion is more pronounced. It well known that for inverse ferrites, the resultant moment is that of the divalent metal ion existing on the B-sites. The magnetic moments of the few remaining Fe^{3+} ions on the A-sites are no longer able to align all the moments of the B ions ant parallel to them, since this is opposed by the B-B interaction which remains unaffected. The B sub lattices will then divide itself into two sub lattices (B$^{'}$ and B$^{''}$), the magnetization of which make an angle with each other differing from 0° to 180°. This results in a decrease in the net magnetic moment of the spinel lattice as $\mu = \mu_B - \mu_A$. Another reason for the decrease of the effective moment, Fig (5.c) is the difference between the moments of Cu^{2+} (2.02 B.M) and both Fe^{2+} and Fe^{3+} ions (5.52 and 5.92 B.M) [29]

134

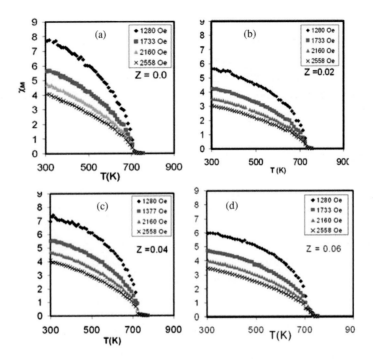

Fig.(4:a-d) The dependence of the molar magnetic susceptibility on the absolute temperature as a function of magnetic field intensity for the samples $Ni_{0.7}Zn_{0.3}Cu_zY_{0.01}Fe_{1.99-z}O_4$; $0 \leq z \leq 0.06$.

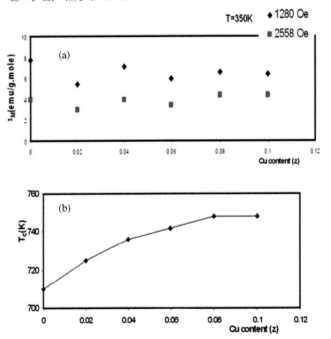

Fig.(5:a, b) The dependence of (a) the molar magnetic susceptibility (b) the Curie temperature on Cu content.

respectively. The spin orbit coupling also plays a significant role where it increases the values of μ more than the theoretical one [30]. The peculiarity appeared at z =0.04, Fig. (5:a, b) agree well with the observation of the splitting of the band υ_1 and the appearance of a new band at ≈ 362 cm^{-1} at z $\geq$ 0.04 in the IR spectra which may be attributed to the pronounced influence of Jahn-Teller effect which is reflected as the increase in the exchange interaction giving a maximum in χ_M. Also the increase of T_C is much slower as shown in Fig.(5.b).

Conclusion

1- $Ni_{0.7}Zn_{0.3}Cu_zY_{0.01}Fe_{1.99-z}O_4$; $0 \leq z \leq 0.1$ crystallized in single phase spinel structure with cubic symmetry.

2- Cu^{2+} substitution increases the grain size, enhances the densification and lowers the porosity values.

3- The four fundamental bands appear in the IR spectra assured the formation of the ferrite in single phase.

4- Jahn-Teller distortion, due to the existence of Cu^{2+} ions on the B-site, is reflected in the IR spectra.

5- The values of χ_M as well as the Curie temperature increases with Cu^{2+} content.

6- The calculated values of the effective magnetic moment decrease with increasing Cu^{2+} concentration.

7- Critical behavior, was observed at z = 0.04.

References

1-W.C.Kim, S.J.Kim, S.W.Lee, C.S.Kim, J.Magn.Magn.Mater.226-230,1418(2001)

2-Z.X.Yue, J.Zhou, X.H.Wang, Z.L.Gui, L.T.Li, Mater Sci. Lett.20, 1327(2001)

3-T.T.Scrinivasan, P.Ravindranathan, J.Appl.Phys.6318, 3789(1988)

4-E.Rezlescu, L.Sacchelarie, IEEE Trans.Magn.36, 3962-3967(2000)

5-T.Nakamura, J.magn.Magn.Mater.168, 285(1997)

6-M.Fujimoto, J.Am.Ceram.Soc.77, 2873(1994)

7-J.J.Shrotri, S.D.Kulkarni, C.E.Despande. J.Mater.Chem.Phys.59, 1-5(1999)

8-I.Z.Rahman, T,T,Ahmed, L.Dowell, J.Meta.Nano.Mater.17, 9-16(2003)

9-W.C.Kim, S.J.Kim, IEEE Trans.Magn.37, 2362-2365(2001)

10-C.W.Kim, J.G.Koh, J.Magn.Magn.Mater.257, 355-368(2003)

11-O.F.Caltun, L.Spinu, IEEE Trans.Magn.37, 2353-2355(1999)

136

12-D.Ravinder, T.Sehsagiri Rao, Crys.Res.Technol.25, 963(1990)

13-M.A.Ahmed, E.Ateia, L.M.Salah and A.A.ELGamal, Mat.Chem. and Physics (2005)

14-M.A.Ahmed, E.Ateia, L.M.Salah and A.A.ELGamal, Phys.Stat.Sol.(a)201,No.13, 3010-3022(2004)

15I.Z.Rahman, T.T.Ahmed, J.Magn.Magn.Mater.290-291, 1576-1579(2005)

16-Z.Kawano, N.Sakurai, S.Kusumi, H.Kishi, J.Magn.Magn.Mater.297, 26-32(2006)

17-R.D.Shanon, Acta crystallography, A32, 751(1976)

18-B.D.Cullity" Element of X-ray diffraction", London, 508(1976)

19-M.A.Ahmed, M.M.EL-Sayed, J.Magn.Magn.Mater.(2006)

20-K.V.S.Badrainath, Phys. Stat.Sol.A 91, k19(1985)

21-C.Prakash, J.S.Bahijal, J.Less Comm.Mat.107, 51(1985)

22-R.Manjula, V.R.Murthy, J.Sobhandari, J.Appl.Phys59(8), 2929(1986)

23-R.D.Waldron, Phy.Rev 99(6), 1727(1955)

24-J.Smit, H.P., Wijn, Ferrite, Cleaver Hume press, London (1959)

25-Y.Yamada, Phys.Rev.Lett.77, 904(1996)

26-J.Kanamori, J.Appl.Phys.31, 145(1960)

27-M.A.Ahmed, R.Seoudi and S.I.EL Dek, J.Molecular Structure, 754(1-3), 41(2005)

28-J.E.Bruke in :W.D.Kingery (Ed.), Ceramic fabrication process, Wily, New York(1958)

29-F.Albert Cotton and G.Wilkinson, "Advanced Inorganic Chemistry" Interscience Publisher, John Wiley&Sons 1966

30-L.F.Bates, "Modern Magnetism" Cambridge University Press 1948

EFFECTS OF RARE EARTH IONS ON THE QUALITY AND THE MAGNETIC PROPERTES OF Ag-FERRITES

M. A. AHMED, N. OKASHA*

Materials Science Lab. (1), Physics Department, Faculty of Science, Cairo University, Giza, Egypt
* Physics Department, Faculty of Girls, Ain Shams University, Cairo, Egypt

Abstract

A set of compounds with the formula $MgAg_{0.4}R_{0.2}Fe_{1.4}O_4$ where R is Lanthanum (La^{3+}), Terbium (Tb^{3+}), and Yttrium (Y^{3+}), were prepared by the flash combustion technique. The effect of rare earths ions on some properties of silver doped was investigated. The obtained data indicated that, by introducing a relatively small amount of rare earths ions instead of Fe_2O_3 ions, an important modification of both the structure and the magnetic properties can be obtained. The Curie temperature, effective magnetic moment and exchange interaction are affected by these substitutions. The effect of rare earth ions were explained both by their partial diffusion in the spinel lattice and by the formation of the crystalline secondary phase on the grain boundaries.
Keywords: Silver ferrites; rare earths elements; *X-ray* analyses; magnetic properties.

Introduction

The spinel ferrite unit cell is based on a close packed oxygen lattice metal cations reside on 8 of the 64 tetrahedral sites (A sites) and 16 of the 32 octahedral sites (B sites). The distribution and valence of the metal ions on these sites determines the materials magnetic and electrical properties. For many years, researchers have studied the magnetic properties of spinel ferrites and the effects of substatutional metals on their magnetic and electronic properties [1].

The properties of spinel ferrites depend on cation distribution that may be suitably modified by the addition of impurity cations and, controlling the preparative on conditions, particularly, the heat treatment. The compound silver- magnesium ferrite is partially inverse spinel and the cation distribution in it is strongly dependent on the factors mentioned because of the high diffusibility of Mg^{2+} ions which have strong preference to occupy B sites and partially occupy A sites [2]. While Ag^{1+} ions, on the other hand, occupy only A sites due to stronger site preference.

The effect of substituting trivalent ions for iron in ferrites has been investigated [3-5]. Patil et al [6] studied the effect of substitution of La for iron in Cu- La ferrite and they detected two phases of spinel. The most important factors that affect the different properties of ferrites are the radius and the magnetic moment of the substituted ion.

The addition of the rare earth ions to the ferrite samples produces a change in the electrical, magnetic as well as the structural properties depending on the type and the amount of rare earth used. The rare earth ions (R_2O_3) were introduced in the starting ferrite powder prior to the calcining, because each of the substitutive can enhance one or two properties. But in some time can deleteriously affect the others.

CP 998, Modern Trends in Physics Research
Third International Conference MTPR-08
edited by L. El Nadi

138

It is known that rare earth atoms play an important role in determining the magnetocrystalline anisotropy in the 4f- 3d intermetalic compounds [7, 8]. Still now, some researches have been carried out about their influence on ferromagnetic oxides [9, 10].

The rare earth ions can be divided into two categories [11], the first one in which R_2O_3 ions with radius very close to that of the iron Fe ion, can enter into the spinel lattice, while the second one which have ionic radius larger than Fe ions, can't occupy the tetrahedral or octahedral sites [12] during the sintering process. Some of these ions will diffuse to the grain boundaries and form an isolating ultra- thin layer around the grains, leading to the formation of secondary phases.

In the engineering of advanced functional materials, doping with different rare earth ions is a well known straight forward and versatile way to tune their physical properties not only because of the lanthanide contraction which induced monotonic change of ionic radius, but also because of the different stable oxidation states and the periodical variation in magnetic moments coming from the sequential filling of electrons in their 4f shells [13].

The structure, electrical and magnetic properties of these ferrites are very sensitive to oxygen content and the particular rare earth [14]. It is known that the rare earth atoms play an important role in determining the magneto-crystalline anisotropy in the 4f- 3d intermetalic compounds [15- 17].

Accordingly, to study the effect of the ionic radius we substituted Fe^{3+} ions in $MgAg_{0.4}R_{0.2}Fe_{1.4}O_4$ with Y^{3+}, Tb^{3+} and La^{3+}, both are nonmagnetic and their ionic radius are quit different ($r_Y=0.93°A$, $r_{Tb}=1°A$, $r_{La}=1.15°A$).

The objective of the work is to investigate the rare earth ions influence on the quality and physical properties of Ag- Mg ferrite.

Experimental

Samples with the chemical formula $MgAg_{0.4}R_{0.2}Fe_{1.4}O_4$; (R = La, Y, and Tb), were prepared by flash combustion technique [18] as in the figure.

Flash method of ferrites

The stoichiometric proportions of corresponding metal nitrate solutions were mixed thoroughly for few minutes. Urea was added as an igniter. The mixture was then introduced into the electrically heated furnace (Lenton, UAF 16/5 England) at temperature 500°C for 30 min. The mixture turns into a foamy and highly porous (fluffy in nature) precursors mass in less than 5 min. The foamy powder was collected and then powdered pressed into pellet form using uniaxial press with pressure of 1.9×10^8 N/ m^2 to final sintering that is performed at 1150°C for 1h with heating rate 2°C/min. All the above samples were cooled to room temperature with the same rate as that of heating. The obtained powder was characterized by XRD and magnetization measurements. The lattice constant of the sintered samples was calculated along with the other properties- the magnetic susceptibility (χ_M), Curie temperature T_C were evaluated using Faraday's method where a very small quantity of sample was inserted at the point of maximum gradient (maximum force). The measurements of the magnetic susceptibility were performed at different temperatures (300- 700K) and as a function of magnetic field intensity (833, 1280, 1733, 2160 Oe). The temperature of the samples was measured using T- type thermocouple with junction near the sample to avoid the temperature gradient.

Results and Discussion

Figure (1) shows XRD pattern of the investigated samples $MgAg_{0.4}R_{0.2}Fe_{1.4}O_4$.

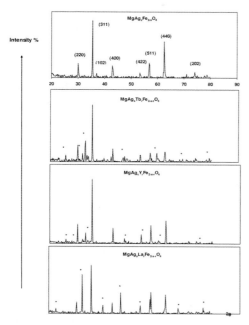

Fig. (1): XRD pattern for $MgAg_{0.4}R_{0.2}Fe_{1.4}O_4$; R: Y^{3+}, Tb^{3+}, and La^{3+}.

It is clear that, all substituted samples besides the spinel phase as major phase, the metallic Ag particle with cubic phase and the crystalline secondary phases were appeared in all cases. Secondary phases were identified by JCPDS cards as R_2O_3. The values of the lattice constant (a) are calculated and plotted in Fig. (2a) and listed in Table (1).

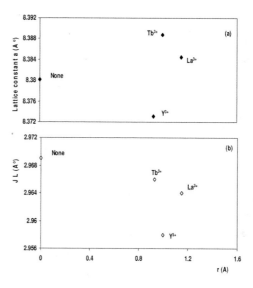

Fig. (2: a,b): a- Variation of lattice constant with different ionic radius of R^{3+} ions. b- Variations of jump length with different ionic radius of R^{3+} ions.

As the ionic radius of the rare earth ions is large compared to that of Fe^{3+} ions, internal stresses must take place in B- sites where the rare earth ions have to occupy these sites. Moreover, the increase in the lattice constant for R^{3+} ion with small radius can explain the partial incorporation of these ions in the spinel lattice. For R^{3+} ions with larger radius, a plausible explanation for decrease in the lattice constant is that the secondary phases produced on the surface of ferrite grains during the sintering process induce lattice distortions in the internal grain region. These lattice distortions which correspond to a lattice contraction might be attributed to the compression induced by the discrepancy in the thermal expansion coefficients between bulk and intergranular material.

The values of the calculated bulk density D, X- ray density D_x, relative density D_r ($D_r = D/D_x$) and the porosity P% corresponding to each substituted R^{3+} ion are given in Table (1).

Table 1: The calculated values of lattice constant a (A°), X-ray density (D_x), Grain size (L), porosity P% and jump length (JL) as a function of ionic radius of rare earth ions (r).

R^{3+}	r (A°)	a (A°)	D_x (g/cm³)	P%	Grain size	
					L (nm)	JL (°A)
non	0	8.3801	4.49	5.26	146.8	2.969
Y^{3+}	0.93	8.3731	5.43	19.05	190.5	2.966
Tb^{3+}	1	8.3888	5.15	19.04	171.2	2.958
La^{3+}	1.15	8.3844	5.35	21.95	180.2	2.964

As it is expected, samples of high relative density have low porosity. Also, the increase of porosity is due to the existence of R^{3+} on the grain boundary with Ag ions leaving cation vacancies.

Figure (2b) shows the variation of jump length (L) as a function of ionic radius of the rare earth used (r). It is clear that, the values of L which calculated from the relation L= a $\sqrt{2}/4$ [19], decreases with increasing ionic radius up to Tb (r =1A) and then increase. These results are explained on the assumption that the substitution of nonmagnetic Ag^{1+} ions instead of Fe^{3+} ions at A sites. This substitution increase the jump length of electrons at B sites and decreases that at A sites by migration of Fe^{3+} ions from A to B sites. Therefore, the effective field at A sites decreases and that of the B sites increases, indicating ferromagnetic order.

The molar magnetic susceptibility (χ_M) of the investigated samples as a function of absolute temperature is illustrated in Fig. (3: a-d). The data indicates that the Curie- Weiss constant (θ) vary slowly with increasing the field intensity and ionic radius. Also, all R^{3+} ions tend to flatten the χ_M - T curve. The figure shows that for some substituted specimens and for pure Mg-Ag ferrite. Besides, the magnetic susceptibility decreases with increasing temperature. This may be attributed to the substitution of nonmagnetic R^{3+} ions instead of Fe^{3+} ions.

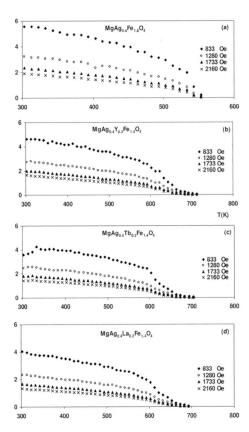

Fig. (3: a-d): Effect of different rare earth types on (χ_M).

According to the fact that the magnetic behavior of ferrites is governed by the iron- iron interaction (the spin coupling of the 3d electrons), an evaluation of this interaction can be obtained by analyzing the Curie temperature or the magnetization. The effect of partial substitution of iron with rare earth ions on Curie temperature T_C was summarized in Fig. (4a).

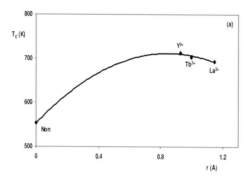

Fig. (4a): Effect of different ionic radius of R^{3+} ions on the Curie temperature T_C.

It was clear that, the R^{3+} substitution decrease the T_C, also, the Curie temperature of pure Mg- Ag ferrite is lower. Taking into consideration that R^{3+} ions substitute Fe^{3+} ions, the change of T_C can be associated with a modification of the A- B exchange interaction strength. Moreover, the decreasing of T_C for R substituted is due to the negative iron- iron interaction is that strangest and R- Fe interaction (3d- 4f coupling) has a minor influence at high temperature. In other words, the substitution effect of R^{3+} ions on the A- B magnetic interaction on the investigated ferrite depending on the ionic radius of R^{3+}, meaning; the R^{3+} ions with smaller ionic radius (Y= 0.93°A, Tb=1°A) can inter the spinel lattice (owing to high Tc) while, the R^{3+} with high ionic radius (La=1.15°A) will diffuse to the grain boundaries with Ag metallic particle (Ag= 1.26°A) and form

an isolating ultra- thin layer around grains (owing low T_C).

Figure (4b) illustrated the relation between the magnetic susceptibility (χ_M) at different ionic radius (r) as a function of different magnetic field intensity. It is clear that, (χ_M) decreases with increasing ionic radius.

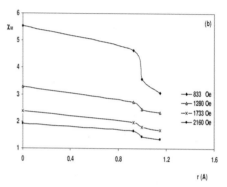

Fig. (4b): The relation between the magnetic susceptibility χ_M and the ionic radius of R^{3+} ions.

This decrease as the applied field increases. This behavior could be explained as follows; it is known that the magnetization is ferrimagnetic materials results from the rotation of the magnetic moments inside the domain wall. Kersten [20] supposed that the domain wall tend to be trapped by nonmagnetic inclusions, precipitates and voids. This trapping of the domain walls hinders their motion. Figure (4c)

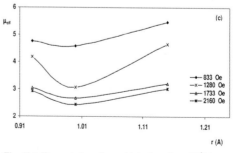

Fig. 4(c): The variation of μ_{eff} with ionic radius of R^{3+} ions.

142

shows the decreasing in the effective magnetic moment μ_{eff} with the increasing of the ionic radius at different magnetic field intensity. This was expected due to the dilution of the magnetic lattice as well as the preference of Ag^{1+} ions to the tetrahedral sites, thus decreasing the A sites magnetization (M_A). Moreover, the sample with R=Tb has the lowest value of effective magnetic moment relative to those of after ions. This could be attributed to the different cation distribution of Ag^{1+}, R^{3+} ions over A and B sites as mentioned before.

Conclusions

1- The ionic radius of R^{3+} cation play's an important role in the investigated ferrite.

2- Substitution of high valency cations namely Y^{3+}, La^{3+} and Tb^{3+} in Ag- Mg ferrites causes significant changed in structure and magnetic properties due to the effect of their ionic radius.

3- The substitution of iron ions by R^{3+} ions provide clearly improved temperature characteristics of the magnetic properties.

4- Rare earth ions facilitate the formation of crystalline secondary phases on the grain boundaries which fill intergranular voids and inhibit the grain growth by a pinning mechanism.

References

[1] V. G. Harris, N. C. Koon, *IEEE* Transactions on magnetism, **31** (1995)6.

[2] L. M. Letyuk, G. I. Zuravlev, "Chemistry and Technology of ferrites" (in Russian) (Khimiya, Leningard) p. **256** (1983).

[3] A. A. Sattar, A. H. Wafik, K. M. El- Shokrofy, M. M. El- Tabby, J. Phys. Stat. Sol. (a), **171** (1999)563.

[4] A. A. Sattar, A. H. Wafik, H. M. El- Sayed, J. Phys. Stat. Sol. (a), **186** (2001)415.

[5] E. Rezlescu, N. Rezlescu, C. Pasnicu, M. L. Crous, P. D. Pora, J. Crys. Res. Technol., **31** (1996)3.

[6] S. A. Patil, S. M. Otari, V. C. Mahajan, M. G. Patil, A. B. Patil, M. K. Soudagar, B. L. Patil and Sawant, Solid State Commun., **78** (1991)39.

[7] Yang Ying- chang, Kong Lin- shu, Sun Shu- he, Gu Dong- mei, J. Appl. Phys., **63** (1988)3702.

[8] Li Hong- shuo, Hu Bo- ping, Coey J. M. D., Solid State Commun., **66** (1988)133.

[9] N. Rezlesu, E. Rezlesu, C. Pasnicu, M. L. Craus, J. Phys. 6 (1994)5707.

[10] N. Rezlescu, E. Rezlescu, P, D. Popa, L. Rezlescu, J. Alloys Compounds 257 (1998)657.

[11] E. Rezlescu, N. Rezlescu, C. Pasnicu, M. L. Crous, P. D. Pora, J. Crys. Res. Technol., **31 (3)** (1996)343.

[12] N. Rezlescu, E. Rezlescu, C. Pasnicu, M. L. Crous, J. Phys. Condensed Matter. 6 (1994)5707.

[13] Biao Zhou, Ya- Wen Zhang, Chun-Sheng Liao, Chun- Hua Yan, Liang-Yao Chen, Song- you Wong, J. Mag. Mag. Mater., **7** (2002)272.

[14] A. Maignan, C. Martin, D. Pelloquin, N. Nguyen, B. Roveau, J. Solid State Chem., **142** (1999)247.

[15] Y. Y- chang, K. L. Shu, Z. Y- bo, S. Hong and P. X. di, J. Physique Coll. **49** C8 (1988)543.

[16] L. H. Shuo, H. Bo- Ping and Coey JMD, Solid State Commun., **66** (1988)133.

[17] Y. Y. Chang, K. L. Shu, S. S- he and G. Dong- mei, J. Appl. Phys., **63** (1988)3702.

[18] N. Balagopal, K. G. K. Warrier, A. D. Damodharan, J. Mater. Sci. Lett., **10** (1991)1116.

[19] H. Takagi, S. Ucemda, H. Eisaki, S. Tanaka, J. Appl. Phys., **63** (1988)4009.

[20] M. Kersten, Grundiagen einer Theorie der ferromanetischen Hysterese und der Koerzitivkraft, S. Hirzel, Leipzig; reprinted J. W. Edwards, Ann Arbor (1943).

II. LASERS, CHEMICAL PHYSICS & DEVICES
II-1 KEYNOTE AND PLENARY PAPERS

VACUUM TECHNOLOGY AND STANDARDIZATION-AN UPDATE

H. M. AKRAM[1] AND H. RASHID[2]

[1]National Institute of Vacuum Science and Technology (NINVAST), P.O. Box No.3125, QAU, Islamabad, Pakistan

[2]Centre for High Energy Physics (CHEP), University of the Punjab, Lahore-54590, Pakistan

Vacuum technology has been vital for the progress in almost every field of modern industrial & scientific research and technological developments. Research in this field is therefore important for the rapid progress in other sophisticated technologies. The modern society require precise know-how of vacuum metrology for its complex and sophisticated manufacturing processes and research activities. Accuracy in vacuum measurements is therefore an essential need for every application. The required accuracy is achieved with the help of well-calibrated vacuum gauges and this is possible only, if there exist proper vacuum standards of required range and accuracy. In this paper, a brief review of recently developed different vacuum standards, namely Standard Mercury Manometer, Standard Volume Expansion System and Standard Orifice Flow System will be presented, employed for the calibration of low, medium and high vacuum gauges respectively. Our recently developed standards are simple in design, least in vibration & degassing rate with desired accuracy, ease of operation and cost effective.

1 Introduction

1.1 Vacuum Technology

Vacuum has proved to be a blessing for the mankind. It had a very interesting history. In the beginning, it had been a common topic of philosophical debate. It was argued in ancient Greek times that when air or other gases have been removed from a certain volume, there remains nothing and "how can nothing be something?"[1]. But over the period, with time-honored vacuum experimentation, our understanding of this phenomena changed. Many experiments regarding the production, measurement and justification of vacuum [2] were performed in this field which kept on progressing during the past centuries. The American Vacuum Society in 1958 defined vacuum as "At Standard Temperature and Pressure (S.T.P) any system having particle density less than about 2.5×10^{19} moecules/cm^3 is called a system under vacuum" [3]. The tools used to produce measure and maintain various levels of vacuum, comprise of what may be termed as "Vacuum Science and Technology". In the 20th century vacuum science and technology has made tremendous progress and contributed significantly towards the furthering of scientific research and manufacturing. Furthermore, because of its association of research in physics, the range of applications has extended to other important sectors of industrial activity including metallurgy, mechanical, electrical, electronics, mechatronics, chemical engineering and other such disciplines, making an incalculable contribution. Therefore, now it can be said that Vacuum is nothing, but everything to us!

1.2 Vacuum Metrology

The science of vacuum measurement is called Vacuum metrology. This metrology due to its specific significance has always been the back bone of vacuum science and technology. Principally this requires the development of devices which can accurately measure low pressure of widely differing magnitudes, from 10^3 mbar to 10^{-16} mar and beyond. Low pressure measurement of broad range is relatively difficult in the sense that there is neither a single effect nor a perfect gauge that can be used to measure the whole vacuum span from atmospheric to extreme ultra high vacuum. Therefore, many gauges have to be employed for measuring vacuum, depending on desired range [4]. From the vacuum physics point of view, this is because of some logical reasons: in the low vacuum range, the gas flow is viscous; mutual interactions of the molecules with one another establish the nature of the flow. In medium vacuum range, a region of Knudsen flow occurs, in which there is a transition from viscous to the molecular flow. For high and ultra high vacuum region, there exists molecular flow, characterized by the molecules which can move freely with practically no mutual interference. Therefore

CP 998, Modern Trends in Physics Research
Third International Conference MTPR-08
edited by L. El Nadi

146

while going from one pressure range to another, the vacuum physics changes, causing the application of different physical principles for pressure measurement, and consequently the application of various gauges. These gauges fall into different main groups: mechanical phenomena gauges, transport phenomena gauges, ionization phenomena gauges and analyzing phenomena gauges [5]. Furthermore, degree of vacuum varies linearly with molecular density, thus defining different vacuum levels and corresponding gauge regions: pressure exertion region, thermal conductivity region, ionization region, enhanced ionization region, and gas analyzing region. For understanding, various parameters are shown in Fig. 1 [6].

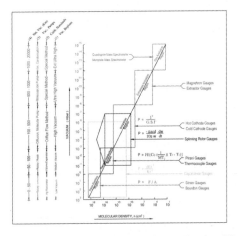

Fig.1: Molecular density verses degree of vacuum [6].

1.3 Vacuum Standardization

Vacuum gauge is a key to the sealed vacuum system which demands that it should be very accurate in its performance to offer precise vacuum measurements for variety of special purposes. The required accuracy can be achieved with the help of some key master system that is designated as having the highest metrological exclusivity. In such a system, the pressure is deduced directly from the involved standard physical quantities (mass, length, time etc.) uniquely with the best accuracy, high precision, finest resolution and highest reliability, known as the primary vacuum standard. It can be believed that it is established or widely recognized as a novel model of authority or excellence [7]. The unambiguous development of such standard systems for the truthful and consistent calibration

of vacuum gauges is called vacuum standardization. This standardization is main area of expertise of vacuum metrology, which is the core of vacuum technology.

2 Primary Vacuum Standards

Briefly, in order to precisely calibrate the vacuum gauges in the needful range for various vacuum applications, three primary vacuum standard systems are recognized. Mercury Manometer Systems (MMS) are mainly used for low vacuum standardization. For medium vacuum standardization, Volume Expansion Systems (VES) are usually employed whereas Orifice Flow Systems (OFS) are suitable for high vacuum standardization. Typical physical principles and particular scientific techniques are employed for such vacuum standard systems. These techniques are primarily correlated with the design parameters of the standards.

2.1 Mercury Manometer System

2.1,1 Principle

As illustrated in Fig. 2: a pressure applied to the left side mercury surface of U-tube displaces it, which generates a differential pressure P determined by its density ρ, the displaced height Δh, and the local acceleration due to gravity g. When the applied pressure and displaced liquid are in equilibrium, allowing for a non-zero reference pressure P_{ref} on right side of the column, absolute pressure P is given by Eq. (1) [8].

$$P = \rho \, g \, \Delta h + P_{ref} \qquad (1)$$

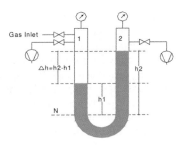

Fig. 2: Simple Model of a Liquid Manometer [8].

Mercury Manometer Systems are normally used for the calibration of low vacuum gauges. These may be as simple as a plastic U tube next to a vertical ruler. However, as the demand for precision increases so does the complication and cost. [7]. Difference between various types of manometers arises critically in the techniques used to measure the differential height.

2.1.2 Ultrasonic Interferometer Manometer (UIM) - KRISS

The technique employed by KRISS for its three columns UIM is rather different. Height difference due to the difference in pressure between mercury columns is measured using ultrasonic interferometry [9]. A small pulse with a carrier frequency of 10 MHz is sent to the transducer. This signal is reflected at the surface of the mercury column and returns to the transducer. This procedure continues until all the ultrasonic energy is lost from the signal by absorption and diffraction. The echoes received by the transducer are amplified, and then the phase sensitivity is detected by employing two detectors with a 90° phase shift between their respective reference signals. The amplitude of the phase sensitive detected signals is proportional to the amplitude of the received signal and to the phase shift between the measured and reference signals. When the liquid-column length changes, phase of each received echo also changes. From these observations, liquid differential height is determined. The electronic circuits for excitation of the transducers are attached to the bottom of three mercury columns. Figure 3 shows the photograph of the UIM [10].

Fig. 3: Photograph of UIM at KRISS [10]

2.1.3 Ultrasonic Interferometer Manometer (UIM) - NIST

NIST has also developed a similar primary vacuum standard, for the calibration of gauges with four liquid-column manometers [11]. Although the principle at the back of primary standard is simple, the manometers are operationally complicated, requiring accurate determination of different parameters: temperature, density, gravity, speed of sound, and ultimately column height. Similar to KRISS an ultrasonic interferometric technique is used in this standard system. In this system custom electronics and software implement a pulse-echo technique that uses the change in phase between sent and received radio-frequency acoustic signals to determine the length of a mercury or oil column. For thermally stabilized controlled environment, these manometers are housed in a double-walled room in which the air temperature is typically controlled to within fine limits. NIST system is shown in Fig. 4.

Fig. 4: Photograph of UIM at NIST [11]

2.1.4 Mercury Manometer System (MMS) – Pakistan

This standard mercury manometer system shown in Fig. 5 [12], has recently been established in Pakistan for the calibration of low vacuum gauges. It consists of a cistern that is a small stainless steel container used as mercury reservoir as well as its first column, connected with a long glass tube used as second column. The whole apparatus is mounted on MS-base plate having 2-leveling meter fitted on its upper side and 3-leveling screws installed in its lower side. This arrangement facilitates to align the

148

Fig. 5: New Hg Manometer [12]

instrument properly for accurate observations. The system is then placed on a massive MS table, which considerably reduces the vibration at the mercury surface for the determination of its exact position. All MS-parts of the system are Ni-plated. A micrometer is fitted on the cistern in order to measure mercury height h1 in the cistern. An accurately ruled scale is attached with the glass tube for measuring height of Hg h2 in this tube. This is done with the help of a scale sliding needle which indicates the height of Hg surface on the ruler scale. Large bore of cistern and glass tube is used to reduce the effect of capillarity. The Hg column measuring system is covered with acrylic sheet for thermally stable environment while the vacuum system essentially required is connected with this system at its lower side. This standard manometer has been developed according to proper design philosophy with many good points like ease of operation, compactness in fabrication, correctness in observation and especially cost effectiveness of the device.

2.2 Volume Expansion System (VES)

2.2.1 Principle

Volume expansion system is also known as static, Knudsen, or series expansion system. It is used for the calibration of medium vacuum gauges. Figure 6 shows that in such a standard system, a small volume V1 of gas at a known high reference pressure P1 and temperature T1, is expanded into a larger evacuated volume V2 at temperature T2, thus generating a lower pressure P2 that can be calculated from the known reference pressure (relatively high), measured volume ratio and in accordance with Boyle's law at constant temperature is given by [8]

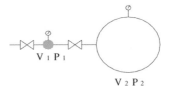

Before Expansion

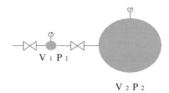

After Expansion

Fig. 6: Principle of Volume Expansion [8]

$$P_2 = \{V_1/(V_1+V_2)\}\,(T_2/T_1)\,P_1 \qquad \text{(A)}$$

$$P_2 = R\,(T_2/T_1)\,P_1$$

where R is the volume ratio of the volumes involved. These devices are simple in concept. Volume expansion systems have been built and are in regular use in international standards laboratories for the accurate generation of pressures down to especially medium vacuum range.

2.2.2 Static Expansion System (SES) - KRISS

The KRISS SES consists of a gas reservoir (D) and three vacuum chambers (A, B, and C) having different volumes. Different vacuum pumping systems consisting of turbo- molecular pumps backed by rotary pumps are used for chambers evacuation. Ionization gauges and some capacitance diaphragm gauges (CDGs) are used for measuring the pressure and volume ratio, respectively. A Quartz Bourdon Gauge (QBG) is used for measuring the reference pressure of chamber A while the ionization gauge is used to measure the ultimate pressure of chamber C. For temperature measurement, three fine quality thermometers are used. Volume of chambers A, B, and C was found to be 73 ml, and 8.4 l, and 67 l, respectively. Volume ratio of chambers A+B+C to chamber A and chambers A+B to chamber A was 1061.61 and 116.79, respectively, while their standard deviation wase 0.9777 and 0.0359, respectively. Figure 7 [10] shows a photograph of the KRISS static expansion system.

Fig. 7: Photograph of SES at KRISS [10]

2.2.3 Static Expansion System (SES) - Mexico

A static expansion system is also developed as a primary vacuum standard at the Centro Nacional de Metrología (CENAM), the Mexican National Metrology Institute, Mexico. This system was developed in a project within the framework of a technical collaboration between Germany and Mexico. The vacuum group from Germany helped CENAM in the development of this primary standard. This

Fig. 8: Photograph of Mexican SES [13]

newly established system covers a reasonable measuring range. Photograph of this system is shown in Fig. 8.

2.2.4 Volume Expansion System- Pakistan

A 4-stage primary medium vacuum standard system, for the generation of calibration pressures, in the range 10^0–10^{-4} mbar has also been developed in Pakistan [14]. It is based on volume expansion method, whereby the range is extended to lower pressures by multiple expansions. The standard system consists of a reference volume Vr and four chambers of volume V1, V2, V3 and V4 (calibration chamber) that help achieve a pressure reduction by a factor of about 10^{-5} in the main calibration chamber after a four-step expansion. For generated pressure, Eq.-(A) for this particular standard takes the following form [15].

$$P = \left[\left(\frac{V_r}{V_r + V_1}\right)\left(\frac{V_1}{V_1 + V_2}\right)\left(\frac{V_2}{V_2 + V_3}\right)\left(\frac{V_3}{V_3 + V_4}\right)\right]P_r$$

The whole designing of the subject standard system is based on Ultra High Vacuum (UHV) technique. In order to avoid degassing of the system, special consideration is given to the choice of the materials used and other components along with their surface finish. Therefore, reference volume, different chambers and other vacuum accessories all are made of Stainless Steel (SS), with properly finished internal surfaces. In order to further reduce degassing, vacuum chambers are made of cylindrical shape, in order to get the inner wall surface area to volume ratio minimum. The actual photograph of the complete system is shown in Fig. 9.

Fig. 9: Photograph of VES at Pakistan [14].

In the simple and specific design of this standard system, due thought has been given to make use

150

of the minimum number of vacuum pumps, gauges, electronic devices and other vacuum accessories without compromising on the accuracy of measurements. All this makes the standard system user friendly with easy operation, vibration less, specifically free from out gassing problems in the defined range and especially cost effective.

2.3 Orifice Flow System

2.3.1 Principle

Orifice flow system is also known as dynamic expansion system or conductance-limited expander. Simple model of the same is illustrated in Fig. 10. The basic determinations

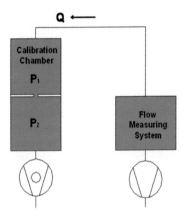

Fig. 10: Simple Model of Orifice Flow System.

for the measurement of generated pressure in the orifice flow standard system are mainly three-fold: To calculate the conductance C of the orifice in the plate partitioning the chamber, to measure the flow rate Q to the calibration chamber and to determine the pressure ratio R of the system with calibration gas. If a gas is allowed to flow in the calibration chamber from the top, a pressure drop will develop across the

$$P_1 - P_2 = \frac{Q}{C}$$
$$P_1 = \frac{Q}{C(1-R)} \qquad \textbf{(B)}$$

orifice, this pressure difference is directly proportional to flow rate and inversely to the conductance C of the orifice [16].

Where pressure P1 and P2 are in the upper and the lower chamber respectively, R= (P1/P2) which is separately and specially measured for the developed system and does not rely upon the actual measurement of absolute pressures (P1 and P2) and P is the pressure generated by the standard system. Q is calculated by the pressure drop ΔP per unit time either in a constant volume V or pressure drop ΔP per unit time in a constant pressure P [17].

2.3.2 Dynamic Expansion System (DES)-KRISS

The KRISS Ultra High Vacuum (UHV) calibration system is basically a two stage flow divider system, as shown in Fig. 11 [10]. It consists of two dynamic calibration systems: one for high vacuum and the other for ultrahigh vacuum. A calibrated CDG and a SRG is used to measure the change in pressure in the gas chamber. The High Vacuum System (HVS) was evacuated using a turbo molecular pump with a pumping speed of 345 l/s for nitrogen, and the UHV system was evacuated using a closed loop helium refrigerator-type cryopump with a nitrogen pumping speed of 1500 l/s. Bellows were used to connect the system to the pumps to avoid any vibration from the pumps. The cylindrical chambers have an internal diameter of 20 cm and heights of 30 and 31 cm for the upper and lower chambers, respectively. Round disk baffles with a diameter of 17 cm are installed at a distance of 25 cm from the orifice plate, for the Maxwellian gas distribution in chambers.

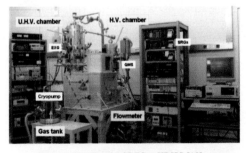

Fig. 11: Photograph of DES at KRISS [10]

2.3.3 Orifice Flow System-NIST

NIST orifice flow standard system is used for the calibration of high vacuum gauges. Similar to other such standards, in this system a known gas flow enters from the top of the upper vacuum chamber, passes through an orifice of known conductance in the centre of the orifice plate, and finally exits from the bottom of the lower vacuum chamber. The conductance of the orifice can be calculated from its known diameter, which in turn allows the pressure drop to be calculated truthfully from the conductance and flow rate according to relation of the type given in Eq.-B. NIST OFS is shown in Fig. 12 [11].

Fig. 12: Photograph of OFS at NIST [11]

2.3.4 Orifice Flow System-Pakistan

Fig. 13: Photograph of OFS at Pakistan [14]

The orifice flow standard system shown in Fig. 13 [14] mainly consists of two parts, one is flow measuring system having separate pumping station and the other is high vacuum calibration chamber. This chamber is made of stainless steel (SS-316) with I.D 305 mm and length 900 mm. An orifice plate of the same material and slightly less diameter is very finely welded at 400 mm below the chamber top end, thus partitioning this long chamber into two chambers namely upper vacuum chamber and lower vacuum chamber having length 400 mm and 500 mm respectively. A turbo molecular pump 551 l/s backed by rotary pump 16 l/s is connected with the lower vacuum chamber. Each chamber has S.S baffle plate of 25mm diameter mounted 350 mm above & below the orifice plate to develop Maxwellian velocity distribution. The orifice plate is 5.0 mm thick with 304mm diameter. The nominal orifice diameter is 9.850 mm having 0.200 mm inner sharp knife edge. The entire assembly is mounted on a massive steel table to damp vibrations due to pumping system used.

Designing, fabrication, instrumentation, and measuring techniques of this standard system all are in need of refinement to fully meet the simple devise goals and fulfill the scientific requirements. In order to reduce the uncertainties, due thought has been given to the relative positions of the gas inlet, the gauges, the orifice and the other effects. More over the system is fully refined, very simple, more compact, user friendly, accurate and especially cost effective.

Summary

In view of growing importance of vacuum science and technology, new developments are being devised from time to time. Three innovative vacuum standards are step forward in the same direction. A selective work of few recent developments on these standards is briefly presented in this work. All these developments meet the desired standards and designed based on particular needs. Our user friendly standards are simple in design with proper deliberation, in the course of compact built-up.

References

1 E:\Vacuum-Wikipedia, the free encyclopedia. Htm.

2 Niels Marquardt "Introduction to the Principles of Vacuum Physics" Institute for Accelerator Physics and Synchrotron

152

Radiation, University of Dortmund, Germany.

3 Jeremiah Mans, Vacuum Physics and Technology, Physics 4052 (2006).

4 John H., Moore; Christopher Davis, Michael A. Copland and Sandra Greer (2002). Building Scientific Apparatus. Boulder, CO: Westview Press. ISBN 0-8133-4007-1.

5 ISO 3529: Vacuum Technology-Vocabulary, Part 3: Vacuum Gauges 9(1981).

6 H. M. Akram, Int. Vac. Workshop on Renewable Energy Technologies, Islamabad, Pakistan, (2003).

7 W. R. Bryant "Physical Methods of Chemistry" Vol.6, 2nd Edition, Chap.2 (by C.R. Tilford). (1992).

8 J. M. Lafferty, "Foundations of Vacuum Sci. & Tech.", John Wiley & Sons Inc. (1998).

9 S. S. Hong, Y. H. Shin, and K. H. Chung, J. Kor. Vac. Soc. 5, 181 (1996).

10 S. S. Hong, Y. H. Shin, and K. H. Chung, J. Vac. Sci. Technol. A 24, 5 (2006).

11 http://www.cstl.nist.gov836/836.06/ pages/P&V%20Group%20facilities.htm

12 H. M. Akram, M. Maqsood, and Haris Rashid, Rev. Sci. Instrum. 78, 7(2007).

13 J C Torres-Guzman, L A Santander1 and K Jousten, Metrologia 42, S157–S160 (2005).

14 NINVAST-CHEP preprint 13/2, (2008) and 26/4 (2008).

15 J. H Leck, Total & Partial Pressure Measurement in Vacuum Systems, 5,128 (1989).

16 P.D. Levine and J. R. Sweda, J. Vac. Sci. Technol. A 12 (4) (1994).

17 Poulter K.F, Review Article J. Phys, E- Sci, instr:10:11(1977).

THE INTERACTION OF HIGH DENSITY SHORT PULSE LASER WITH MATTER

LOTFIA EL NADI[1, 2] and MAGDY M. OMAR[1]

[1] *Laser Physics Laboratory, Physics Department, Faculty of Science, Cairo University, Giza, Egypt*
[2] *IC-SAS of HDSP Lasers, National Institute of Laser Enhanced Science, Cairo University, Giza, Egypt*
lotfianadi@gmail.com

Abstract

The study of the interaction of High Density Short Pulse (HDSP) lasers with matter is an important rapidly expanding branch of Physics. Since 1985 these lasers have been developed to generate very short pulses with typical high performance parameters: peak powers up to hundreds of TW, pulse duration less than 20 fs, pulse energy more than 2J, repetition rate up to 10 Hz and wavelength of 800nm that could be lowered through higher harmonic generation. When such photons are properly focused on a target, creation of simultaneous unprecedented conditions in the laboratories within very short time takes place, namely: brightness up to 10^{20} W/cm^2, electric fields up to 10^{11} V/cm, magnetic fields up to 10^9 gauss, temperatures of the order of 10^{12} °K, pressures ~10^9 bars and acceleration of up to 10^{26} cm/s^2. These conditions could definitely initiate severe nonlinearities within the exposed materials. The main objective of this paper is to review the studies on particle generation, particularly neutrons, by the interaction of high intensity lasers with solid or gas targets. Then propose their possible applications.

Introduction

The development of terawatt table-top lasers [1] have opened up opportunities for extensive research of interactions of high intensity (I = 10^{14}–10^{19} W/cm^2) short laser pulses (τ = 30 fs - 100 ps) with solid and gas targets. The physics of this system is considerably different from conventional interactions of longer pulses since a short time scale hydrodynamic motion is no longer a dominant factor and high density plasmas are produced. The non-linear character of the interaction is apparent [2] and a considerable part of the energy is transferred to a group of very fast particles [3]. Such lasers when focused can create quasi-instantaneous plasmas on the targets generating a medium formed of free ions as well as electrons. Inside this plasma, the transverse electric laser field can be turned into longitudinal plasma electron oscillations, known as plasma waves, which are indeed suitable for electron acceleration [4]. Additionally, due to the high laser intensity, strong quasi-static electric fields can be induced, which can accelerate ions and possibly protons [5].

This approach has unique features: a) particle bunches originating from the small laser focal volume, b) they are evoked in sub-ps laser pulses which suggest bunch duration of the same order of magnitude, c) the high electric field gradients can induce particle acceleration to high energies through significantly short distances [6]. However for these conditions to continue, high repetition rate lasers, maintaining the same focal intensities are required in order to reach valuable applications.

Neutron generation [7] is also possible through fusion of the low energy ions which are not sufficiently energetic to escape the plasma, providing confinement conditions.

CP 998, Modern Trends in Physics Research
Third International Conference MTPR-08
edited by L. El Nadi

154

In this paper we shall introduce the basic properties of high density lasers. As well as a review on recent experimental results of neutron generation encountered during the interaction of high density laser with solid or gas targets. We shall then propose their possible applications.

1. High Power Lasers

An increase of on-target laser intensities can be obtained a priori in different ways. The first one consists in decreasing the focal spot size, but we are finally constrained by beam diffraction. Another possibility would be the increase in energy but this implies bulkier amplifiers...

The easy way to circumvent these problems consists in stretching in time the laser pulse before its amplification and in allowing for its final recompression. Fig.1 summarizes the various stages of such Chirped Pulse Amplification technique:

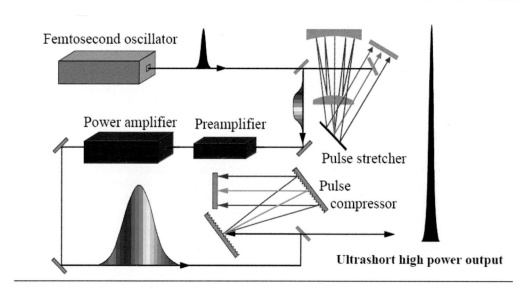

Figure 1. Chirped pulse amplification CPA after reference [1]

1. An initial large band pulse ($_{-}$ 10^{-9} J, 100 fs) delivered by an oscillator (titanium-sapphire) is stretched (one says also chirped) in time by a dispersive optical system, in general made up of one afocal and two diffraction gratings laid out in an anti-parallel configuration: the various pulse wavelengths are diffracted at different angles and follow different optical paths. A second passage in the stretcher eliminates the transverse space shift introduced during the first passage, producing a pulse 10^3–10^4 time longer than the initial pulse and with the spectral components temporally ordered, the long wavelengths before the shorter ones.

The peak pulse intensity is thus lowered below the damage threshold of the amplifier medium.

2. The "chirped" pulse is amplified in several steps by a factor 10^8 – 10^{10} in amplifiers with adequate band-width, remaining below the damage threshold of the crossed optical mediums.

3. Then a dispersive system, symmetrical to the stretcher, compresses the beam. The shorter wavelengths follow an optical path shorter than the longer wavelengths in order to compensate for the relative delays introduced by the stretching. Focusing the obtained pulse with energy of a few joules and duration lower than 1 ps, in a focal spot of ten microns diameter, leads to an on-target intensity higher than 10^{18} W/cm2 could be realized.

4. Multi-pass Ti: Sapphire amplifier is now commercially available to deliver 10 Hz pulses of energy up to 8000 mJ as shown in Fig. 2 and time duration of ~20 fs thus reaching power of ~400 TW in a beam diameter of 5.5 cm. This beam when focused to 50 μm diameter would provide brightness of approximately 20 X 10^{20} W/cm^2.

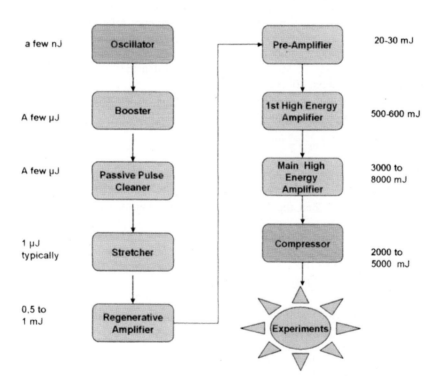

Figure 2. Multi- pass Ti: Sapphire Amplifier (www.amplitude-technologies.com)

2. Laser-Matter Interaction in the Ultra High Intensity UHI Regime

The use of Ti:Sapphire as amplifier medium to deliver pulse ranging between 1 – 2 J, with a wavelength of 815 nm, in 30 – 40 fs of installations is nowadays becoming current fact and the continuous search for increasing the interaction power is already leading soon to petawatt systems, where the intensity can approach or exceed the 10^{21} W/cm^2. Such projects are launched at LOA and CEA in France, at RAL in UK, at GSI in Germany, at ILE in Japan, at INRS in Canada, at CUOS and UNL in USA.

The development of such lasers is mainly justified by the extreme physical conditions which they make possible. Indeed the electric field corresponding to a linearly polarized laser pulse is

$$E \ [Vm^{-1}] = 2.7 \times 10^{12} \ I18^{1/2}$$

where I18 is the laser intensity expressed in units of $10^{18} W/cm^2$. For I18 = 1, the laser field corresponds to approximately 5 times the field binding in an electron to an hydrogen atom

$$EH \ [V \ m^{-1}] = 5 \times 10^{11} \ Vm-1.$$

Any atom subject to such a field is thus ionized practically instantaneously.

Immediately after ionization the electron oscillation velocity in the laser electric field is very close to the speed of light c, and plasma thus created becomes relativistic. In this relativistic regime, we observe:

• The conversion of a large fraction ($\approx 0.1 - 0.5$) of the laser energy into very energetic particles (until tens and even hundreds of MeV). The currents and densities of current in the target can reach extreme values, of about 10^7 A and 10^{12} Acm^{-2}.

• The laser wave propagation beyond the critical density n_c, as a consequence of the reduction of the effective plasma electron frequency ω_{pe} due to the relativistic increase in the electron mass,

$$\omega'_{pe} = \omega_{pe} / \ \gamma^{1/2}$$

where γ is the Lorentz factor. This phenomenon is called induced transparency [8]

The refraction index ($n < n_c$) is also modified, supporting the collimated propagation of the laser pulse on distances longer than the Rayleigh length, a phenomenon called relativistic self-focusing.

• The electromagnetic pressure of such UHI laser pulse is very high and can exceed 1 Gbar:

$$P_{rad} = 2I0/c \approx 600 \ I18 \ Mbar$$

By comparing this expression with the thermal pressure of plasma with a mass density ρ and of temperature T:

$$P_{Th} \ [Mbar] \approx 480 \ \rho \ [gcm-3] \ T[keV]$$

it is seen that a pulse with I18 = 1 is able to contain the thermal expansion of an ionized solid with a temperature of 1 keV and thus to ensure its confinement. The modification of the target surface and the creation of shock waves take place for $P_{rad} > P_{Th}$ [9].

Short pulse duration ($\approx ps$) limits the hydrodynamic expansion of the target surface and the interaction conditions are very different from those obtained in the nanosecond regime, where the phenomena of absorption take place in a wide region of subcritical and critical plasma. In the ultra-intense regime, the interaction zone has dimensions often smaller than the laser wavelength λ_0 and is much denser than in the nanosecond case: the interaction almost occurs with a solid. In practice, however, the density profile is determined by the pedestal of the UHI pulse, which can last few nanoseconds. For an intensity of I = 10^{18}–$10^{19} W/cm^2$, a temporal contrast of 1: 10^8 is sufficient to produce a coronal plasma with a gradient length

$$L_{grad} = n_e \ / \ (dn_e/dz) \approx \ \lambda_0$$

3. Applications of Laser-Matter Interaction in the Relativistic Regime

The development of UHI laser systems, able to produce new physical phenomena, allowed the development of original research in a large number of fields. The results described in this paper are limited however to the interaction of intense and short pulse laser with solid and gas targets. The study of the intense laser-generated relativistic electron currents in matter among other applications were previously reported by our group [10]

This review of research is more deeply related to the Fast Ignition concept in the inertial confinement fusion ICF context, which will be described below.

4. Fast Ignition

Fast ignition is an alternative approach to Inertial Confinement Fusion. This approach, suggested in 1994 [11], benefits of UHI lasers technology. Before entering in the Fast Ignition FI.

Specifities, let us place ourself in the more general context of thermo nuclear Fusion TNF and, in particular, of the Inertial Confinement Fusion ICF.

4.1 Thermonuclear Fusion TNF

One of the current major scientific challenges consists in finding a new source of energy, able to compensate the future exhaustion of fossil fuels (oil, gas), and, at the same time, to avoid the problems of nuclear waste reprocessing, resulting from the fission of heavy atoms. A possible way is the controlled thermonuclear fusion of light nuclei. The effective reaction is

$$D_2 + T_3 \ \text{------>} \ He_4 \,(3.5 \text{ MeV}) + n_1 \,(14.1 \text{ MeV})$$

The goal is to confine Deuterium-Tritium plasma (DT) during a sufficient long time with a rather high density and with a thermonuclear temperature (10^8 K $\approx$ 10 keV)
in order to satisfy the Lawson criterion:

$$n_e \tau \ > 10^{14} \,\text{cm}^{-3} \,\text{s}$$

where τ is the length of time of plasma confinement and n_e its electronic density. Two ways are dealt with in parallel: magnetic confinement fusion, where one exploits the confinement time, and inertial confinement fusion (ICF), where the plasma electronic density is dealt with. We are interested to explain this second way where the compression and the heating of a small mass of over dense DT are carried out using very powerful laser beams. The implosion is determined by the target hydrodynamics and the inertia of the medium.

4.2 Inertial Confinement Fusion

Fusion could be obtained according to the following steps:
1. Focusing light beams on the target.
2. Ionization of the surface of a microballon ablator which contains DT
3. Ablation due to the increase in the laser heating
4. Because of the conservation of the momentum, propagation of a centripetal shock wave towards the interior of the target.

158

5. Ignition of the nuclear fuel in a central hot spot which satisfies the conditions of the necessary temperature and pressure
6. Combustion of the DT gradually carried out by the propagation of the particle which maintain the reaction (thermonuclear burn wave).

Two different approaches are being studied to carry out the implosion:

The direct attack consists in imploding the microballon of DT by laser beams directly focused on the target. This technique is very sensitive to the inhomogeneities of irradiation at the Origin of hydrodynamic instabilities (Rayleigh-Taylor, Richtmeyer-Meshkov...)

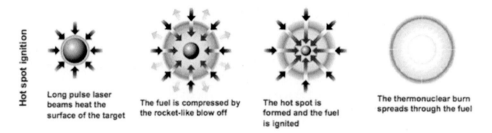

Figure 3. The scenario of direct implosion [13]

In the scenario of indirect attack, the implosion is ensured by the X-radiation emitted by the interior walls of high atomic number cavity, on which the laser beams, are focused. The microballon of DT is inside this cavity. The principal constraints of this approach relate to the X-radiation: the conversion rate of laser energy into X-radiation must be as high as possible and the thermalisation of the radiation in the cavity must be optimized to obtain an isotropic compression. Another uncertainty concerns the laser beams propagation in the cavity: to avoid plasma filling out the cavity entries, the cavity is filled with a light gas which is ionized by the lasers. The produced plasma, transparent to the laser, constitutes a medium favorable for the growth of parametric instabilities, potentially detrimental to the laser beams propagation.

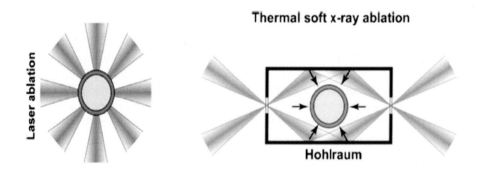

Figure 4. Scenario of Indirect Implosion [14]

The lasers under consideration for implosion have an intensity peak of the order of 10^{15} W/cm^2, pulse duration 100 fs down to 10's nanoseconds and a wavelength of 0.35 μm

4.3 Cone Guided Fast Ignition

Re-entrant conical laser target geometries have been studied for several years in the context of cone-guided fast ignition inertial fusion.

Cu – Cone   **DT microballon inside high Z number cavity**

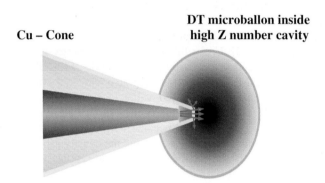

Figure 5. Novel micro- structured Cu conical target, IR fs Laser and DT micro balloon [16]

The use of sharp-tipped cone targets has been proposed as a means to increase high-intensity short-pulse laser target coupling [12]. Recently, significant improvements in both laser-acceleration of protons and material heating, have been achieved using novel micro-structured conical targets as shown in figure 5 [13, 14], in experiments at the LULI 100TW and LANL Trident lasers. Rassuchi et al reviewed the fundamental laser-cone interaction physics, and its dependence on the target geometry and the pre-formed plasma conditions observed experimentally, and Fillipo et al simulated the experiments with the aid of collisional PIC Particle In Cell simulations. They find that in the absence of pre-formed plasma, the laser pulse propagates to the cone wall, where strong (~10 MG) quasi-static magnetic fields can be formed, which appear to confine electrons (of up to 100 keV) and enhance the heating of the cone material. In the presence of pre-formed plasma inside the cone, the laser pulse is largely absorbed before reaching the wall and magnetic confinement is predicted. The generation of hot (MeV) electrons is increased and acceleration of protons by the Target Normal Sheath Acceleration TNSA mechanism can be enhanced with suitable target design with a larger shot-to-shot variability. The prospect for using such micro-structured targets for ion and x-ray sources for potential future medical applications is expected.

Two projects aiming at the demonstration of the ICF concept are currently being developed:

A) The National Ignition Facility (NIF) at the Lawrence Livermore National Laboratory [15] (LLNL) in the United States with 48 groups of 4 beams delivering energy higher than 1.5 MJ,

B) The laser MégaJoule of CEA/DAM in Bordeaux with 60 groups of 4 beams delivering energy between 1.6 and 2 MJ [16]

4.4 Simulations of Fast Ignition with Cone-Guided Targets

In the pioneering cone-guided targets for fast ignition experiments at Osaka University it was reported by Kodama et al that the fuel core was heated up to ~ 0.8 [keV] [17, 18], but efficient heating mechanisms and achievement of such high temperature were not clarified. To estimate scheme performance of the fast ignition, H. Sakagami [19] et al considered the following:

160

1) Overall fluid dynamics of the implosion,

2) laser-plasma interaction and fast electron generation, and

3) Energy deposition of fast electrons within the core.

It is, however, impossible to simulate all these phenomena with a single simulation code due to divergence of both space and time scales and one must simulate each phenomenon with individual codes and integrate them holistically.

To attack this challenging problem H. Sakagami *et al* promoted the Fast Ignition Integrated Interconnecting code (FI3) project [20]. Under this project, the Arbitrary Lagrangian Eulerian hydro code (PINOCO) [21], the collective Particle-in-Cell code (FISCOF1) [22], and the relativistic Fokker-Planck code (FIBMET) [23] were integrated with data exchanges. The result obtained was far from the experimentally reported value of 0.096 keV. The group of H. Sakagami [24] then introduced the density gap at the contact surface between the cone tip and the imploded plasma. The period of core heating became longer and the core was heated by 0.162 [keV],~69% higher increment compared with ignoring the density gap effect.

Further factors are still needed to verify the obtained experimental results.

5. Application of Neutron Sources originating from laser matter interactions

Adding to the highly important field of Fast Ignition, the generated neutrons could be greatly utilized in several applications. The following fields that could only be verified using such initiated neutron sources.

5.1 Plasma diagnostics

Deuterium material irradiated by UHI lasers can produce neutrons by the ions accelerated at the target surface, according to the reaction D + D – 3He + n. These neutrons can be used to infer plasma properties (ion temperature, plasma evolution). Photoneutrons are also produced by the interaction of the radiation described above with the target atoms [25].

5.2 Astrophysics studies

Laser-matter interactions with intensities exceeding 10^{20} W/cm2 could allow to reproduce Astrophysical conditions in the laboratory. We can thus imagine studying nuclear reaction rates in the dense matter, the physics of metals at ultra-high pressures (phase transition, metallization and hydrogen crystallization), or the physical mechanisms controlling the supernovas, stars and nebulas [26].

6. Conclusion

This review article is meant to enlighten the present status of the achievements in the field of HDSP Lasers and their interaction with matter. Particular emphasis is devoted to the generation of neutrons as well as the utilization of these interactions in the field of Fast Ignition FI. The importance of this field consists the current major scientific challenges needed to find a new source of energy, able to compensate the future exhaustion of fossil fuels (oil, gas), and, at the same time, to avoid the problems of nuclear waste reprocessing, resulting from the fission of heavy atoms. It also clarify the international projects which are launched at LOA and CEA/DAM in France, at RAL in UK, at HIPER in Europe, at GSI in Germany, at ILE in Japan, at INRS in Canada, at CUOS, (NIF) at (LLNL) and UNL in USA.

The persue of research in this field is expected to invoke new information about unprecedented fundamental facts that are yet to be discovered.

References

[1] D. Strickland and G. Mourou, Optics Communications, 56(3), 219 (1985).

[2] R.W. Lee et al., J. Opt. Soc. Am. B**20**, 770 (2003).

[3] O.L. Landen et al. J. Quant. Spectrosc. Radiat. Transf. **71**, 465 (2001).

[4] E. Nardi et al. Phys. Rev. E**57**, 4693 (1998).

[5] D. Riley, et al. Plasma Phys. Control. Fusion **47**, B491 (2005).

[6] O.L. Landen et al. Rev. Sci. Instr. **72**, 627 (2001).

[7] S.H. Glenzer et al., Phys. Rev. Lett. **17**, 175002-1 (2003).

[8] G. Gregori et al. Phys. Rev. E**67**, 026412 (2003).

[9] R. Thiele et al. J. Phys. A**39**, 4365 (2006).

[10] Lotfia M. El Nadi et al., WSP Conf. Proc. 7599, 19 (2007)

[11] J. Meyer-ter-Vehn, Nuclear Fusion, 22(4),561 (1982)

[12] Y. Sentoku et al, Phys. Plasmas 11, 3083 (2004)

[13] J. Rassuchine et al, J. Phys. Conf. Ser. 112, 022050 (2008)

[14] K. Flippo et al, Phys. Plasmas 15, 056709 (2008)

[15] J.D. Lindl et al., Physics of Plasmas, Vol. 11, 339 (2004)

[16] Jean-Pierre Chièze, CLEFS CEA - No. 49, 75 (2004) & *P. Bétrémieux, J.-J. Dupas, F. Signol*, Laser Mégajoule Project ICALEPCS, 1 (2007) *CEA/DAMFrance*

[17] R. Kodama et al., Nature 412, 798 (2001)

[18] R. Kodama et al., Nature 418, 933 (2002)

[19] H. Sakagami et al. Elsevier Int. Conf. Proc.2001, Koyoto, 380 (2002)

[20] H. Nagatomo et al.. Elsevier Int. Conf. Proc.2001, Koyoto, 140 (2002)

[21] T. Sakaguchi et al. Computer Science, 1 (2005)

[22] T. Johzaki et al. Fusion Sci. Technol. 43, 428 (2003)

[23] H. Sakagami and K. Mima, Laser and Particle Beams **22**, 41 (2004)

[24] H. Sakagami, et al., Laser and Particle Beams **24**, 191 (2006)

[25] D. Batani, et al. Phys. Rev. Lett., 94(5):055004 (2005)

[26] Bruce A. Remington, R. Paul Drake, Dmitri D. Ryutov, Rev. Mod. Phy. 78,755 (2006)

ELASTIC BACKSCATTER LIDAR SIGNAL TO NOISE RATIO IMPROVEMENT FOR DAYLIGHT OPERATIONS: POLARIZATION SELECTION AND AUTOMATION

YASSER Y. HASSEBO*

Math/Engineering Dept., LaGuardia Community College of the City University of New York
31-10 Thomson Ave., Long Island City, NY 11101, USA

KHALED ELSAYED

Department of Physics, Faculty of Science, Cairo University, Egypt

Signal-to-Noise Ratio (SNR) improvements is one of the important issue in lidar measurements, particularly for lidar daytime operations. Skylight background noise precincts lidar daytime operations and disturbs the measurement sensitivity. In the past, polarization selective lidar systems have been used mostly for separating and analyzing polarization of lidar returns for a variety of purposes. A polarization discrimination technique was proposed to maximize lidar detected SNR taking advantage of the natural polarization properties of scattered skylight radiation to track and minimize detected sky background noise (BGS). In our previous work this tracking technique was achieved by rotating, manually, a combination of polarizer and analyzer on both the lidar transmitter and receiver subsystems, respectively. Minimum BGS take place at polarization orientation that follows the solar azimuth angle, even for high aerosol loading. In this article, we report a design to automate the polarization discrimination technique by real time tracking of the azimuth angle to attain the maximum lidar SNR. Using an appropriate control system, it would then be possible to track the minimum BGS by rotating the detector analyzer and the transmission polarizer simultaneously, achieving the same manually obtained results. Analytical results for New York City are summarized and an approach for applying the proposed design globally is investigated.

Keywords: Lidar SNR Remote Sensing, Skylight noise, Polarization, Azimuth Angle, Control System

1. Introduction

Polarization selective lidar systems have, formerly, been used mostly for separating and analyzing polarization of lidar returns, for a variety of purposes, including examination of multiple scattering effects and for differentiating between different atmospheric scatterers and aerosols. [1-6] For instance, Polarization Diversity Lidars (PDL is a lidar with two channels to detect two polarizations) [7-8] are famous lidars to measure and detect clouds. Mie

---

*yhassebo@lagcc.cuny.edu

scattering is the basic theory to distinguish between cloud phases (liquid and solid) where the backscattering from non-spherical (e.g., crystal phase) particles changes the polarization strongly, but the spherical (water droplets) particles do not.[9] Both spherical and non-spherical cloud particles have a degree of depolarization ($\delta = I_\perp / I_{||}$) due to the multiple scattering effects, where $I_\perp, I_{||}$ are respectively the perpendicular and the parallel intensity components for the incident light. It is well known that the degree of depolarization in non-spherical cloud particles is greater than the degree of depolarization of spherical particles depolarization ($\delta_{NS} > \delta_S$). Previously, we succeeded in extending the polarization lidar approach to improve lidar Signal-to-Noise (SNR). [13-16] In our efforts, among others, to improve lidar SNR, we devised a *manual* polarization selective scheme to reduce the sky background signal (BGS). This approach led to improvements in SNR up to 300% and attainable lidar ranges improvement above 30%, which are important considerations in daylight lidar operations. The principles of operation for the polarization discrimination technique are well-documented [13-16] and are reviewed briefly below.

The approach discussed in our polarization selective scheme is based on the fact that most of the energy in linearly polarized elastically backscattered lidar signals retains the transmitted polarization [1, 6], while the received sky background power observed by the lidar receiver shows polarization characteristics that depend on both (1) the scattering angle θ_{sc} between the direction of the lidar and the direct sunlight, and (2) the orientation of the detector polarization relative to the scattering plane. In particular, the sky background signal (BGS) is minimized in the plane parallel to the scattering plane, while the difference between the in-plane component and the perpendicular components (i.e., degree of polarization) depends solely on the scattering angle. For a vertically pointing lidar, the scattering angle θ_{sc} is the same as solar zenith angle θ_s. The degree of polarization of sky background signal observed by the lidar is largest for solar zenith angles near $\theta_s \approx 90^o$ and smallest at solar noon. [10-12] The essence of the approach (previously reported) is therefore to first determine, manually, the parallel component of the detected sky background signal (BGS) with a polarizing analyzer on the receiver, thus minimizing the detected BGS. This parallel component in a scattering plane makes an angle equal to the azimuth angle with respect to the reference axis. Simultaneously we orient manually the polarization of the outgoing lidar signal so that the polarization of the received lidar backscatter signal is aligned with the receiver polarizing analyzer. This ensures unhindered passage of the primary lidar backscatter returns, while at the same time minimizing the received sky background signal (BGS), and thus maximizing both SNR and attainable lidar ranges.

The system geometry and measurements approach for the polarization discrimination scheme is well-documented [13-16] in our previous publications. Section 2 introduces Polarization selective scheme globalization. Diurnal variations in BGS as functions of different solar angles are given and the SNR improvement is shown to be consistent with the results predicted from the measured degree of linear polarization, with maximum improvement restricted to the early morning and late afternoon. Automated control system will present in Section 3, where the proposed controller instruments and the model description will be discussed. Conclusions and discussion are presented in Section 4

2. Polarization Selective Scheme Globalization

Solar Zenith Angle Impact on SNR

The SNR improvement factor (G_{imp}) is plotted as a function of the local time, Figure 1a, and the solar zenith angle, Figure 1b. Since the solar zenith angle retraces itself as the sun passes through solar noon, it would be expected that the improvement factor (G_{imp}) would be symmetric before and after the solar noon and depend solely on the solar zenith angle. This symmetry is observed in Figures 1a and 1b for measurements made on 19 February 2005 and is supported by the relatively small changes in optical depth (AOD) values obtained from a collocated shadow band radiometer, (morning $\tau = 0.08$, afternoon $\tau = 0.11$)

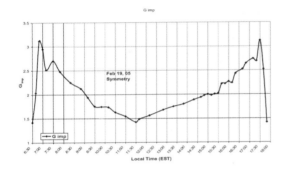

Figure 1(a). G_{imp} in detection wavelength of 532 nm verses local time (NYC EST) on Feb 19, 05

Figure 1(b). G_{imp} in detection wavelength of 532 nm verses solar zenith angle on Feb 19, 05

Solar Azimuth Angle Impact On SNR

While the magnitude of the SNR improvement factor is to some extend diminished due to scattering and depolarization, it is still important to confirm if the scattering plane defining the maximum and minimum polarization states has changed. Within the single scattering theory, the polarization orientation at which the minimum BGS occurs should equal the azimuth angle of the sun (see previous papers [13-16]). To validate this result, the polarizer rotation angle was tracked (by rotating the detector analyzer) over several seasons since February 2004 and compared with the azimuth angle calculated using the U.S. Naval Observatory standard solar position calculator [21] (14 April 2005). As expected, the polarizer rotation angle needed to achieve a minimum BGS closely tracks the azimuth angle as shown in Figure 2.

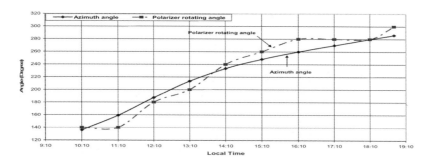

Figure 2. Comparison between solar azimuth angle and angle of polarization rotation needed to achieve minimum P_b: 14 April 2005

While it is intuitive that the maximum noise suppression should occur when the receiver polarization is parallel to the scattering plane in the single scattering regime, we have also examined the orientation of the scattering plane for the case of multiple scattering. However, we confirmed that even for high optical depth (multiple scattering regime $\tau_{aer} = 0.5$), the maximum noise improvement factor occurs when the differential azimuth angle is zero (i.e. the scattering plane and the observation plane are the same). [13-16]

This relationship is significant since it allows us to design an automated approach that makes use of a pre-calculated solar azimuth angle as a function of time and date to automatically rotate and set both the transmitted lidar polarization and the detector polarizer at the orientations needed to minimize BGS. With an appropriate control system, it would then be possible to track the minimum BGS by rotating the detector analyzer and the transmission polarizer simultaneously to maximize the SNR, achieving the same results as would be done manually as described above. An integration of an automated approach is proposed in the following section.

3. Automated Control System

The approach proposed here is a global shared control system that can be used with lidar virtually for all ground-based, in situ probes (airborne), and space-based lidar platforms. In this paper, we concentrate on typical lidar ground based stations. By knowing the longitude, the latitude and the azimuth angles during a given day, an optimization system can be applied to maximize lidar signal-to-noise ratio and corresponding lidar range automatically. We are proposing a design for an automated negative feedback position control system to minimize lidar BGS and maximize the SNR and its attainable range using our polarization discrimination technique which device by us previously. [13-16] The main advantages of this automated control system are: potential for automated data collection, fast and accurate lidar operations, applicability to different lidar configurations (vertically pointing and scanning lidars) and for different types of lidar returns (Rayleigh, Mie, Raman, DIAL, Doppler, and florescence lidars, and globalization. Finally, in the new era of remote sensing including ground-based, in situ probes (airborne), and space-based lidar platforms, this approach can be adopted, with some differences, to many space borne applications. In section 3.1

the typical information exchange between lidar devices is discussed. The control system design process is introduced in section 3.2. In section 3.3 the proposed controller instrument is discussed. Finally, in section 3.4 the control model is described.

A Typical Information Exchange Between Lidar Subsystems

- The research presented in this paper is based on the following lidar parameters, hypothesis and assumptions:
- The experimental results are to be carried out with monostatic (coaxial in the lab and biaxial in the vehicle) elastic (Mie and Rayleigh) scattering lidars, for which the wavelength of backscattered observation is the same as that of the laser
- The lidars used are lidars operating in the Visible spectral range.
- All experimental results shown above were taken during the daytime operations at the CCNY site, USA (longitude 73.94 W, latitude 40.83 N)
- The polarization is assumed to be linear polarization (the polarizer is to pass a single polarization and extinguish the orthogonal polarization)
- The single scattering regime and clear sky conditions were assumed

Figure 3 shows typical information exchanges needed, and which summarizes the interactions used as the basis for the control model operation.

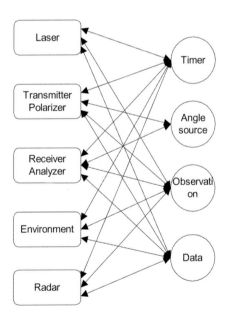

Figure 3. Typical information interactions between lidar devices

Flow for Azimuth Angle and Position Correction

The first step in a control system design is to obtain a configuration, identification of the key components of the proposed lidar system to meet a requirement 'goal''. In this section we introduce the sequential design of an automated negative feedback position control system to improve lidar SNR. This includes the controller goal, the variable parameter to be controlled, proposed hardware to be used in the control system, and finally the flow for the system setup and the SNR experiment.

Controller Goals
1- Minimize the lidar BGS
2- Maximize the Lidar return signal
3- Control system with fast response and accurate results

Components to be control:
Generally, the lidar optical components and the electronic devices are discussed in section two and listed in table I. We fixed most of these components and devices except the components that we desired to control. These components to be controlled are:

- The polarization device at the receiver subsystem
- The polarization device at the transmitter subsystem
- Power-meter at the receiver subsystem

Proposed Controller Instruments

The polarization devices at both subsystems are mounted on the rotation stages. Since we wish to rotate the polarizer at the receiver according to the azimuth angle, we select a rotation stage as the actuator. These stages can be controlled using controller devices as shown in Fig 4 and Fig 5. Programmable logic control (PLC) can be also used as a controller. Also since the microprocessor calculation speed is fast compared to the rate of change of the azimuth angle and the input signal we can consider a microprocessor as a good position controller model with very accurate measurements. A well-known closed loop position control model is the PICOMETER closed loop driver model 8751-C and/ or 8753. This model and its communications adapter cables, and the setup [22] are shown in Fig. 5 and 6.

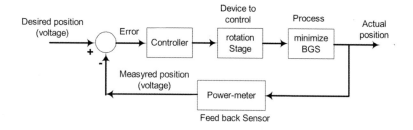

Figure 4. Block diagram of a negative feedback position control system to minimize lidar BGS

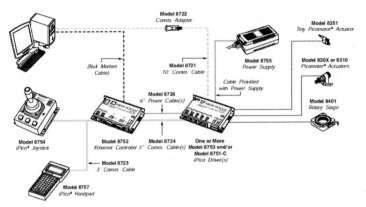

Figure 5: The Intelligent Picomotor network can be configured for different motion control applications [22] (User's Guide: Intelligent Picomotor Control Modules)

(a)   (b)

Figure 6. (a) Model 8751-C Closed- Loop Driver With (b) Model 8310 Closed-Loop Picomotor [22]

Model Description

We have developed an instrument control model design suitable for simulating a sequential theoretical design of an automated negative feedback position control system to minimize lidar sky background signal (BGS) and maximize the SNR and its attainable range. The model can be described in four stages. The first stage deals with creating a source data pool (such as date, time and the corresponding azimuth angle, and location), and then prepares the system to start. The second stage explains how to minimizing BGS at the receiver subsystem. The third stage describes how to maintain a maximum lidar return using a polarizer at the lidar transmitter subsystem. Finally the fourth stage illustrates data collection and processing. The flow chart of this proposed design is presented in Figure 7.

Model flow

Stage 1: Creating a source data pool
1- Get the lidar lab longitude and latitude
2- Create a data pool for the azimuth angle for this position according to date and time (10 minuets step)
3- Reset control system timer

Stage 2: Minimizing BGS at the receiver
4- Block the lidar transmitted beam
5- Get the azimuth angle from the data pool

6- Rotate the polarization device at the receiver subsystem according the azimuth angle

7- Measures the BGS (use it as an offset)

Stage 3: Maximizing lidar return signal

8- Unblock the transmitted beam

9- Measure the lidar return signal

10- Rotate the polarization device at the transmitter subsystem to maximize the lidar return signal (use a "For loop" supported with Power-meter and/or Labview interface)

Stage 4: Data collection/ processing

11- Start collecting/saving lidar data

12- Stop saving after 8 mins

13- Check control system timer to start the second period of measurement exactly after 10 mins from the previous period

14- Repeat steps 4 to 13

170

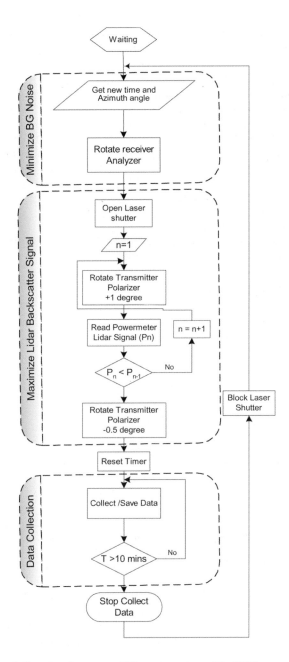

Figure 7: Flow chart for automated lidar system setup and the SNR improvement

4. Conclusion and Discussion

A polarization discrimination technique was used to maximize lidar detected SNR taking advantage of the natural properties of the scattered skylight radiation to track and minimize detected sky background noise (BGS). This tracking technique was achieved in the previous work by rotating, manually, a combination of polarizer and analyzer on both the lidar transmitter and receiver subsystems, respectively. Lidar elastic backscatter measurements at 532 nm, carried out continuously, but manually, during daylight hours, and showed a factor of $\sqrt{10}$ improvement in signal-to-noise ratio and the attainable lidar range up to 34% over conventional un-polarized schemes. . In this article, a design for an automated negative feedback position control system to minimize lidar sky background signal (BGS) and maximize the SNR and its attainable range was presented and the same factor of improvement of SNR was established. This approach can be employed for any place and time. This can be achieved by knowing the longitude, the latitude and the azimuth angles during a given day in a certain location. The main advantages of this automated control system are: potential for automated data collection, fast and accurate lidar operations, applicability to different lidar configurations (vertically pointing and scanning lidar), and for different types of lidar returns (Rayleigh, Mie, Raman, DIAL, Doppler, and florescence lidars), and globalization, as well as to lidars operated on aircraft or space platforms, with some differences, such as the A-Train.

Acknowledgements

This work is partly supported by a 2007-2008 Professional Development Grant administered by the Educational Development Initiative Team (EDIT) of LaGuardia Community Collage.

References

1. R. M. Schotland, K. Sassen, and R. J. Stone, "Observations by lidar of linear depolarization ratios by hydrometeors," *J. Appl. Meteorol.* **10,** 1011–1017, (1971)
2. K. Sassen, "Depolarization of laser light backscattered by artificial clouds," *Appl. Mete.* **13,** 923–933 (1974)
3. C. M. R. Platt, "Lidar observation of a mixed-phase altostratus cloud," *J. Appl. Meteorol.* **16,** 339–345 (1977)
4. K. Sassen, "Scattering of polarized laser light by water droplet, mixed-phase and ice crystal clouds. 2. Angular depolarization and multiple scatter behavior," *J. Atmos. Sci.* **36,** 852–861 (1979)
5. C. M. R. Platt, "Transmission and reflectivity of ice clouds by active probing," in *Clouds, Their Formation, Optical Properties, and Effects*, P. V. Hobbs, ed. Academic, San Diego, Calif., 407–436 (1981)
6. Kokkinos, D. S., Ahmed, S. A. "Atmospheric depolarization of lidar backscatter signals" *Lasers '88; Proceedings of the International Conference*, Lake Tahoe, NV, A90-30956 12-36, McLean, VA, STS Press, 538-545 (1989)
7. G.P.Gobbi, "Polarization lidar returns from aerosols and thin clouds: a framework for the analysis," *Appl. Opt.* **37,** 5505-5508 (1998)
8. N. Roy, G. Roy, L. R. Bissonnette, and J. Simard, "Measurement of the azimuthal dependence of cross-polarized lidar returns and its relation to optical depth," *Appl. Opt.* **43,** 2777-2785 (2004)
9. J. Hansen, and L. Travis, "Light Scattering in Planetary Atmospheres, "Space *Science R.* **16,** 527-610 (1974)
10. Takashi Fhjii and T. Fukuchi. *Laser Remote Sensing*, Taylor and Francis Group (2005)

11. Sassen, K. "Advanced in polarization diversity lidar for cloud remote sensing." *Proc. IEEE* **82**: 1907-1914 (1994).

12. Sassen, H. Z. K., et al. "Simulated polarization diversity lidar returns from water and precipitating mixed phase clouds." *Appl. Opt.* **31**: 2914-2923 (1992).

13. Yasser Y. Hassebo, Barry Gross, Min Oo, Fred Moshary, and Samir Ahmed "Polarization discrimination technique to maximize lidar signal-to-noise ratio for daylight operations" *Appl. Opt.* **45,** 5521-5531 (2006)

14. Yasser Y. Hassebo, B. Gross, F. Moshary, Y. Zhao, S. Ahmed "Polarization discrimination technique to maximize LIDAR signal-to-noise ratio" in *Polarization Science and Remote Sensing II*, Joseph A. Shaw, J. Scott Tyo, eds., *Proc. SPIE* **5888**, 93-101 (2005)

15. Yasser Y. Hassebo, Barry M. Gross, Min M. Oo, Fred Moshary, Samir A. Ahmed "Impact on lidar system parameters of polarization selection / tracking scheme to reduce daylight noise" *in Lidar Technologies, Techniques, and Measurements for Atmospheric Remote Sensing,* Upendra N. Singh, ed., *Proc. SPIE* **5984,** 53-64 (2005)

16. S. Ahmed, Y. Hassebo, B. Gross, M. Oo, F. Moshary, "Examination of Reductions in Detected Skylight Background Signal Attainable in Elastic Backscatter Lidar Systems Using Polarization Selection", in *23rd International Laser Radar Conference (ILRC),* Japan (2006)

17. Agishev, R. R. and a. A. Comeron "Spatial filtering efficiency of monostatic biaxial lidar: analysis and applications." *App. Opt.* **41**: 7516-7521 (2002).

18. Yasser Hassebo, Ravil Agishev, F. Moshary, S. Ahmed, Optimization of biaxial Raman lidar receivers to the overlap factor effect, in *the Third Annual NOAA CREST Symposium* Hampton Virginia, USA, April (2004)

19. Yasser Hassebo, Khaled El Sayed, "The Impact of Receiver Aperture Design and Telescope Properties on lidar Signal-to-Noise Ratio Improvements", in *AIP Conference Proceedings* **888**, 207-212 (2007)

20. Welton, E., J. Campble, et al. First Annual Report: The Micro-pulse Lidar Worldwide Observational Network, Project Report (2001).

21. Solar Calculator Webpage: http://aa.usno.navy.mil/data/docs/AltAz.html

22. User's Guide: Intelligent Picomotor Control Modules

ANALYSIS OF PLASMA PRODUCED BY LASERABLATION FOR COPPER

A. I. REFAIE

amal_hazzaa@yahoo.com

I. EL GHAZALI, S. H. ALLAM and Th. M. EL SHERBINI

Laboratory of Lasers and new Materials
Department of Physics, Faculty of Science, Cairo University
Giza, Cairo, Egypt

A time-resolved diagnostic technique was used to investigate the emission spectra from the copper plasma produced by the high power Q-switched Nd : YAG laser that generates 670 mJ pulses in 6 ns of duration at a frequency 10 Hz and 1064 nm pulse-laser ablation in air. Spectroscopic measurements were devoted to determine the plasma temperature by using Boltzmann plot for the spectral lines of Cu I which are free from self-absorption. Electron number density was also deduced from the stark broadening measurements for different delay times (0.5-10 μs) under the assumption of local thermodynamical equilibrium (LTE). Branching ratios for the some experimental relative transition probabilities have been determined. Calculations with a relativistic Hartree-Fock wavefunctions have been carried out for Cu II in order to place the experimental data on an absolute scale. The Results are compared with that measured and with the available data in literature.

Key words: atomic data, laser ablation, transition probabilities.

1. Introduction

In recent years, due to the interest of studying peculiar structure of closed shells, Ni-like ions have been widely applied in the laboratory and in astronomical plasma [1-7]. The data will also be applicable to solar system and spectroscopy of the interstellar medium. In addition, Laser Induced Plasma Spectroscopy (LIPS) is an alternative elemental analysis technology based on the optical emission spectra of the plasma produced by the interaction of high-power laser with gas, solid and liquid media. The LIPS is based on analysis of line emission from the laser-induced plasma, obtained by focusing a pulsed laser beam onto the sample. The physical and chemical properties of the sample can affect the produced plasma composition. Plasma characteristics and analytical performances of plasmas produced by laser strongly depend on the experimental parameters; the laser beam properties, the effect on surrounding atmosphere on the laser matter interaction and the plasma emission, the properties of the material to analyze [8]. Moreover, the plume typically produced in laser-ablation processes is also of considerable interest in plasma physics. Thus, intensive theoretical and experimental investigations are still in progress in order to obtain accurate knowledge of all the physical processes involved in the phenomenon. The

CP 998, Modern Trends in Physics Research
Third International Conference MTPR-08
edited by L. El Nadi

optical emission spectroscopy (OES) is the tool by which the plasma can be diagnosed. The diagnostics of the plasma can be done through the measurements of electron number density and excitation temperature. The ultraviolet and visible spectra of plasmas produced by N_2-laser radiation focused onto a copper target have been studied [9] in air as well as in vacuum have been recorded photographically. Characteristics of laser produced Cu plasma using spectroscopy have been also studied [10] by using a CCD camera, and a Langmuir single probe. A pulsed Nd: YAG laser of 52 mJ, 335 nm, and pulse duration 7 ns has been used for generating high density plasma in vacuum and argon buffer gas. A time-resolved diagnostic technique using the emission spectra from the plasmas produced by 1064 nm, 10 ns of metal Cu target has been investigated [11]. The spectral line-broadening mechanism was analyzed according to the obtained spectra of the excited atoms.

The aim of the work is the determination of the excitation temperature using Boltzmann plot of Cu I emission lines at different delay times. This has been done from the measurement of the intensity of its optically thin spectral lines assuming that the population of the energy levels follows the Boltzmann distribution law. The electron density measurement has been also performed using a high resolution spectroscopic technique. In addition, the experimental transition probabilities for 30 transition lines in Cu II in the wavelength range (202 – 225 nm). The absolute transition probabilities are obtained by normalizing our relative intensity values on the transition probability value of the 224.7 nm line from $3d^9 4p$ (^2D) $^3p_2$ to $3d^9 4s$ (^2D) $^3D_3$ transition in Cu II [12]. The calculated theoretical values for the transition probabilities are also included for comparison. The results show a fairly good agreement with the data in references [13-15].

2. Experimental set up

A Q-switched Nd: YAG laser type Brilliant B from Quantel is used to produce plasma by irradiating copper titanium dioxide (5% Cu + 95% TiO_2) target at the fundamental wavelength 1064 nm. The energy per pulse 670 mJ was measured at the target surface and the pulse duration is 6 ns FWHM. This corresponds to a power density of 5.6 x 10^{10} W/cm^2 for a laser focal spot radius of 0.25 mm. The laser was focused on the target using 10 cm quartz lens. However, the target was fixed on an x-y precession translational stage at a distance of 9.5 cm to avoid breakdown in air as shown in figure 1. The emission spectrum has been recorded using SE 200 Echelle spectrograph (Catalina Corp.), equipped with ICCD camera (Andor model iStar DH734-18F). The instrumental bandwidth of the system has been measured by using an Oriel low pressure Hg-lamp used for wavelength calibration and found to be 0.12 ± 0.02 nm. The image and the spectrum of the mercury lamp are shown in figure 1.

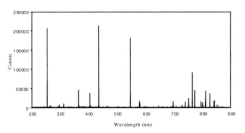

Fig. 1. Emission line spectrum of Hg lamp use for wavelength calibration.

The spectra have been acquired at different delays (1-10 μs) after the laser pulse, using constant opening gate of 2 μs of ICCD camera. The whole acquisition and operation of laser are controlled by a PC program (KestrelSpec Version 3.96). A time resolved diagnostic technique has been used to investigate the emission spectra from the plasma produced by the pulsed-laser ablation of a copper titanium dioxide target in air.

3. Results and discussion

3.1. Plasma parameters

The electron number density and the excitation temperature are considered the most important parameters used to characterize the state of the plasma. The line width of the measured spectral lines is noticeably broadened above the natural line width. Broadening mechanism in laser plasma are; Stark broadening, Doppler broadening and resonance broadening from collisions between like neutral species on strong resonance line [16-17]. In the hottest region of the laser plasma, the Stark effect leads to a broadening of the atomic and ionic emission lines as well as to a shift in the line-center wavelength. Stark broadening results from collisions of electrons (and ions) and leads to a shift in the energy levels of the emitting atom.

In case of well-isolated lines in neutral and singly ionized atoms, Stark broadening is predominantly occurred by electron collision. So the half width of these lines can be computed. The full half width $\Delta\lambda_{1/2}$ of a Stark broadened line from neutral atoms is given by [16, 18]

$$\Delta\lambda_{1/2}(\overset{o}{A}) = 2w(N_e/10^{16}) + 3.5A(N_e/10^{16})^{1/4}[1 - \tfrac{3}{4}N_D^{-1/3}]w(N_e/10^{16})$$ (1)

where $\Delta\lambda_{1/2}$ is the full width half maximum (FWHM) of the line, N_e is the electron density in cm^{-3} and w is the electron impact width parameter or Stark width parameter, A is the ionic –impact broadening parameter and N_D is the number of particles in the Debye sphere, given by[19]

$$N_D = 1.72\times10^9 \, [T_e(ev)]^{3/2} / [N_e(cm^{-3})]^{1/2}$$ (2)

The electron density measurement is evaluated using a high resolution spectroscopic technique which depends on the measurement of the Stark broadening of some optical lines [20]. The spectral line 465.1 nm of Cu I is used for measuring N_e. The Stark width parameter w of this line was taken from reference [9]. The contribution is almost entirely due to electron impact, and

therefore the half width of the Stark-broadened transition can be estimated by only the first term in equation 1. The FWHM $\Delta\lambda_{1/2}$ of the Stark-broadened of the line is given by

$$N_e = (\frac{\Delta\lambda_{1/2}}{2w})\times10^{16}$$ (3)

The Cu I spectral line at 465.1 nm arises from the upper lying energy level, then the line is not subjected by self absorption. The observed line shape is corrected by subtracting the contribution of the instrumental width, and then the Stark FWHM of the spectral line 465.1 nm of Cu I is determined after fitting the line with Voigt function at a delay time 3μs. The calculated electron density using equation 3 of the Cu plasma produced by Nd : YAG laser 1064 nm in air at different delay times(1 – 10 μs) are in the range of $(3.16\times10^{16} - 0.95 \times10^{16}$ cm$^{-3})$ for the range of obtained excitation temperatures (2.29 - 1.07 eV), and $\Delta E \sim 2$ eV. This means that, in principle, LTE conditions should prevail over the major part of plasma duration. The emitted spectral line intensity I_{mn} is a measure of the population of the corresponding energy level in the plasma. If the plasma is in LTE, thus according to Boltzmann's law [21].

$$N(T)/U(T) = \lambda_{mn}I_{mn}/g_{mn}A_{mn} \exp(E_m/kT)$$ (4)

where I_{mn}, g_i, A_{mn} and E_m are the intensity, the statistical weight, the transition probability and the excited energy level, T and k are the temperature and the Boltzmann constant respectively. $N(T)$ is the total number density and $U(T)$ is the partition function. ln (I λ /g A) versus E_{exc} has been plotted, then the so called Boltzmann plot is obtained at different delay times (1, 2, 3, 5, 7, 8, 10 μs) by using Cu I spectral lines at wavelengths 458.69 nm (3d^{9}4s5s ^{4}D$_{5/2}$-3d^9 4s4p ^{4}F$_{7/2}$), 465.11 nm (3d^{9}4s5s ^{4}D$_{7/2}$-3d^{9}4s4p ^{4}F$_{9/2}$), 515.32 nm (3d^{10}4d ^{2}D$_{3/2}$-3d^{10}4p ^{2}p$_{1/2}$), 521.82 nm (3d^{10}4d ^{2}D$_{5/2}$-3d^{10}4p ^{2}p$_{3/2}$). The atomic data of these selected lines are available in the literatures and their emission intensities are quite detectable. Atomic data of these lines are given in table 1[22, 23].

Table 1. Spectroscopic data for neutral copper lines

λ_{mn}(nm)	$E_{m\,(eV)}$	$E_{n\,(eV)}$	g_m	g_n	A_{mn}(s^{-1})
458.69	7.8051	5.1027	6	8	2.569E+07
465.11	7.7375	5.0724	8	10	4.194E+07
515.32	6.1915	3.7861	4	2	1.034E+08
521.82	6.1924	3.8169	6	4	1.221E+08

These selected spectral lines are checked to be sensitive for temperature changes in the plasma and are not influenced by self absorption, i.e., the plasma is optically thin for the lines used. The variation of the plasma temperature with different delay times has been shown in figure 2. The results indicate that the excitation temperature at the initial stage is very high and decrease quickly with time. After the end of the laser pulse, the atoms in plasma are excited while leaving the target surface and traveling in the direction of the laser source. Moreover, the impact excitation after the end of the laser pulse is the dominant process in the plasma when it is cooling. As well as higher laser irradiance gives rise to more target heating, melting and vaporization.

on branching ratio method [24-27]. The calibration of the system for measurements of the spectral intensity in the range between 200-350 nm has been carried out using standard Euler carbon arc in air. Theoretical calculations have been performed for Cu II by using relativistic Hartree-Fock (RHF) method [28]. Energy levels, wavelengths, transition probabilities have been calculated for the configurations $3d^{10}$, $3d^8\,4s^2$, $3d^9\,4s$, $3d^9\,4p$, $3d^9\,4d$, $3d^9\,4f$, $3d^9\,5s$, $3d^9\,5p$, $3d^9\,5d$, $3d^9\,5f$ and $3d^9\,5g$. The calculations are carried out with RHF in LS coupling and the relativistic correction is included also in the calculations. The interpretation of the configuration level structures are made by least-squares fit of the observed levels. The experimental data has been placed on an absolute scale with the theoretical values for the transition probabilities for 30 transition lines in the wavelength range (202 – 225 nm) for Cu II. The Results are also compared with the available data in the literature as listed in table 2. The theoretical values for the transition probabilities are also included for comparison. The results show a fairly good agreement with the data in references [13-15].

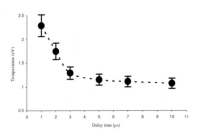

Fig. 2. Variation of the plasma temperature with different delay times (1 – 10 µs).

3.2. Determination of the transition probabilities of Cu II

For the determination of transition probabilities of Cu II spectral lines experimentally, the emitted spectral line should not affected by self absorption and well isolated. The determination of transition probability experimentally is based

177

Table 2. Transition probabilities (A_{ji}) for Cu II lines in the wavelength range (202-225) nm.

Transition Levels		λ (nm)	A_{ji} This Work ($\times 10^8$ sec^{-1})		A_{ji} Previous Work ($\times 10^8$ sec^{-1})		
Lower	Upper		Theo.	Exp.	Ref.[13]	Ref.[14]	Ref.[15]
3d^9 (^{2}D)4d ^{1}P$_1$	3d^9 (^{2}D) 4p ^{3}P$_1$	202.713	2.087	3.24	2.163	1.43	
3d^9 (^{2}D) 4d ^{3}P$_1$	3d^9 (^{2}D) 4p ^{3}P$_2$	202.99	1.418	3.28	1.418	1.01	
3d^9 (^{2}D) 4d ^{3}P$_2$	3d^9 (^{2}D) 4p ^{3}P$_2$	203.10	6.418	2.02	6.018	5.27	
3d^9 (^{2}D) 4p ^{3}D$_1$	3d^9 (^{2}D) 4s ^{3}D$_1$	203.58	4.856	3.81	4.398		3.87
3d^9 (^{2}D) 4p ^{1}F$_3$	3d^9 (^{2}D) 4s ^{3}D$_2$	203.71	1.515	1.55	1.317		1.33
3d^9 (^{2}D) 4p ^{3}D$_3$	3d^9 (^{2}D) 4s ^{3}D$_3$	204.38	1.981	1.67	1.794		2.14
3d^9 (^{2}D) 4d ^{3}F$_2$	3d^9 (^{2}D) 4p ^{3}F$_2$	205.42	2.510	2.20	1.917		
3d^9 (^{2}D) 4p ^{3}D$_2$	3d^9 (^{2}D) 4s ^{3}D$_2$	205.49	2.230	2.16	2.100		1.69
3d^9 (^{2}D) 4d ^{3}D$_2$	3d^9 (^{2}D) 4p ^{3}P$_1$	206.24	1.873	2.12	1.955		
3d^9 (^{2}D) 4d ^{3}S$_1$	3d^9 (^{2}D) 4p ^{3}P$_2$	207.86	7.110	10.3	6.810	6.55	
3d^9 (^{2}D) 4d ^{3}F$_3$	3d^9 (^{2}D) 4p ^{3}F$_3$	208.79	3.057	1.42	2.893		
3d^9 (^{2}D) 4d ^{3}P$_1$	3d^9 (^{2}D) 4p ^{3}P$_1$	209.36	2.202	3.40	2.151	2.36	
3d^9 (^{2}D) 4p ^{1}D$_2$	3d^9 (^{2}D) 4s ^{1}D$_2$	212.29	3.136	2.71	2.938		2.76
3d^9 (^{2}D) 4d ^{1}G$_4$	3d^9 (^{2}D) 4p ^{3}D$_3$	212.51	3.872	1.39	2.811	1.90	
3d^9 (^{2}D) 4p ^{3}F$_2$	3d^9 (^{2}D) 4s ^{3}D$_2$	212.60	1.806	2.52	1.658		1.75
3d^9 (^{2}D) 4p ^{3}F$_4$	3d^9 (^{2}D) 4s ^{3}D$_3$	213.59	5.093	1.80	4.671		4.45
3d^9 (^{2}D) 4d ^{3}S$_1$	3d^9 (^{2}D) 4p ^{3}P$_1$	214.54	1.557	3.45	1.244		1.18
3d^9 (^{2}D) 4d ^{3}F$_3$	3d^9 (^{2}D) 4p ^{3}F$_2$	215.18	3.844	1.48	2.542		
3d^9 (^{2}D) 4d ^{3}G$_3$	3d^9 (^{2}D) 4p ^{3}D$_2$	216.13	3.930	1.70	2.940		
3d^9 (^{2}D) 4p ^{3}F$_3$	3d^9 (^{2}D) 4s ^{3}D$_2$	219.22	3.538	2.30	3.266		3.00
3d^9 (^{2}D) 4d ^{3}F$_4$	3d^9 (^{2}D) 4p ^{3}D$_3$	219.56	3.577	1.55	3.643	3.93	
3d^9 (^{2}D) 4p ^{3}F$_2$	3d^9 (^{2}D) 4p ^{3}D$_1$	220.05	3.492	3.43	2.349		
3d^9 (^{2}D) 4p ^{3}D$_3$	3d^9 (^{2}D) 4p ^{3}D$_3$	220.98	1.784	2.79	1.675		
3d^9 (^{2}D) 4p ^{3}D$_2$	3d^9 (^{2}D) 4s ^{1}D$_2$	221.02	1.802	3.92	1.578		1.38
3d^9 (^{2}D) 4d ^{1}F$_3$	3d^9 (^{2}D) 4p ^{1}D$_2$	221.51	5.404	2.46	4.954		
3d^9 (^{2}D) 4d ^{1}D$_2$	3d^9 (^{2}D) 4p ^{3}D$_1$	221.85	1.672	4.01	2.359		
3d^9 (^{2}D) 4d ^{3}F$_2$	3d^9 (^{2}D) 4p ^{1}P$_1$	222.46	1.135	2.97	1.340		
3d^9 (^{2}D) 4d ^{3}D$_2$	3d^9 (^{2}D) 4p ^{3}D$_2$	222.67	3.496	5.23	3.137		
3d^9 (^{2}D) 4d ^{3}F$_4$	3d^9 (^{2}D) 4p ^{1}F$_3$	224.89	1.995	1.46	1.342	0.91	
3d^9 (^{2}D) 4d ^{3}D$_1$	3d^9 (^{2}D) 4p ^{1}P$_1$	225.49	1.719	4.09	1.669		

Conclusion

A Copper plasma is produced using high power Q-switched Nd:YAG laser at 1064 nm, energy 750 mJ / pulse, pulse duration 6 ns and repetition rate 10 Hz. The characteristics of the formed plasma are studied. Excitation temperature of the core of the plasma plume is estimated for Cu I in air using Boltzmann's equation. It is found that, the core of the plume has the excitation temperature in the range of (1.07 – 2.29 eV) in air at different delay times (1 – 10 μs). The results indicate that the excitation temperature at the initial stage is very high and decreases quickly with time. After the end of the laser pulse, the atoms in plasma are excited while leaving the target surface and traveling in the direction of the laser source. Moreover, the impact excitation after the end of the laser pulse is the dominant process in the plasma when it is cooling. As well as higher laser irradiance gives rise to more target heating, melting and vaporization, resulting in a higher vapor density, velocity and temperature in the plume.

The results show a fairly good agreement with the data in literatures.

References

[1] W. T. Y. Mohamed, Improved LIBS limit of detection of Be, Mg, Si, Mn, Fe and Cu in aluminum alloy samples using a portable Echelle spectrometer with ICCD camera, *Optics and Laser Technology*, 40 (2008) 30–38.

178

[2] H. E. Bauer, F. Leis and K. Niemax, Laser induced breakdown spectrometry with an échelle spectrometer and intensified charge coupled device detection, *Spectrochim Acta Part B*, 53 (1998) 1815–1825.

[3] R. E. Russo and X. L. Mao, Chemical analysis by laser ablation. (San Diego: Academic Press 1998) pp. 375-380.

[4] J. T. Davies and J. M. Vaughan, A new tabulation of the Voigt profile, *Astrophys. J.*, 137 (1963) 1302–1317.

[5] H. O. Di Rocco, Systematic trends and relevant atomic parameters for Stark line shifts and widths, *Spectrosc. Lett.*, 23 (1990) 283–292.

[6] C. Colon, G. Hatem, E. Verdugo and P. Ruis, Campos, J. Measurement of the Stark broadening and shift parameters for several ultraviolet lines of singly ionized aluminum, *J. Appl. Phys.*, 73 (1993) 4752–4758.

[7] A. N. Mostovych, L. Y. Chan, K. J. Kearney, D. Garren, C. A. Iglesias, M. Klapisch and F. J. Rogers, Opacity of dense, cold, and strongly coupled plasma, *Phys. Rev. Lett.*, 75 (1995) 1530–1533.

[8] V. Margetic, A. Pakulev, A. Stockhaus, M. Bolshov, K. Niemax and R. Hergenroder, A comparison of nanosecond and femtosecond laser induced plasma spectroscopy of brass samples, *Spectrochim. Acta Part B*, 55 (2000) 1771-1785.

[9] V. Henc-Bartolic, Z. Andreic, M. Stubicar and H.-J. Kunze, Nitrogen laser beam interaction with copper surface, *FIZIKA A*, 7 (1998) 4, 205– 212.

[10] M. A. Hafez, M. A. Khedr, F. F. Elaksher and Y. E. Gamal, Characteristics of Cu plasma produced by a laser interaction with a solid target, *Plasma Sources Sci. Technol*, 12 (2003) 185–198.

[11] B. Y.Man, Q. L. Dong, A. H. Liu, X. Q.Wei, Q. G. Zhang, J .L. He and X. Twang, Line-broadening analysis of plasma emission produced by laser ablation of metal Cu, *J. Opt. A: Pure Appl. Opt.*, 6 (2004) 17–21.

[12] J. E. Sansonettia... and W. C. Martin, Handbook of Basic Atomic Spectroscopic Data, *J. Phys. Chem. Ref. Data* 34 (2005) 1679-1683.

[13] R. L. Kuruczs, Atomic spectral line data base CD- ROM 23.

[14] M. Ortiz *et al.*, Radiative parameters for some transitions arising from the 3d94d and 3d84s2 electronic configurations in Cu II spectrum, *J. Phys. B: At. Mol. Opt. Phys.*, 40(2007)167-176.

[15] E. Biemont *et al.*, Core-polarization effects in Cu II, *Physica Scripta,* 61 (2000) 567-580.

[16] H. R. Griem, Spectral line broadening by plasmas (Academic press, New York, 1974).

[17] G. Bekefi, in principles of laser plasmas (Wiley Interscience, New York, 1976) Ch. 13, pp. 549.

[18] H. R. Griem, Plasma spectroscopy (New York, McGraw-Hill, 1964).

[19] F. F. Chen, Introduction to plasma Physics (Plenum Press, New York, 1974) Ch. 5.

[20] S. C. Wilks, W. L. Kruer, W. B. Mori, Odd harmonic generation of ultra-intense laser pulses reflected from an overdense plasma, *IEEE Trans. Plasma Sci.,* 21 (1993) 120-124.

[21] A. D. Sappey, T. K. Gamble, Laser-fluorescence diagnostics for condensation in laser-ablated copper plasmas, *Appl. Phys,* B 53 (1991) 353-361.

[22] K. Fu, M. Jogwich, M. Knebel, K. Wiesemann, Atomic transition probabilities and lifetimes for the Cu I system, At. Data Nucl. Data Tables, 61 (1995) 1-30.

[23] WWW.NIST.gov.

[24] L. J. Radziemski, D. A. Cremers, Laser induced plasmas and applications (New York: Marcel Dekker, 1989).

[25] V. Detalle, M. Sabsabi, L. St-Onge, A. Hamel, and R. Héon, Influence of Er:YAG and Nd:YAG Wavelengths on laser-induced breakdown spectroscopy measurements under air or helium atmosphere, *Applied Optics*, 42 (2003) 5971-5977

[26] B. Martinez, F. Blanco, J. Campos, Application of laser produced plasma to the study of 5s5d-5p2 configuration interaction in In II, *J. Physc. B: At. Mol. Opt. Phys.*, 29 (1996) 2997-3007.

[27] F. S. Ferrero, J. Manrique, M. Zwegers and J. Campos, Determination of transition probabilities of 3d^8 4p-3d^8 4s lines of Ni II by emission of laser produced plasmas, *J. Phys. B: At. Mol. Opt. Phys.*, 30 (1997) 893-903.

[28] Cowan R. D., The theory of atomic structure and spectra, (University of California press, Berkeley, 1981).

II-2 INVITED LECTURE PAPERS

PREPARATION OF GaN NANOSTRUCTURES BY LASER ABLATION OF Ga METAL

LOTFIA EL NADI*, MAGDY M. OMAR, GALILA A. MEHENA, HUSSIEN M. A. MONIEM

Physics Dept., Laser Communication Lab., Faculty of Science
Cairo University, Giza, EGYPT
*corresponding author lotfianadi@gmail.com

ABSTRACT: In the present study, GaN nanodots (0D) and nanowires (1D) nanostructures were prepared on stainless steal substrates applying laser ablation technique. The target of Ga metal mixed with $NaNO_2$ was introduced in a central bore of a graphite rod of a confined geometry set up. The laser beam was normally focused onto the central bore and the ablated plume of Ga metal was deposited on stainless steal substrate lying below the graphite rod in an atmosphere of slow flow of nitrogen gas with or without ammonia vapor. The pulsed N_2 laser beam having a wavelength of 337 ± 2 nm, pulse duration 15 ± 1 ns and energy per pulse of 15 ± 1 m J, could be focused on the central bore by a cylindrical quartz lens to a spot of dimensions 500 X 700 μm^2 t providing target irradiance of 0.2-0.3 GW/cm^2 per pulse. The ablated plum was collected after several thousand laser shots. The morphology and structure of the formed nanostructures were investigated by Scanning electron microscope and Energy Dispersive X-Ray Spectroscopy. The growth mechanism is most likely by Solid-Liquid-Vapor phase during the laser ablation processes. The role of the carbon, the $NaNO_2$ and the flowing gas on the growth of Nanostructures of GaN are discussed.

Keywords: Nanostructures, Laser Ablation, GaN, SEM, EDX, Growth Mechanism

1. INTRODUCTION

Once it was said small is beautiful, but we say small is smart. Since the last seven years a tremendous success has been achieved by several groups to grow 0 dimension (0D) and 1 dimension (1D) nanostructures applying chemical procedures. Important properties of such nanosized materials, are revolutionizing their application in electronic, photonic and medical fields. New properties other than that of the bulk were reported especially for semiconductors and laser semiconductors forming an era of nano-lasers and nano-optoelectronic devices [1-4].

GaN semiconductors in bulk are used as blue emitters and are already used in DVD systems and other microelectronic equipments on commercial scale [5]. The physical properties of (1D) nanostructures based on GaN compounds have been the subject of several studies recently [6-9]. The growth techniques of such interesting material have been reported by Li Yang *et.al.* [10]. They applied a two-step growth technology. The growth methods based on vapor-liquid-solid (VLS) mechanism [11] are chemical vapor deposition (CVD) method [12], or direct reaction of metal gallium with nitride gases [13], or by catalytic methods [14]. Bae *et.al* synthesized GaN nano belts by thermal reaction of gallium metal with ammonia using iron and boron oxide as catalysts [15].

Here we attempted two methods in which we applied laser ablation technique on gallium metal, under two different gas environment and catalysts. Zero dimensional GaN nano-dots and one dimensional GaN nanowires were proved to grow during each method providing different structural properties. Speculation of growth origin is discussed for both methods.

2. EXPERIMENT

The laser ablation process was performed using N_2 laser with λ 337 ± 2 nm, pulse duration τ 15 ± 1 ns and energy per pulse of 15 ± 1 m J, for up to a total focused laser power 200-500 GW/cm^2 on the target. The plasma plumes produced during the two methods; ambient flowing nitrogen gas only or with ammonia gas jet, were each allowed to deposit directly on cleaned stainless steel (Fe 91 %,Ni 4.5 %,Cr 4.5 %) substrates. The confined geometry initiated by our group and described before [16] is schematically represented in Figure 1.

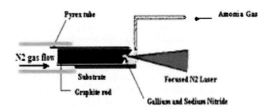

Figure 1: Schematic representation of the
experimental set up (drawn not to scale)

THE FIRST METHOD: gallium metal was introduced in the 10 mm long 3 mm diameter graphite rod. Carbon from graphite plays the role of the catalyst. Nitrogen gas flow was flushing the target set up during the ablation process to provide the nitrogen needed as vapor and to prevent the formation of Ga Oxide, when in contact with air oxygen.

THE SECOND METHOD: gallium metal was mixed with $NaNO_2$ in 1:1 ratio by % weight and then introduced in the central pore of the graphite target. Ammonia gas jet was synthesized on the spot during laser irradiation using the low rate flow of NaOH solution dropping from a separating funnel on solid NH_3Cl. The rate of ammonia gas jet flow on the target surface was adjusted to flow regularly during the experiment as shown in figure 1. The nitrogen gas flow through the target set up was also kept the same as in the first method.

The stainless steel substrates with the grown nanostructures were examined using Scanning Electron microscope (SEM) type Joel JXA-840 A, which was provided by Energy Dispersive X-Ray Spectrometer (EDX) INCA-X-Oxford Instrument micro analyzer. The morphologies of the micrographs were studied and analyzed to determine the type and size of the formed nanostructures. Interesting ideas about the origin and the growth of GaN nanostructures could then be speculated. The GaN growth was proved by X-ray diffraction using type Phil. Co- Kα line of $\lambda= 1.71°$.

3. RESULTS AND DISCUSSION

Figure 2(a) shows a typical SEM image in which appears a central 1.45 µm average diameter GaN condensed droplet on the surface of the stainless steel substrate. The solidified droplet is surrounded by dots of average diameter 40 nm and short rods average diameter 80 nm scattered in a way that might suggest that they originated from the impinged laser ablated plume on the substrate surface. This micrograph has been obtained from the samples prepared by the first method for total accumulated laser power density of ≈ 500 GW/cm^2.

A high density of parallel grown nanowires can be recognized in Fig. 2(b) having average length varying between 5 µm to 15 µm and average diameter of 300 nm and average density of 6.6 x 10^7 cm^{-2}. There is a clear formation of GaN crystallite at the tip of some wires and at some of them, it is noticed that dots are formed successively along the wire. These conditions for such growth were through the first method of preparation but at lower accumulated laser power density of ≈ 250 GW/cm^2.

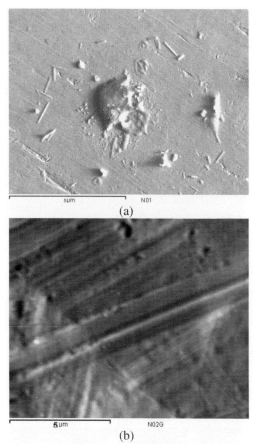

(a)

(b)

Figure 2: SEM image of GaN deposited on SS substrate, and prepared by first method (a) GaN droplets ablated by laser power density of 500 GW/cm^2 (b) GaN wires ablated by laser power density of 250 G W/cm^2

Considering the GaN grown by the second method in presence of ammonia gas the growth of nanodots exceeded any other structure as clear from Fig. 3(a). The morphology revealed the existence of both nanodots and nanowires, the average diameter of the dots 82±20 nm, the rods have interesting wurzite crystal shapes and are confirmed to have the GaN structure from the measured XRD pattern. The average width of the top surface is 160 ±40nm and the average length is 1.77±0.5μm. Figure 3(b) represents the size of nanodots confirming the average diameter mentioned above and confirming the high density of the formation of the nanodots than the rods being 7.93×10^8 cm^{-2}. These two micrographs were obtained by accumulated power 200 GW /cm^2 laser power density in ablation of Ga metal mixture with Sodium Nitride NaNO$_2$ in presence of graphite and both N$_2$ and NH$_3$ gas, for the deposited plum

In the micro graph 3c the GaN nanodots and nanowires are formed in scattered orientations under the same conditions of Figs. 3(a) and 3(b) but applying higher accumulated laser power density reaching up to 400 GW/cm^2.

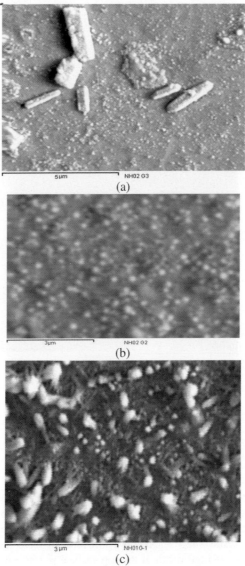

(a)

(b)

(c)

Figure 3: SEM image of GaN deposited on SS substrate, and prepared by second method. (a) GaN nanodots and nanowires ablated by laser power density of 500 GW/cm^2 (b) GaN nanodots and nanowires ablated by laser power density of 200 GW/cm^2, (c) GaN nanodots and nanowires ablated by laser power density of 400 GW/cm^2.

Selective area SEM and EDX measurements are shown in figures 4a and 4b. The spectrum shows the presence of GaN with peaks of Ga Lα and Kα lines. The unlabeled peaks correspond to the Fe Kα line from the substrate and other low intensity constituents. The lines corresponding to the sputtered thin Au layer, used to get better contrast in the SEM measurements, are also unlabeled. The qualitative results obtained automatically after subtraction of the background is shown in the upper right of the figures. It emphasizes the prevailing %weight of GaN.

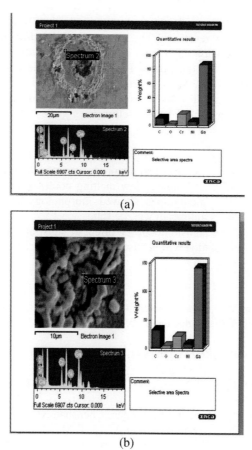

(a)

(b)

Figure 4: Selective area spectra of SEM and EDX measurements. Both (a) and (b) are different areas of the same sample. The L$_\alpha$ and K$_\alpha$ lines of Ga are clear, and K$_\alpha$ of N lines at low energy overlaps with C and O, and percentage by weight is 90% for (a) and 150% for (b).

The reactants existing in the plume and deposited on the substrate during the growth of GaN using method 1 could be represented by

$$\text{reactants} \quad \text{Ga}_{\text{liquid}} + \text{C}_{\text{solid}} + \text{N}_{2\text{gas}} \rightarrow \text{GaN}_{\text{solid}} + \text{CN}_{\text{gas}} \tag{1}$$

The CN gas can continuously be dissolved in the liquid Ga metal forming rolling droplets such as:

186

$$\text{Proucts(1)} \quad CN_{gas} + Ga_{liquid} \rightarrow GaN_{solid} + C_{solid} \tag{2}$$

In case of the second method of growth of GaN the reactants in the ablated plume deposited on the substrate could be represented as forming tiny droplets of GaN.

$$\text{Reactants} \quad 2Ga_{liquid} + NH_{3\,gas} + NaNO_{2\,solid} \rightarrow 2GaN_{solid} + NaOH_{vapor} + H_2O_{vapor} \tag{3}$$

In addition reactants similar to that in method 1 are also involved. Droplets from the reaction 3 and the rolling droplets from reaction 1 are summed up and thus allowing time to grow the wurzite crystallites. Such suggestion might help in explaining the fact that nano rods are exceeding nano dots in the first method. While quite dense formations of droplets are formed during method 2. The absence of sodium lines in the EDX spectrum indicates that the formed sodium hydroxide vaporizes away from the substrate. The sodium nitride could be considered as a catalyst helping to provide more products of GaN as a clear from reaction 3.

4. CONCLUSION:

One might state that GaN nanodots and nanorods were successfully sensitized through laser ablation of Ga metal as liquid phase (having low melting point) in the presence of nitrogen rich gases or solid catalysts. SEM micrographs showed morphology of 0D as well as 1D GaN nanostructures as well as GaN crystallites in wurzite state.

The detailed physical mechanism about the effect of substrate and possibility of growth by lower power density UV laser ablation will be discussed in another paper.

References:

[1]   D.K.T. Ng, L.S. Tan, M.H. Hong, Current Applied Physics 6 (2006) 403.
[2]   Y. Cui, L.J. Lauhon, M.S. Gudiksen, J. Wang, C.M. Lieber, Appl. Phys. Lett. 78 (2001) 2214.
[3]   B. Johlsson, M.T. Bjork, M.H. Magnusson, K. Deppert, L. Samuelson, Appl. Phys. Lett.79 (2001) 3335.
[4]   P.A. Smith, C.D. Nordquist, T.N. Jackson, T.S. Mayer, Appl. Phys. Lett.77(2000) 1399.
[5]   http://sony.inc.
[6]   Li Yang, Xing Zhan, Ru Huang, Guoyan Zhan, Chenshan Xe, Physica E28 (2005) 237.
[7]   G.S. Cheng, L.D. Zhang, Y. Zhu, G.T. Fei, L. Li, C.M. Mo, Y.Q. Mao, Appl. Phys. Lett. 75 (1999) 2456.
[8]   C.C. Tang, S.S. Fan, H.Y. Dang, P. Li. Y.M. Liu, Appl. Phys. Lett. 77 (2000) 1961.
[9]   C.C. Chen, C.C. Yeh, Adv. Mater.12(2000) 738.
[10]   G.S. Cheng, S.H. Chen, X.G. Zhu, Y.Q. Mao, L.D. Zhang, Mater. Sci. Eng. A286 (2000) 165.
[11]   X.F. Duang, C.M. Lieber, J. Am. Chem. Soc.122 (2000) 188.
[12]   Y.U. Peng, T.X. Zhau, N. Wang, F.Y. Zheng, S.L. Liao, S.W. Shi, S.C. Lee, T.S. Lee Chem. Phys. Lett. 327 (2000) 263.
[13]   M. He, I. Minus, P.Z. Hou, S.N. Mohammed, *et al.*, Appl. Phys. Lett. 77 (2000) 3731.
[14]   J.Y. Li, Z.Y. Qiao, X.L. Chen, Y.G. Cao, Y.C. Lan, C.Y. Wang, Appl. Phys. A. 71 (2000) 587.
[15]   Seung Young Bae, Hee Won Seo, Jeunhee Park, Hyunik Yang, Ju Chul Park, Sou Young Lee, Appl. Phys. Lett. 81(2002)126.
[16]   Lotfia M.Elnadi, Galila Mehena, Magdy M.Omar, Hussein A. Moniem, Fakiha H.A.Taieb, Faried A.Rahiem, AIP Conf. Proc. 888 (2007) 167.

LASER-INDUCED BREAKDOWN SPECTROSCOPY TECHNIQUE IN IDENTIFICATION OF ANCIENT CERAMICS BODIES AND GLAZES

KHALED ELSAYED*, HISHAM IMAM **, FATMA MADKOUR***, GALILA MEHEINA**, YOSR GAMAL*

* Physics Dept, Faculty of Science, Cairo University, Egypt

** NILES, Cairo University, Egypt

*** Conservation Dept, Faculty of Fine Arts, Minia University, Egypt

In this paper we report a study on Laser Induced Breakdown Spectroscopy (LIBS) as a promising non-destructive technique for the identification of the colored glazes, and clay's bodies of Fatimid ceramics ancient artifacts. The scientific examination of ceramics may be helpful in unraveling the history of ancient shards, particularly as the process of its production such as firing condition and temperatures. The analysis of pottery, ceramic bodies and glazed coatings is required in order to structure the conservation or restoration of a piece. Revealing the technical skills of ancient potters has been one of the most important issues for gaining a deep insight of bygone culture and also it is required in order to structure the conservation or restoration of a piece of art. LIBS measurements were carried out by focusing a Nd-YAG laser at 1064 nm with pulse width of 10 ns and 50 mJ pulse energy on the surface of the sample by a 100-mm focal length lens. The plasma emission was collected by telescopic system and transferred through a fiber to Echelle spectrometer attached to an ICCD camera. The focal spot diameter is found to be in the range of 100-150 μm. which is small enough to consider this technique as a non-destructive technique. LIBS technique clarified that each piece of archaeological objects has its own finger print. X-ray diffraction (XRD) analysis was carried out on these archaeological ceramic body samples to study raw materials such as clays, which allowed the investigation of the crystal structure and showed the changes in its structure through firing process. This provided information on the ceramic characteristic and composition of the ceramic bodies.

Keywords: LIBS, Ceramics artifacts, analytical techniques, Archeology

Corresponding author: hishamimam@niles.edu.eg

CP 998, Modern Trends in Physics Research
Third International Conference MTPR-08
edited by L. El Nadi
Copyright @ 2011 by World Scientific Publishing Co. 978-981-4317-50-4 / 981-4317-50-0

1. Introduction

Art restoration is important if we are aiming to preserve our history and culture. It involves the repair of damaged areas on a piece of artworks and its maintenance. The collaboration between scientists and art historians is becoming a more frequent occurrence. Together they work to achieve the unambiguous chemical characterization of many different types of archaeological artworks and artifacts such as paintings, pottery, ceramics, textiles etc.

The scientific examination of ceramics may be helpful in unraveling the history of ancient shards, particularly as the process of its production. The analysis of pottery, ceramic bodies and glazed coatings is required in order to structure the conservation or restoration of a piece and can also be useful in its dating or authentication. Revealing the technical skills of ancient potters has been the subject of much research since it has been one of the most important issues for gaining a deep insight of bygone culture and also it is required in order to structure the conservation or restoration of a piece of art [1,2].

There are many analytical techniques which can be used to study the physical and chemical composition of ancient ceramic bodies and coatings. Examples of techniques used in this capacity include, laser-induced breakdown spectroscopy (LIBS) [3-9], Raman spectroscopy, X-ray fluorescence (XRF), X-ray diffraction (XRD), particle-induced X-ray emission (PIXE) and Fourier-transform infrared spectroscopy (FTIR), atomic absorption spectrometry (AAS). In this work LIBS technique was used

for the identification of ancient ceramic bodies and coatings from Fatimid period. In this context, laser – induced breakdown spectroscopy (LIBS) is a potential alternative to other spectroscopic, mass spectrometric, or X-ray techniques used in art conservation and archaeology related applications. It is a practically non-destructive as well as rapid elemental analysis technique with the critical advantage of being applicable in situ, thereby avoiding sampling and sample preparation [8]. When a laser is focused on sample surface, a tiny amount of the material is vaporized, and through further photons absorption, it is heated up until it ionizes. This laser-induced plasma is a micro–source of light that is analyzed by using a spectrometer. The obtained spectra consist of emission lines corresponding to the elements evaporated from the sample surface.

Useful information about ancient technology, materials composition, kinds of surface glazes and evaluation of firing conditions and temperatures, extracted from LIBS technique, will help in restoration and repairing of these objects. LIBS technique clarified that each piece of archaeological objects has its own finger print. The potential of LIBS technique for direct qualitative analysis of ceramics bodies and coatings was evaluated and compared to commonly done by X-ray diffraction spectroscopy which gives more direct and detailed description of the ceramic bodies. There are however clear advantages of LIBS over XRD such as non-destructive, non-sample preparation and fast analysis and most importantly it has the potential to be a portable system [10,11].

2. Methodology

2.1. Description Archaeological ceramic samples from Fatimid period

The Islamic period is characterized above all in the decorative arts by the extraordinary development of glazed ceramics. At first the development was slow and the products of the earliest Islamic periods in the 7th century cannot be distinguished from what went before. But the established production of rich and sophisticated kinds of pottery across the Islamic world by the 9th century indicates that an industrial revolution of considerable proportion had taken place. It introduced glazed ceramic production to places that previously had never made it, and knew of it only as a rare and exotic import. [12]. Theses ceramic objects have been employed as storage vessels for transport, jars, bowls, serving dishes, jugs, cups, and many other uses.

The Fatimid period (359-567A.H / 969 -1171 A.D) can be regarded as a renaissance for the arts in Egypt. For pottery production, it was a golden age and there were three major types: vessels covered with a colored monochrome glaze, luster painted wares and the so–called North African polychrome-painted in-glaze wares and this type was produced in the North African i.e. Morocco, Algeria, Tunisia and Libya [13].
The first sample (sample no. 1) is a fragment of a monochrome glazed that covered with a pale green glaze and the clay body is buff earthenware. The second sample (sample no.2) is a fragment of polychrome glazed ware (dark yellow and green) and the body is red earthenware. The third sample in this period (sample no. 3) is a fragment of a monochrome glazed ware coated with a turquoise glaze and decorated

with incised decoration and the body is buff earthenware. Figure 1 shows these samples.

Figure 1. Ceramics samples from Fatimid Period.

The samples were collected from Al-Fustat excavation in Old Cairo, the first capital of Islamic Egypt (641 A.D-21 A.H), which continued occupation at least the first century of the Fatimid period (969- 1171 A.D) [14].

2.2 Experimental Set up

Figure.2 shows the typical experimental set-up for Laser Induced breakdown Spectroscopy technique (1-3). The plasma emission was produced by a Q-Switched Nd:YAG (Continuum, Surlite II) operated at the fundamental wavelength of 1064 nm, with a repetition rate of 1 Hz. Individual laser pulses had a pulse length of approximately 10 ns. The energy of these pulses was adjusted for pulse energies of 50 mJ, using a beam spilter.. The laser beam was focused on the sample surface using a 5-cm focal length lens as show in Figure. 2. The emission from the plasma was collected by a 200 μm diameter, one-meter length wide band fused-silica optical fiber, connected at its other side to a spectral analysis system. This system consisted of an echellel spectrometer (PI - Echelle, Princeton Instruments, USA) with a gateable, Intensified Charge Coupled Device (ICCD I-MAX, Princeton Instruments) attached to the spectrometer. The emission signal was corrected by subtracting the dark signal of the detector through the LIBS software. A gate width of 10000 ns and delay of 1000-ns were chosen for maximizing spectral line intensity while maintaining a good temporal resolution. The gate width and delay time have been chosen after performing an optimization experiment for maximizing spectral line intensity. The choice of gate width and delay time for the spectroscopic data acquisition is accomplished by computer-controlled system. The analysis of the emission spectra was accomplished using the commercial software (GRAMS/AI v.8.0, Thermo-electron co.). The reproducibility of measurements strongly depends on the experimental conditions. The average energy of laser pulse could be precisely adjusted to maintain a pulse-to-pulse variation of approximately 2%. To improve LIBS precision, spectra from several laser shots have to be averaged in order to reduce statistical error due to laser shot-to-shot fluctuation. We reproduced the measurements at five locations on the sample surface in order to avoid problems linked to sample heterogeneity. Fifty laser shots were fired at each location and saved in separated files and the average (average of 250 spectra) was computed and saved to serve as the library spectrum.

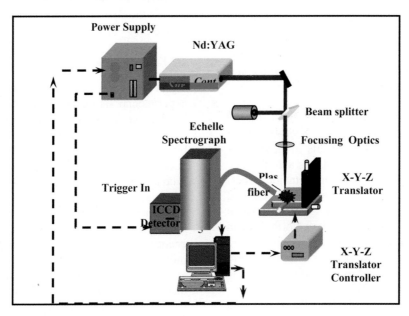

Figure 2. LIBS experimental setup

3. Results and discussion

3.1 XRD Results

Mineralogical analysis by X-ray diffraction to ceramic bodies provides useful information about their chemical composition and assessing firing temperatures. Some mineral phases may disappear and new ones form during firing process, as each phase is stable within a definite interval of temperature: thus, recognition of a suitable number of phases may allow one to determine a temperature range encompassing the original firing temperature. However, one should take into account that phase transformations in fired clays may also occur during cooling [15]. The results are summarized in table 1

Sample no. 1

Beside quartz, gehlenite is exist as a minor compound which appears from 800 °C and its presence is associated with the existence of carbonates [16]. Hematite, geothite, diopside, and wollastonite exist as traces indicating that the firing temperature for this sample was between also 800-900 °C.

Sample no. 2

This sample contains quartz as a major compound, tridymite and hematite as traces. The existence of these compounds together indicates that the firing temperature ranging from 900 to1000 °C. [17]

Sample no. 3

In this sample the major compound is calcite which forms through the beginning of the firing process and decomposes at about 870 °C. Dolomite and quartz are existed as minor compounds and goethite as a trace. The existence of calcite and dolomite indicate that the clay used was calcareous clay and the firing temperature was around between 800-850 °C. [16]

Table (1): XRD results of ceramic bodies.

No. of Sample	Color of body Sample	Compounds		
		Major	Minor	Traces
1 F	Buff	Quartz SiO_2	Gehlenite $Ca_2Al_2SiO_7$	- Wollastonite $CaSiO_3$ - Diopside - $Ca(Mg,Al)(Si,Al)_2O_6$ - Geothite $FeOOH$ -Hematite Fe_2O_3
2 F	Red	Quartz SiO_2	_	- Hematite Fe_2O_3 - Tridymite SiO_2
3 F	Buff	Calcite $CaCO_3$	- Dolomite $CaMg(CO_3)_2$ - Quartz SiO_2	- Geothite $FeOOH$

3.2 LIBS Results

Different experiments have been made to investigate the element composition of colored glazes, and clay's bodies of Fatimid ceramics samples. The main difficulty in the quantitative LIBS analysis of glaze and clay is related to the sample inhomogeneity when single pulse acquisition is applied. This could be partially overcome by averaging over different analytical lines of the elements acquired in different spectral parts, i.e. sample points. Another analytical difficulty is due to interference of the layer beneath the analyzed one, which might be also ablated by the laser. Although the laser pulse energy was kept low, it was noticed that in all LIBS measurements on glaze, it is even more difficult to distinguish the glaze from body because all the major element of clay such as Ca, Al, Si, Mg are also present in the glaze as will explain later. A significant amount of

192

tin and lead, present only in the glaze, was observed. For accurate evaluation of the elements, specifically the plasma parameters has to be kept as constant as possible. By measuring the electron density and plasma temperature on line, the plasma reproducibility can be assessed. Electron temperature T_e was estimated using very simple formula based on two lines Boltzmann plots [3]. In most cases, we used the two Ca lines at 452.69 nm (E = 43 933 cm^{-1}) and 430.77 nm (E = 38 417cm^{-1}) for relevant spectral segments (see Figure.4). Ionic temperature T_{ion} was estimated using the Saha-Boltzman relation, and our estimate was based on two Mg lines, namely the ionic line at 280.2 nm (E= 35 669 cm^{-1}), and the neutral atom line at 285.21 nm (E=35 051 cm^{-1}), for the relevant spectral segments (see Figure. 5). Electron density, Ne, was calculated using the standard relation, which can be found in [3, 4]. An equality expression related to the electron density was used to estimate whether local thermodynamic equilibrium LTE is likely to prevail in our measurements. The value $N_e = 4.7x10^{15}$cm^{-3}, which according to [3, 4] is needed for LTE to exist, is well below the values encountered in our experiments. This suggests that the analytical measurements in our study were most likely carried out under LTE conditions. Note that during all experiments the value for plasma temperature (in LTE all 'temperatures' are assumed to be equal, i.e. $T_e\sim\_T_{ion}\sim\_T_{plasma}$.) was kept as close as possible to $T_{plasma} = 9100 + 300$ K. The electron density was adjusted to a value of $N_e = 1.7$ x $10^{17}+1.2$ x 10^{17} cm^{-3}. It is important to remember that the parameters addressed above may dramatically change with the matrix composition, and that in order to keep the plasma conditions for different matrices comparable, one should basically adjust the laser pulse energy, being the parameter easiest to control. To get a feeling for the range of variation to be expected from the various sample matrices, the plasma temperature and electron density were recorded for glazes and bodies. For example, the ablation thresholds for the ceramic glaze is lower than that of ceramic body [12, 13], and thus one would expect lower temperatures for glaze.

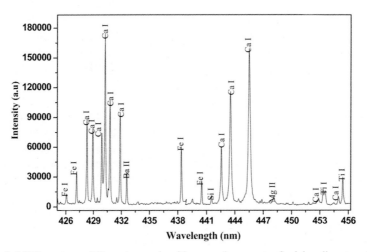

Figure3 LIBS spectrum of Ceramic sample with zoomed segments of calcium lines to calculate T_e

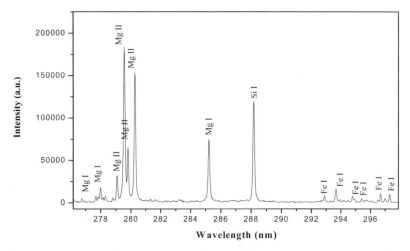

Figure. 4. LIBS spectrum of Ceramic sample with zoomed segments of calcium lines to calculate T_{ion}

The actual values of the plasma temperature, derived from the measurements according to the standard relations addressed above (based on Ca and Mg spectral lines), are 8800 and 13700 K for ceramic glaze and ceramic body, respectively. The electron densities for ceramic glaze and ceramic body were 1.2 and 2.1 x 10^{17} cm^{-3}, respectively (calculated using the Ca line mentioned above).

LIBS technique has been demonstrated to be quite useful tool for the investigation of elemental composition of ancient ceramic and coatings. It provides non-destructive (micro-destructive) and analysis of both the surface glazes and the ceramic bodies. The optical microscope picture for a glaze surface sample (as an example) shows a laser spot diameter on the surface of about 70 μm diameter as illustrated in Figure. (5). Three types of Fatimid ceramic objects were analyzed by LIBS technique. Selected spectra from clay bodies of samples no. 1, 2 are shown in Figures. (6, 7). As shown from the characteristic emission lines in this spectrum, the clay bodies of these two samples contain a relatively high contains of iron (Fe). This indicates that the clays used in these samples were Iron clay. Iron-rich minerals, with the iron primarily in the form of hematite Fe2O3, goethite (FeOOH) have

been found in the clay bodies of sample 1 and 2. Also Figure. 8 shows that the two samples contain high ratio of silicon Si as confirmed by the results obtained by XRD (see table 1). Spectra from clay body of sample number 3 is shown in Figure. 9 As seen by the characteristic emission lines in these spectra that the clay body of sample no. 3 contains high ratio of calcium, this indicates that the clay used was calcareous clay. These results show that the ceramic objects are often made from local clay sources.

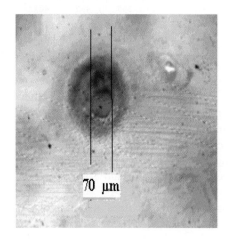

Figure. 5. Image obtained with OM of Famtimid glazed sample after on laser shot.

194

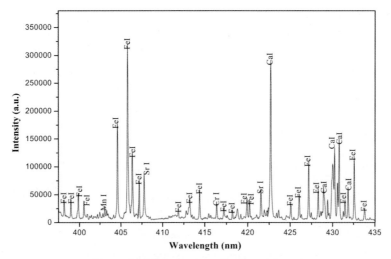

Figure. 6.   LIBS spectrum of body sample no.1

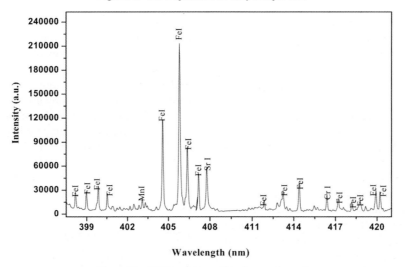

Figure. 7.   LIBS spectrum of body sample no.2

Glaze is silica based mixture often containing pigment applied on the surface of a ceramic vessel and often fire over 1200 °C producing a decorative, impermeable, glassy coating [8]. Selected spectra from glaze samples of sample 1, 2, and 3 are shown in Figure (10, 11, 12). As seen by the characteristic emission lines in these spectra, the glazes of the samples 1 and 2 contain lead while the glaze of sample 3 contains lead and sodium. This indicated that flux, used in ingredients of these glazes to decrease the firing temperature, was lead in samples 1 and 2 and lead and sodium in sample 3. In this case lead may be added in the form of red lead oxide, Pb_3O_4 while sodium added in the form of sodium oxide or sodium carbonate, Na2O, Na_2CO_3. [18].

Normally glazes are transparent and colorless but can be given color by the addition of metal oxides [18].

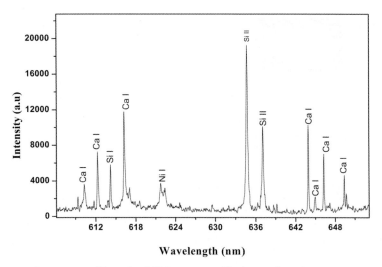

Figure.8.   LIBS spectrum of body sample no.2

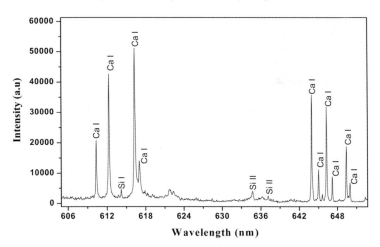

Figure.9.   LIBS spectrum of body sample no.3

The analysis of the LIBS spectra obtained from pale green glaze of sample no. 1, and yellow glaze of sample no 2. (see Figure.10 and Figure 11) showed that glaze of these two sample contains Fe. Iron-riched minerals with iron primarily in the form of hematite Fe_2O_3 and other oxides have been extensively used as pigments in these glazes. Upon firing iron in the form ferric oxide Fe_2O_3 in lead glaze undergo transformation to ferrous oxide Fe_2O which give glaze

sample no 1 this pale green tint in the reduction condition while give the yellow color in glaze sample no. 2 in oxidation condition.

The spectra from glaze surface of sample number 2, 3 shown Figure (11,12) show that the glaze of the two sample contain substantial amounts of Cu, this indicate that the source of turquoise blue glaze in sample no3 is copper oxide, which used in different Islamic periods to produce this color in lead-alkaline glaze where sodium is presented beside lead in this sample. While the strong green color in sample no.2 is obtained by using copper in lead glaze. Magnesium and calcium may be presented in glaze of sample no 3 as impurities in sand.

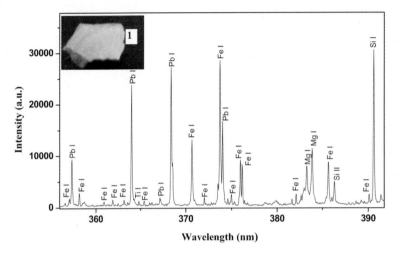

Figure.10.   LIBS spectrum of glaze sample no.1

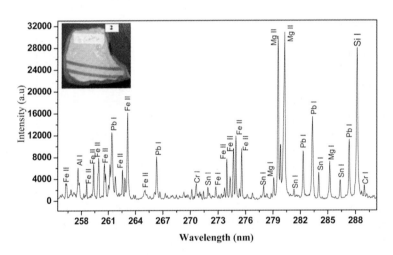

Figure.11.   LIBS spectrum of glaze sample no.2

The spectra from glaze of sample no 3 show that the glaze of this sample contains tin, Sn, as opacifier agent to obtain the opaque white glaze in this sample, Tin oxide was used in the production of Islamic opaque glazes from the ninth century ad, and subsequently in enamels applied to Islamic glasse from the 12th century ad onwards [19]. Tin opacifiers were used first introduced in Basra, Iraq in the first of the eighth century A.D and were developed later in Iraq and Egypt and subsequently this technology spread to the rest of the Islamic world and also Europe [20].

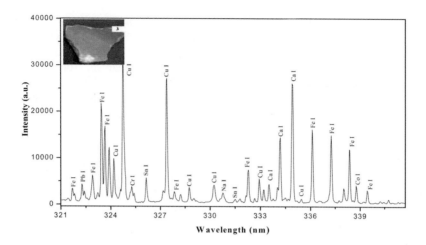

Figure.12.   LIBS spectrum of glaze sample no.3

The spectra of clay boy and glaze of all sample show that both bodies and glazes contain Al. Aluminum oxide in the form of alumina (Al_2O_3) was used in glaze recipes as a bonding material and to prevent the glaze flow off the ceramic through firing process [21].

Comparing the data obtained from XRD and LIBS for clay bodies, a good agreement is observed. The result obtained by both techniques indicates that silicon/quartz is the predominant mineral and it is observed in samples no.1, 2 but calcium/calcite is major element in the sample no. 3 and silicon/quarts is minor element. The quartz was added in the form of sand to the original clays as degreaser in order to facilitate the mould of the pastes.

4. Conclusions

The information extracted from LIBS results suggested that LIBS can be employed to determine/confirm the date of certain types of pottery. The scientific examination of ceramics may be helpful in unraveling the history of ancient shards, particularly as the process of its production such as firing condition and temperatures. The analysis of pottery, ceramic bodies and glazed coatings is required in order to structure the conservation or restoration of a piece. Revealing the technical skills of ancient potters has been one of the most important issues for gaining a deep insight of bygone culture and also it is required in order to structure the conservation or restoration of a piece archaeology sample. The results obtained, once compared to other analytical techniques such as XRD demonstrates, that LIBS technique is promising for fast non-destructive analysis of ancient ceramics. An important aspect of LIBS is the speed of analysis, which can allow the quick examination of a large number of samples in the field, at a museum, or at an excavation site, which leading to determine characterization and classification of archaeological findings.

References

1-Barclay, K.: "Scientific analysis of archaeological ceramics". Handbook of resources. The Information Press, Great Britain, Oxford (2001).

198

2-Zoppi, A., Lofrumento, C., Castellucci, E.M. and Migliorini, M.G.: "Micro-Raman technique for phase analysis on archaeological ceramics" Spectroscopy Europe, Firenze, Italy (2002).

3- Muller, K. and Stege H.: "Evaluation of the analytical potential of Laser Induced Breakdown Spectrometry (LIBS) for the analysis of historical glasses" Archaeometry, Vol.45, Issue 3, August, pp. 421-433 (2003).

4- Fotakis, C.: "Laser technology for the preservation of cultural heritage: diagnostic and restoration applications" Colloquia du service de Physique Theorique et Mathematique (2003).

5- Rebollar, E., Oujja, M., Gaspard, S. and Castillejo, M.: "Cleaning and microanalysis of artworks using lasers" (2006).

6- Marco, L., Klaus, D., Johan, M.: "Controlled laser cleaning of artworks with low resolution LIBS and linear correlation analysis" (2005).

7-Nikolov, I. P., Popmintchev, T., Todorov, T., Buchvarov, C., Surtchev, M. and Tzaneva, S.: "Laser restoration of ceramic artifacts with archaeological value", Applied Physics A Material Science and Processing, Vol.79, Issue 4-6, pp.1111-1115 (2004).

8- Melessanaki, K., Mateo, M., Ferrence, S.C., Betancourt, P. and Anglos, D., The application of LIBS for the analysis of archaeological ceramic and metal artifacts , Applied Surface science 197-198, (2002) 156-163.

9- Tornari,V., Zafiropulos, A., Bonarou, N.A., Vainos, C. and Fotakis, C., Modern technology in artwork conservation: a laser-based approach for process control and evaluation, Optics and Lasers in Engineering 34 (2000) 309-326.

10-Melessanaki, K., Angols, D., Chluveraki, S. and Ferrence, S.S.: "A new transportable analytical instrument for fast elemental analysis of archaeological objects" In: 4th Symposium on Archaeometry, Hellenic Society for Archaeometry, Athens, Greece, May 28-31(2003).

11- Lazic, V., Colao, F., Fantoni, R., Palucci, A., Spizzichino, V., Borgia, I., Bswnetti, B., Sgamellotti, A..: "Characterization of luster and pigment composition in ancient pottery by laser induced fluorescence and breakdown spectroscopy" Journal of cultural heritage (2003).

12- Watson, O., Ceramic from Islamic Lands, Thames and Hudson Ltd., London, U.K. 2004.

13-Fehervari, G.: "Pottery of the Islamic World in the Tareq Rajab Museum", Printed in the State of Kuwait (1998)

14- UNESCO Division of Cultural Heritage, Supreme Council of Antiquities, Egypt, Archaeological discoveries, 2004.

15- Mirti, P.: "Recent advances in the study of ancient ceramic bodies & coatings" In: Fourth Euro-Ceramics, Vol. 14, The Cultural Ceramics Heritage, 3rd European Meeting on Ancient Ceramics, C.N.R. IRTEC, Faenza, Italy, pp.13-22 (1995)

16- Guiraum, A., Barrios, J. & Flores Ales, V., A group of Roman Ceramics building materials from the settlements of Astigi and Arva in the province of Seville (Spain), In: Fourth Euro-Ceramics, Vol. 14,The Cultural Ceramics Heritage, 3rd European Meeting on Ancient Ceramics, C.N.R. IRTEC, Faenza, Italy, pp.363-370 (1995).

17- Thierrin – Michel, G.: "Production of Italic wine amphora: some technical aspects" In: Fourth Euro-Ceramics, Vol. 14, The Cultural Ceramics Heritage, 3rd European Meeting on Ancient Ceramics, C.N.R. IRTEC, Faenza, Italy, pp.173-183 (1995).

18-Al-Hassan, A.Y. and Hill, D.R.: "Islamic technology, an illustrated history" Cambridge University Press, Great Britain (1986).

19 -Tite, M.S., Pradell, T. and Shortland, A., Discovery, Production and use of Tin- Based Opacifiers in Glasses, Enamels and Glazes from the Late Iron Age Onwards: A Reassessment, Archaeometry, Vol.50, Issue 1,pp. 67-84, February 2008.

20- Mason, R.B. and Tite, M.S., "The beginning of tin-opacification of pottery glazes" Archaeometry, 39, 1, 41-58 (1997).

21- Downi, C.A.: "Glaze analysis of 15th to 17th century Islamic ceramics from the collection of the Arthur M. Sackler Gallery" Queens University, Kingston, Ontario, Canada (2002).

THERMAL STABILITY AND INFRARED-TO-VISIBLE UPCONVERSION EMISSIONS OF Er^{3+}/Yb^{3+} CO-DOPED $70GeO_2$–$20PbO$–$10K_2O$ GLASSES

SAMAH M. AHMED[a], I. SHALTOUT[b], Y. BADR[a]

[a] *National Institute of Laser Enhanced Sciences (NILES), Cairo University, Cairo, Egypt*
[b] *Faculty of Science, Al Azhar University, Physics Department, Cairo, Egypt*

Er^{3+}/Yb^{3+} co-doped potassium-lead-germanate ($70GeO_2$–$20PbO$–$10K_2O$) glasses with a fixed concentration of Er^{3+} ions (0.5 mol. %) and different concentrations of Yb^{3+} ions (0, 0.5, 1.5, and 2.5 mol. %), have been synthesized by the conventional melting and quenching method. The structure and vibrational modes of the glass network were investigated by the infrared absorption and Raman spectroscopy. The thermal behavior of all glass samples was investigated by the differential thermal analysis. Infrared-to-visible frequency upconversion process was investigated in all glasses. Intense green and red upconversion emission bands centered at around 532, 546, and 655 nm were observed, under excitation at 980 nm of diode laser at room temperature. The dependence of these emissions on the excitation power was investigated.

1. Introduction

Trivalent rare earth ions $(RE)^{3+}$-doped glasses are excellent active media for photonic devices such as lasers and optical fiber amplifiers. [1, 2] Recently, there has been a great deal of interest in studying of the conversion of near infrared radiation into visible light through frequency upconversion in $(RE)^{3+}$-doped glasses to develop visible-wavelength lasers pumped by near infrared diode lasers. [3–6] The main interest in solid-state, visible-wavelength lasers arises from potential applications including color displays, optical data storage, biomedical diagnostics, sensors, lasers, optoelectronics and undersea optical communications. [7]

Er^{3+} ion is one of the most efficient $(RE)^{3+}$ ions for obtaining frequency upconversion because it can be optically excited with powerful near infrared diode lasers, yielding blue, green, and red emissions [5] as well as, infrared emission for optical amplification at the third communication window (at 1.5 μm). [8, 9] The sensitization of Er^{3+}-doped glasses with Yb^{3+} ions is a well-known method for increasing the optical pumping efficiency through the efficient energy transfer from Yb^{3+} to Er^{3+} ions. Pulsed and continuous wave laser actions have been demonstrated in Er^{3+}/Yb^{3+}-codoped glass lasers. [5, 6]

Glasses as host materials for $(RE)^{3+}$ ions are attractive due to: (i) they are transparent over a wide optical range and a relatively large amount of rare earth ions can be easily and homogeneously incorporated into their matrices, [7] (ii) low fabrication cost and ease of fabrication into complex shapes including fibers, [10] and (iii) in glass hosts, emission of $(RE)^{3+}$ ions consist of broad bands which is interesting in view of tuneability of lasers. [3]

Heavy metal oxide (HMO) glasses are desirable hosts for optically active ions. Among these HMO glasses, lead-germanate glasses are of growing interest because they combine higher mechanical strength, higher chemical durability, and better thermal stability with good transmission in the infrared region up to 4.5 μm. [2, 5, 7] Moreover, they have high refractive index (~2) and relatively low maximum phonon energy (800-900 cm^{-1}). These properties enhance the radiative transitions in doping ions and increase the quantum efficiency of luminescence in these optical media. [2, 11] In addition, the lower optical loss of germanate glasses in the IR (0.02–0.05 dB/km at 2.89 μm for GeO_2) makes them more promising for practical use such as long optical fibers. [12]

The quest for glasses that combine desirable optical properties, good thermal stability, and that can be readily manufactured has led to development of many oxide glass formulations. [10] In this work we report on the effect of Yb^{3+} ions on different optical and physical properties of the Er^{3+}-doped potassium-lead-germanate glasses.

2. Experimental

The glasses used in this work were synthesized by the conventional melting and quenching method. The starting materials are: potassium carbonate (K_2CO_3), lead oxide (PbO), and high purity germanium oxide GeO_2 (Alfa, 99.9999 %). The rare earth ions Er^{3+} and Yb^{3+} were introduced as (Er_2O_3) and (Yb_2O_3) (Alfa, 99.99%). The glass hosts with composition (in mol. %): $70GeO_2$-$20PbO$-$10K_2O$ (labeled as $G_{70}P_{20}K_{10}$), were doped with 0.5 mol. % of concentration of Er_2O_3 and different concentrations of Yb_2O_3 (0, 0.5, 1.5, and 2.5

mol. %). Batches of about 8 gm of the raw materials were carefully weighted and mixed in an agate mortar for good homogenization. The well-mixed powders were melted in a covered platinum crucible in an electric furnace (Carbolite furnace, RHF 14/9) at a temperature of 1250ºC for 30 min. The melts were then poured and rapidly quenched by pressing between two stainless steel plates kept at room temperature. In this work, the glass samples have not been subjected to any annealing process. The obtained glass samples, for the naked eyes, are bubble-free, homogenous, clear and transparent. The amorphous nature of the prepared glasses has been confirmed by the x-ray powder diffraction using CuK$_\alpha$ radiation (Scintag, advanced diffraction system) over 2θ range of 3-70. The differential thermal analysis (DTA) measurements were performed with a differential thermal analyzer (Setaram Labsys TG-DSC16) at a heating rate of 10°C /min under Ar atmosphere in the temperature range of 50-750°C. Some pieces of the glasses were polished carefully for optical measurements.

Raman scattering spectrum of the undoped glass sample was recorded with a FT-Raman spectrometer (Nicolet 670), in the spectral range of 100 - 1200 cm^{-1} at room temperature. The IR spectra of all samples pressed into KBr pellets were obtained with a FT-IR spectrometer (Bruker, IFS 66/S) in the spectral range of 400-1200 cm^{-1} at room temperature. Absorption spectra of glass samples were obtained with a Shimadzu (UV-3101PC) UV/VIS/NIR spectrophotometer in the 350-1700 nm spectral range. The upconversion emission spectra were obtained upon excitation of 980 nm laser diode with a power of 1 W. The emitted light from the samples was analyzed with a monochromator (Spex 750m), and the light signal detected by a photomultiplier and finally recorded and amplified by a lock-in amplifier (SR 510). All measurements were taken at room temperature.

3. Results & discussion

3.1. Thermal Stability of the Glasses

The differential thermal analysis (DTA) curve of the undoped G$_{70}$P$_{20}$K$_{10}$ glass sample, in the temperature range of 300-750°C at a heating rate of 10°C / min., is shown in Fig. (1).

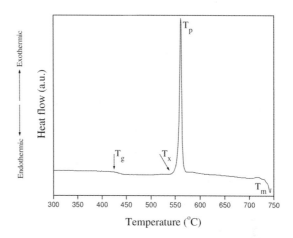

Figure 1. The DTA curve of the undoped G$_{70}$P$_{20}$K$_{10}$ glass sample.

The sample initially underwent a broad endothermic transformation corresponds to the glass transition (onset at T$_g$), [13] at which the glass begins to behave as a viscoelastic solid on heating. [14] Then an exothermic transformation corresponds to successive crystallization of the oxide glassy matrix (onset at T$_x$), and finally an endothermic transformation (at T$_m$) represents the melting of the crystalline phase. [13]

The thermal stability factor, ΔT = T$_x$ - T$_g$, can be used to represent the stability against crystallization of the glasses. The larger ΔT is, the stronger the inhabitation to nucleation and crystallization processes, and consequently, the larger the thermal stability of the glass. [15] In glass fiber technology, it is desirable for a glass host to have ΔT as large as possible, since fiber drawing is a reheating process and any crystallization during the process will increase the scattering loss of the fiber and then degrade the optical properties. [16, 17] The thermal stability factor of the present undoped G$_{70}$P$_{20}$K$_{10}$ glass sample is ΔT = 126 °C which is better than those of tellurite (118°C) and fluoride (105°C) glasses. [16]

In order to investigate the effect of Yb^{3+} ion concentration on the thermal stability of the glasses, the DTA curves of all glass samples were obtained and shown together in Fig. (2). It can be seen that, the onset of crystallization (T$_x$) and the temperature of the crystallization peak (T$_p$) are obviously shifted towards higher temperatures with increasing of Yb^{3+} ion concentration. The thermal parameters T$_g$, T$_x$, T$_P$ and ΔT of all glass samples, collected from Fig. (2), are

202

summarized in Table (1) and shown as a function of Yb³⁺ ion concentration in Fig. (3).

Table (1). The thermal parameters of undoped and Er³⁺/Yb³⁺ co-doped $G_{70}P_{20}K_{10}$ glasses.

Er³⁺:Yb³⁺ concentration ratios	T_g (°C)	T_x (°C)	T_p (°C)	$\Delta T = T_x - T_g$ (°C)
Undoped	432	558	561	126
1 : 0	433	563	577	130
1 : 1	437	580	597	143
1 : 3	442	627	664	185
1 : 5	454	649	666	195

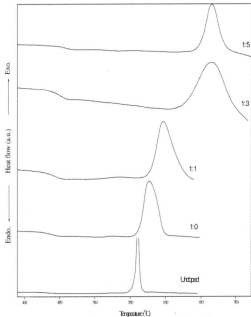

Figure 2. The DTA curves of the undoped and Er³⁺/ Yb³⁺ co-doped $G_{70}P_{20}K_{10}$ glasses.

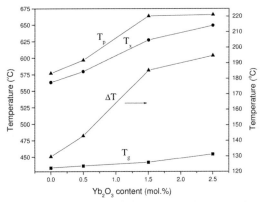

Figure 3. The thermal parameters vs. Yb₂0₃ content (in mol. %).

Clearly, T_g, T_x, T_P and ΔT increase with increasing of Yb³⁺ ion concentration. This in turn suggests that the addition of Yb³⁺ ions is reinforcing the glass network and acting as crystallization inhibitor in these $G_{70}P_{20}K_{10}$ glasses. This modification may be attributed to the large positive charge and large ionic radii of the rare earth ions dopants. They are surrounded by non-bridging oxygens in the glass network, bringing about additional degree of disorder in the glass network, which prevent it from devitrification. [18]

The thermal stability factor of the Er³⁺/Yb³⁺ co-doped $G_{70}P_{20}K_{10}$ glass with 0.5 mol.% of Er₂O₃ and 2.5 mol.% of Yb₂O₃ (sample 1: 5) is relatively high ($\Delta T=195°C$) which makes this glass is much preferable for preform fabrication and crystal-free fiber drawing. Furthermore, the presence of a single glass transition and a single crystallization peak in the DTA curves of all glass samples indicates that there is no phase separation in these glasses on reheating.[14] This suggests that these glasses are homogeneous in structure.

3.2. Raman Spectrum

The Raman spectrum of the undoped $G_{70}P_{20}K_{10}$ glass sample, obtained at room temperature in the spectral range of 100-1200 cm⁻¹, is shown in Fig. (4).

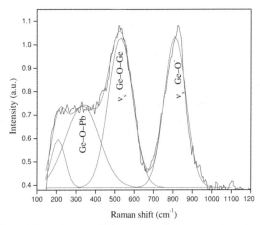

Figure 4. The deconvoluted Raman spectrum of the undoped $G_{70}P_{20}K_{10}$ glass. (v_s refers to symmetric vibration.)

The Raman spectrum is deconvoluted using Gaussian distribution to four bands centered at around 209, 333, 529, and 811 cm⁻¹. By considering the features of Raman scattering spectra for many reported GeO₂-based glasses, the spectrum for the undoped $G_{70}P_{20}K_{10}$

glass is interpreted as follows: The low-frequency band at 209 cm^{-1} could be attributed to vibrations involving the heavy metal atoms Pb. [19] The band at around 333 cm^{-1} is assigned to the vibrations of Ge–O–Pb bridges. [19–21] The intermediate-frequency band at around 529 cm^{-1} is assigned to the symmetric stretching vibrations of Ge–O–Ge bridges connecting [GeO$_4$] tetrahedra. [19, 21, 22] The high-frequency band at around 811 cm^{-1} is assigned to the symmetric stretching mode of the [GeO$_4$] tetrahedron with one non-bridge oxygen (Ge–O$^-$ bond). [22–25]

In pure GeO$_2$ glass, the random network structure is formed by corner-sharing between tetrahedral [GeO$_4$] units. The germanium atom in the centre is covalently bonded to four oxygen atoms and the oxygen atoms in the lattice are predominantly bridging (i.e. linking one tetrahedron to another). The addition of a modifier such as K$_2$O leads to breaking of some oxygen bridges and creation of [GeO$_4$] tetrahedra with non-bridging oxygens, [26, 27] which are responsible for the band at 811 cm^{-1}. In addition, the PbO enters the glass network as an intermediate. At low content (such as in this work, 20 mol %) Pb plays the role of modifiers in the germanate glass structure, increasing the number of non-bridging oxygens in the glass network. [28] When rare earth ions are incorporated in the glass matrix, they are surrounded by non-bridging oxygens of the [GeO$_4$] tetrahedra, and the energy of local vibrational modes coupled to them could be smaller than that of the maximum phonon energy of glass network. [29]

The Raman spectrum indicates that the maximum phonon energy in the undoped G$_{70}$P$_{20}$K$_{10}$ glass is around 811 cm^{-1} which is smaller than those of borate (~ 1350 cm^{-1}), phosphate (~ 1100 cm^{-1}), silicate (~1000 cm^{-1}), some other germanate (~ 900 cm^{-1}) and pure GeO$_2$ glass (~ 865 cm^{-1}) glasses. [29] Therefore, the multiphonon relaxation rate of Er^{3+} ions in this G$_{70}$P$_{20}$K$_{10}$ glass host is the lowest in the above oxide glass systems, since the low maximum phonon energy of the glass host reduces the nonradiative relaxation of the rare earth ions. Consequently, the upconversion luminescence intensity of Er^{3+} ions in this G$_{70}$P$_{20}$K$_{10}$ glass could be the highest among those mentioned above.

3.3. Infrared Absorption Spectra

The FT-IR absorption spectrum of the undoped G$_{70}$P$_{20}$K$_{10}$ glass sample dispersed in KBr pellet, in the spectral range of 400-1200 cm^{-1}, is shown in Fig. (5).

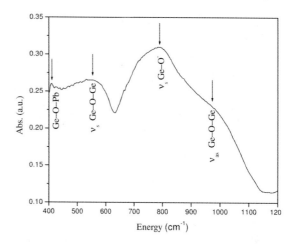

Figure 5. FT-IR absorption spectrum of the undoped G$_{70}$P$_{20}$K$_{10}$ glass. v$_s$ refers to symmetric vibration and v$_{as}$ refers to asymmetric vibration.

The spectrum is dominated by a band centered at around 788 cm^{-1}, with other bands centered at around 410, 550, and a shoulder at 974 cm^{-1}. The band at around 410 cm^{-1} could be attributed to the Ge–O–Pb bridging vibrations, while the band at around 550 cm^{-1} is assigned to the symmetric stretching of oxygen atoms in the Ge–O–Ge bridges. [30] The high-frequency band at 788 cm^{-1} could be assigned to the symmetric stretching mode of Ge–O$^-$ bond with non-bridge oxygen. [24] The shoulder at around 974 cm^{-1} could be assigned to the asymmetric stretching vibration mode of Ge–O–Ge bridges connecting [GeO$_4$] tetrahedra. [22, 31]

Fig. (6) shows the FT-IR absorption spectra of the undoped and the Er^{3+}/Yb^{3+} co-doped G$_{70}$P$_{20}$K$_{10}$ glasses, in the spectral range of 400-1200 cm^{-1}.

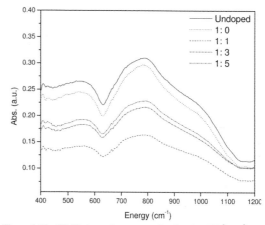

Figure 6. The FT-IR absorption spectra of undoped and Er^{3+}/Yb^{3+} co-doped G$_{70}$P$_{20}$K$_{10}$ glasses with different Er^{3+}:Yb^{3+} concentration ratios.

204

As shown in Fig. (6), the IR spectra of the Er^{3+}/Yb^{3+} co-doped glasses are almost identical with the spectrum of the undoped glass. That is, the increasing of Yb^{3+} ion concentration (from 0 to 2.5 mol. %) in these $G_{70}P_{20}K_{10}$ glasses, does not influence the positions of the IR absorption bands. This in turn indicates a very good dispersion of the Yb^{3+} ions, with no evidence of the formation of Yb^{3+} ion clusters in these $G_{70}P_{20}K_{10}$ glasses. [32]

3.4. Optical Absorption Spectra

The absorption spectra of the Er^{3+}/Yb^{3+} co-doped $G_{70}P_{20}K_{10}$ glasses with different Er^{3+}:Yb^{3+} concentration ratios, in the 350-1700 nm spectral range at room temperature, are shown in Fig. (7).

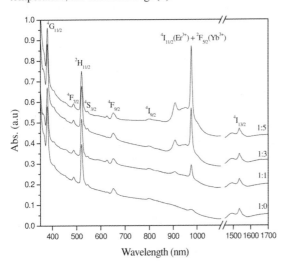

Figure 7. The absorption spectra of Er^{3+}/Yb^{3+} co-doped $G_{70}P_{20}K_{10}$ glasses with different Er^{3+}:Yb^{3+} concentration ratios.

The absorption spectra consist of a number of inhomogeneously broadened bands assigned to transitions occurring within the 4f - shell of the rare earth ions. [2] The broadening of the absorption bands is attributed to the amorphous nature of the glass host. That is, in a glassy matrix, the local environment around the rare earth ions varies from site to site, leading to variations in the ligand field. This results in slight broadening of the energy levels [33-35] Table (2) lists the absorption peaks and the corresponding assigned transitions. [1, 36, 37]

Table (2). The absorption peaks and their corresponding assigned transitions.

Peak position λ (nm)	Average Energy (cm^{-1})	Assigned transition
378	26455	$(Er^{3+})\,^4I_{15/2} \rightarrow {}^4G_{11/2}$
488	20492	$(Er^{3+})\,^4I_{15/2} \rightarrow {}^4F_{7/2}$
522	19157	$(Er^{3+})\,^4I_{15/2} \rightarrow {}^2H_{11/2}$
546	18315	$(Er^{3+})\,^4I_{15/2} \rightarrow {}^4S_{3/2}$
652	15337	$(Er^{3+})\,^4I_{15/2} \rightarrow {}^4F_{9/2}$
800	12500	$(Er^{3+})\,^4I_{15/2} \rightarrow {}^4I_{9/2}$
976	10246	$(Er^{3+})\,^4I_{15/2} \rightarrow {}^4I_{11/2}$ and $(Yb^{3+})\,^2F_{7/2} \rightarrow {}^2F_{5/2}$
1535	6514	$(Er^{3+})\,^4I_{15/2} \rightarrow {}^4I_{13/2}$

The unique absorption band of Yb^{3+} ions (at around 976 nm) corresponds to the $^2F_{7/2} \rightarrow {}^2F_{5/2}$ transition and overlaps that of Er^{3+} ions for the $^4I_{15/2} \rightarrow {}^4I_{11/2}$ transition. Compared to Er^{3+} ions, the absorption of Yb^{3+} near 976 nm is predominant, as shown in Fig. (7), where the band intensity at around 976 nm is strongly increasing with Yb^{3+} ion concentration due to the high absorption cross-section of Yb^{3+} ions at this wavelength. [2, 12, 17, 38-41]

Since there is only one electronic excited state for Yb^{3+} ions (the $^2F_{5/2}$ state), when a diode laser at around 976 nm is used to excite the Yb^{3+} ions, the emission from the $^2F_{5/2}$ state of Yb^{3+} overlaps the absorption band for the $^4I_{15/2} \rightarrow {}^4I_{11/2}$ transition of Er^{3+} ions. This results in an efficient resonant-energy transfer (ET) from Yb^{3+} to Er^{3+} ions in the Yb^{3+}/Er^{3+} co-doped systems. Hence, the addition of Yb^{3+} to Er^{3+}-doped glass can enhance the pumping efficiency of 977 nm diode laser indirectly through the energy transfer (ET) from Yb^{3+} to Er^{3+}. [12, 40, 41]

The measurement of the absorption intensity variations with rare earth ions concentrations is a convenient way of examining the solubility of these ions in the glassy host. [8, 38, 40] Fig. (8) shows the absorption intensity at around 976 nm as a function of Yb^{3+} concentration. Within the analyzed range of Yb^{3+} ion concentrations, the absorption intensity at 976 nm increases linearly with the Yb^{3+} ion content. Hence, it is confirmed that Yb^{3+} ions have a good solubility in this $G_{70}P_{20}K_{10}$ glass. Furthermore, it can be deduced that Er^{3+} ions could have good solubility owing to the similar nature of rare earth ions. [40] Many device applications require a high solubility of dopants to enable small, high power density devices needed for local area network and power laser applications. [10]

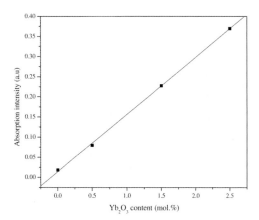

Figure 8. The absorption intensity at 976 nm vs. Yb₂O₃ content.

3.5. Upconversion Emissions

Fig. (9) shows the room temperature upconversion emission spectrum of the Er^{3+}/Yb^{3+} co-doped $G_{70}P_{20}K_{10}$ glass with $Er^{3+}:Yb^{3+}$ concentration ratio of 1: 5, obtained under IR excitation of diode laser at 980 nm with a power of 1W.

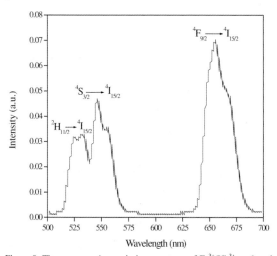

Figure 9. The upconversion emission spectrum of Er^{3+}/Yb^{3+} co-doped $G_{70}P_{20}K_{10}$ glass with $Er^{3+}:Yb^{3+}$ concentration ratio of 1: 5, under excitation at 980 nm of DL with a power of 1W.

The observed emissions correspond to transitions from excited states to the ground state of Er^{3+} ions. The intense green and red emission bands centered at around 532, 546 and 655 nm, are assigned to the transitions $^2H_{11/2}{\rightarrow}^4I_{15/2}$, $^4S_{3/2}{\rightarrow}^4I_{15/2}$ and $^4F_{9/2}{\rightarrow}^4I_{15/2}$, respectively. [30, 40, 42, 43] It was noted that the upconversion emissions

were bright and visible to the naked eye even for 100 mW of the pump laser power at room temperature.

In order to investigate the effect of increasing Yb^{3+} ion concentration on the upconversion emission intensity, the upconversion emission spectra of Er^{3+}/Yb^{3+} co-doped $G_{70}P_{20}K_{10}$ glass samples with fixed a Er^{3+} ion concentration and different Yb^{3+} ion concentrations, shown in Fig. (10), were obtained under 980 nm diode laser excitation with the same laser power (1W). A significant increase in the overall intensity of all bands is observed with increasing of Yb^{3+} ion concentration, indicating that Yb^{3+} ions play an important role in the upconversion emissions.

Fig. (11) shows the integrated intensities I_{up} of the upconversion emission bands as a function of Yb^{3+} concentration. As can be observed, the integrated intensities of the upconversion emission bands increase gradually with increasing of Yb^{3+} ions concentration. This in turn confirms that there exists an effective Yb^{3+} to Er^{3+} energy transfer. [43] Cleary, the red emission intensity (655 nm) increases faster than the green emissions intensities (532 and 546 nm) which indicates that there might be higher populations on the $^4F_{9/2}$ level than that of $^2H_{11/2}$ and $^4S_{3/2}$ levels with increasing of Yb^{3+} ions content. [8]

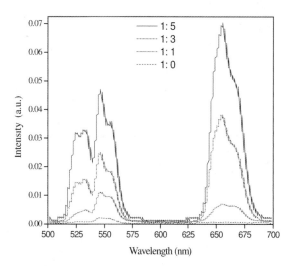

Figure 10. The upconversion emission spectra of Er^{3+}/Yb^{3+} co-doped $G_{70}P_{20}K_{10}$ glasses with different $Er^{3+}:Yb^{3+}$ concentration ratios, under excitation of 980 nm DL with a power of 1 W.

206

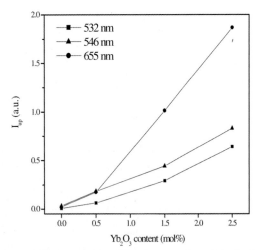

Figure 11. The integrated intensities of upconversion emission bands vs. Yb³⁺ content.

In an upconversion mechanism, the unsaturated upconversion emission intensity I_{UP} is proportional to the mth power of the IR excitation power ($I_{UP} \propto P^m_{IR}$) where m is the number of the IR photons absorbed per visible photon emitted. A plot of $\log(I_{UP})$ vs. $\log(P_{IR})$ yields a straight line with slope m. [2, 5, 30, 42] Figs. (12a), (12b) and (12c) show the corresponding logarithmic plots of the integrated intensities of the 532, 546, and 655nm bands as a function of the laser power, for the Er³⁺/Yb³⁺ co-doped glasses with Er³⁺:Yb³⁺ concentration ratios of 1: 1, 1: 3, and 1: 5, respectively.

The slope values of the linear fits, obtained for the emission bands 532, 546, and 655 nm, are 2.13, 1.76, and 1.94 (for sample 1:1), 2.14, 1.72, and 1.90 (for sample 1: 3), and 2.04, 1.78, and 1.91 (for sample 1:5), respectively. These results indicate that the dependence of the 532, 546, and 655 nm emission bands intensity on the excitation power is nearly quadratic. Hence, it is concluded that a two-photon upconversion process populates the ²H₁₁/₂, ⁴S₃/₂, and ⁴F₉/₂ levels. (i.e., two IR pump photons are absorbed to generate one emitted visible photon).

The possible upconversion emission mechanisms are discussed based on the quadratic pump power dependence and the simplified energy levels of Er³⁺ and Yb³⁺ ions, presented in Fig. (13), as follows:

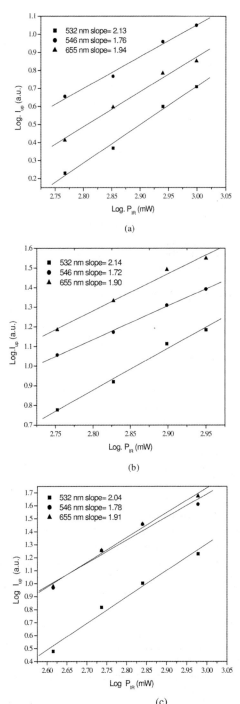

Figure 12. The Log–log plot of upconversion emissions versus ump power for the Er³⁺/Yb³⁺ co-doped glasses with Er³⁺:Yb³⁺ concentration ratios of (a) 1:1, (b) 1:3, and (c) 1:5.

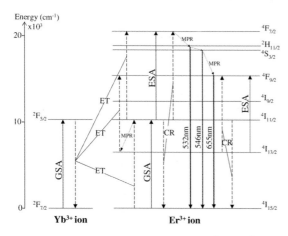

Figure 13. The simplified energy level diagram of Er^{3+} and Yb^{3+} ions doped in $G_{70}P_{20}K_{10}$ glass and the possible upconversion mechanisms.

For the 532 and 546 nm emission bands, the first step involves the excitation to the $^4I_{11/2}$ level of Er^{3+} ion by the following two possible processes. An Er^{3+} ion in its ground state absorbs a 980 nm pump photon and be excited to the $^4I_{11/2}$ level (GSA). And/or an Yb^{3+} ion in its ground state absorbs a 980 nm pump photon and be excited to the $^2F_{5/2}$ level, then transfers its excitation energy nonradiatively to a neighboring ground-state Er^{3+} ion, which is then excited to the $^4I_{11/2}$ level (ET). ET process is dominant in the excitation of $^4I_{11/2}$ level, since the Yb^{3+} ions have a larger absorption cross-section than the Er^{3+} ions at around 980 nm. [2, 5, 6, 30, 43] Many studies have confirmed that the long-lived $^4I_{11/2}$ level is the intermediate level involved in the upconversion mechanisms. [2, 9, 34]

In the second step, the populated $^4I_{11/2}$ level can be excited to the $^4F_{7/2}$ state by excited state absorption of a second pump photon, by energy transfer from an excited Yb^{3+} ion, and/or by cross-relaxation (CR) between two adjacent excited Er^{3+} ion. [2, 5, 6, 30, 43]

The populated $^4F_{7/2}$ level then relaxes rapidly and nonradiatively to the next lower $^2H_{11/2}$ and $^4S_{3/2}$ levels through multiphonon relaxation (MPR) process because of the small energy gap between them (~ 1330 cm^{-1}). The energy gap between the $^2H_{11/2}$ and $^4S_{3/2}$ levels is very small (~840 cm^{-1}), and a rapid thermal equilibrium is established between the thermally coupled $^2H_{11/2}$ and $^4S_{3/2}$ levels. Finally, radiative decay (RD) from $^2H_{11/2}$ and $^4S_{3/2}$ levels to the ground state $^4I_{15/2}$ yields green emissions centered at around 532 and 546 nm, respectively. [2, 5, 6, 30, 43] These processes could be simply visualized as:

$^4I_{15/2}$ (Er^{3+}) + a pump photon (980nm) $\rightarrow$ $^4I_{11/2}$ (Er^{3+})	(GSA)
$^2F_{7/2}$ (Yb^{3+}) + a pump photon (980nm) $\rightarrow$ $^2F_{5/2}$ (Yb^{3+})	(GSA)
$^4I_{15/2}$ (Er^{3+}) + $^2F_{5/2}$ (Yb^{3+}) $\rightarrow$ $^2F_{7/2}$(Yb^{3+}) + $^4I_{11/2}$ (Er^{3+})	(ET)
$^4I_{11/2}$ (Er^{3+}) + a pump photon (980nm) $\rightarrow$ $^4F_{7/2}$(Er^{3+})	(ESA)
$^4I_{11/2}$ (Er^{3+}) + $^2F_{5/2}$ (Yb^{3+}) $\rightarrow$ $^2F_{7/2}$(Yb^{3+}) + $^4F_{7/2}$(Er^{3+})	(ET)
$^4I_{11/2}$ (Er^{3+}) + $^4I_{11/2}$ (Er^{3+}) $\rightarrow$ $^4I_{15/2}$ (Er^{3+}) + $^4F_{7/2}$(Er^{3+})	(CR)
$^4F_{7/2}$ (Er^{3+}) $\rightarrow$ $^2H_{11/2}$(Er^{3+}), $^4S_{3/2}$ (Er^{3+})	(MPR)
$^2H_{11/2}$(Er^{3+}) $\rightarrow$ $^4I_{15/2}$(Er^{3+}) + emitted photon (532nm)	(RD)
$^4S_{3/2}$ (Er^{3+}) $\rightarrow$ $^4I_{15/2}$(Er^{3+}) + emitted photon (546nm)	(RD)

The red emission at 655 nm is originated from the $^4F_{9/2} \rightarrow ^4I_{15/2}$ transition where the population of $^4F_{9/2}$ level is based on the following possible mechanisms: A fast nonradiative decay through multiphonon relaxation from the populated $^4S_{3/2}$ level to $^4F_{9/2}$ level. However, the energy gap between $^4S_{3/2}$ and $^4F_{9/2}$ levels is about 3047 cm^{-1}, and Raman spectrum shows that the maximum phonon energy of the glass network is ~ 811 cm^{-1}. Therefore, about 4 phonons are required to bridge the gap. Hence, the multiphonon relaxation from the $^4S_{3/2}$ state occurs with a small rate. The other possible mechanism involves the population of the $^4I_{13/2}$ state by nonradiative relaxation from the upper $^4I_{11/2}$ level (which populates by means of the processes described previously). Then an Er^{3+} ion in $^4I_{13/2}$ state can be excited to the $^4F_{9/2}$ state by energy transfer from an excited Yb^{3+} ion and/or by excited state absorption of a pump photon. In addition, cross-relaxation (CR) between two adjacent Er^{3+} ions (one being in level $^4I_{13/2}$ and the other is in level $^4I_{11/2}$) may contribute to populate $^4F_{9/2}$ level. Finally, radiative decay from this level to the ground state produces the red emission band at 655 nm. [2, 5, 6, 30, 43] The above processes could be simply visualized as:

$^4S_{3/2}$ (Er^{3+}) $\rightarrow$ $^4F_{9/2}$(Er^{3+})	(MPR)
$^4I_{13/2}$ (Er^{3+}) + a pump photon (980nm) $\rightarrow$ $^4F_{9/2}$(Er^{3+})	(ESA)
$^4I_{13/2}$ (Er^{3+}) + $^4I_{11/2}$ (Er^{3+}) $\rightarrow$ $^4I_{15/2}$ (Er^{3+}) + $^4F_{9/2}$(Er^{3+})	(CR)
$^4I_{13/2}$ (Er^{3+}) + $^2F_{5/2}$ (Yb^{3+}) $\rightarrow$ $^2F_{7/2}$(Yb^{3+}) + $^4F_{9/2}$(Er^{3+})	(ET)
$^4F_{9/2}$ (Er^{3+}) $\rightarrow$ Er^{3+}($^4I_{15/2}$) + emitted photon (655nm)	(RD)

Since Er^{3+} ions concentration is kept fixed in all glass samples, the energy transfer processes from $^2F_{5/2}$ (Yb^{3+}) to $^4I_{15/2}$ (Er^{3+}), $^4I_{11/2}$ (Er^{3+}) and $^4I_{13/2}$ (Er^{3+}) levels are the dominant processes for populating $^2H_{11/2}$, $^4S_{3/2}$, and $^4F_{9/2}$ levels of Er^{3+} ions after infrared excitation with 980 nm diode laser.

The upconversion intensity of the Er^{3+}-singly doped glass is relatively weak, this is due to the transition $^4I_{15/2} \rightarrow ^4I_{11/2}$ at about 980 nm has low absorption cross-section. Thus, Er^{3+} ion has a weak ability to absorb the excitation light, and therefore it has a very low pumping efficiency. Consequently, the population accumulations

208

of $^2H_{11/2}$, $^4S_{3/2}$, and $^4F_{9/2}$ levels of Er^{3+} are difficult and lead to the weak emission bands. On the contrary, Yb^{3+} ions have stronger ability to absorb pumping light than that of Er^{3+} ions. [40] Furthermore, with the addition of Yb^{3+} ions the spatial distance between Yb^{3+} and Er^{3+} ions become shorter and each Er^{3+} surrounded by more than one Yb^{3+} ions and the energy transfer can easily occur from Yb^{3+} to Er^{3+} and this enhances the pumping efficiency. [12] Therefore, the populations in the $^2H_{11/2}$, $^4S_{3/2}$ and $^4F_{9/2}$ levels increase and as a result, the intensity of the emission bands increase. However, the increasing population accumulation rate of the $^4F_{9/2}$ level is faster than that of the $^2H_{11/2}$ and $^4S_{3/2}$ levels with increasing of Yb^{3+} ions content, which results in relatively higher intensity of the red emission than that of the green emissions.

Conclusion

Er^{3+}/Yb^{3+} co-doped 70GeO$_2$–20PbO–10K$_2$O glasses with a fixed Er^{3+} ion concentration and different Yb^{3+} ion concentrations have been synthesized. DTA curves show that, with increasing of Yb^{3+} ion concentration, the thermal parameters T_g, T_x, and T_p are shifted towards higher temperatures and the thermal stability factor ($\Delta T = T_x - T_g$) is improved from 129 to 195 $^{\circ}$C. Raman spectrum of the undoped glass indicates that the maximum phonon energy of the glass network is around 811 cm^{-1}. IR absorption spectra indicate a very good dispersion of the Yb^{3+} ions, with no evidence of the formation of Yb^{3+} ion clusters in these G$_{70}$P$_{20}$K$_{10}$ glasses with increasing of Yb^{3+} ion concentration.

Optical absorption measurements show that, with increasing of Yb^{3+} ion concentration, the absorption intensity around 976 nm increases linearly. Intense green and red upconversion emission bands centered at around 532, 546, and 655 nm were observed, under excitation at 980 nm of diode laser at room temperature, and assigned to the transitions $^2H_{11/2} \rightarrow {}^4I_{15/2}$, $^4S_{3/2} \rightarrow {}^4I_{15/2}$ and $^4F_{9/2} \rightarrow {}^4I_{15/2}$ of Er^{3+} ions, respectively. The quadratic pump power dependence of these emissions indicates that a two-photon upconversion process contributes to population of the emitting levels. It was found that the upconversion emission intensity obviously enhances with increasing of Yb^{3+} content due to the efficient energy transfer from Yb^{3+} to Er^{3+} ions. The obtained results suggest that this Er^{3+}/Yb^{3+} co-doped G$_{70}$P$_{20}$K$_{10}$ glass system could be used as active media for upconversion lasers and optical fiber amplifiers.

References

1. H. Lin, K. Liu, E.Y.B. Pun, T.C. Ma, X. Peng, Q.D. An, J.Y. Yu, S.B. Jiang, Chemical Physics Letters **398**, 146, (2004)
2. M. Ajroud, M. Haouari, H. Ben Ouada, H. Măaref, A. Brenier, B. Champagnon, Materials Science and Engineering C **26** 523, (2006).
3. A. Kanoun, N. Jaba, A. Brenier, Optical Materials **26**, 79, (2004).
4. P.V. dos Santos, M.V.D. Vermelho, E.A. Gouveia, M.T. de Araujo, A.S. Gouveia-Neto, F.C. Cassanjes, S.J.L. Ribeiro, Y. Messaddeq, Journal of Alloys and Compounds **344**, 304, (2002) .
5. R. Balda, J. Fernández, M.A. Arriandiaga, J.M. Fdez-Navarro, Optical Materials **25**, 157, (2004).
6. Hongtao Sun, Shixun Dai, Shiqing Xu, Lei Wen, Lili Hu, Zhonghong Jiang, Materials Letters **58**, 3948, (2004).
7. Hongtao Sun, Liyan Zhang, Meisong Liao, Gang Zhou, Chunlei Yu, Junjie Zhang, Lili Hu, Zhonghong Jiang, Journal of Luminescence **117**, 179, (2006).
8. Tie-Feng Xu, Guang-Po Li, Qiu-Hua Nie, Xiang Shen, Spectrochimica Acta Part A **64** 560, (2006).
9. Jianhu Yang, Nengli Dai, Shixun Dai, Lei Wen, Lili Hu, Zhonghong Jiang, Chemical Physics Letters **376**, 671, (2003).
10. Richard Weber, Tean A. Tangeman, Paul C. Nordine, Richard N. Scheunemann, Kirsten J. Hiera, Chandra S. Ray, Journal of Non-Crystalline Solids **345&346**, 359, (2004).
11. L.R.P. Kassab, W.G. Hora, J.R. Martinelli, F.F. Sene, J. Jakutis, N.U. Wetter, Journal of Non-Crystalline Solids **352**, 3530, (2006).
12. Zhongmin Yang, Shiqing Xu, Lili Hu, Zhonghong Jiang, Journal of Alloys and Compounds **370**, 94, (2004).
13. Robert F. Speyer, "Thermal Analysis of Materials", Marcel Dekker Inc., (1993).
14. James E. Shelby, "Introduction to Glass Science and Technology", The Royal Society of Chemistry, (1997).
15. Meisong Liao, Shunguang Li, Hongtao Sun, Yongzheng Fang, Lili Hu And Junjie Zhang, Materials Letters **60** 1783, (2006).
16. Shiqing Xu, Dawei Fang, Zaixuan Zhang, Zhonghong Jiang, Journal of Solid State Chemistry **178**, 1817, (2005).
17. Qiu-Hua Nie, Yuan Gao, Tie-Feng Xu, Xiang Shen, Spectrochimica Acta Part A **61**, 1939, (2005).
18. V.K. Tikhomirov, A.B. Seddon, D. Furniss, M. Ferrari, Journal of Non-Crystalline Solids **326&327**, 296, (2003).

19. Hongtao Sun, Junjie Yang, Liyan Zhang, Junjie Zhang, Lili Hu, Zhonghong Jiang, Solid State Communications **133,** 753, (2005).

20. Hongtao Sun, Lili Hu, Chunlei Yu, Gang Zhou, Zhongchao Duan, Junjie Zhang, Zhonghong Jiang, Chemical Physics Letters **408,** 179, (2005).

21. G.F. Yang, Q.Y. Zhang, T. Li, D.M. Shi, Z.H. Jiang, Spectrochimica Acta Part A **69,** 41, (2008).

22. T. Fukushima, Y. Benino, T. Fujiwara, V. Dimitrov, T. Komatsu, Journal of Solid State Chemistry **179,** 3949, (2006).

23. A. Céreyon, B. Champagnon, V. Martinez, L. Maksimov, O. Yanush, V.N. Bogdanov, Optical Materials **28,** 1301, (2006).

24. Youngsik Kim, Jason Saienga, Steve W. Martin, Journal of Non-Crystalline Solids **351,** 3716, (2005).

25. Zhongmin Yang, Shiqing Xu, Lili Hu, Zhonghong Jiang, Materials Research Bulletin **39,** 217, (2004).

26. P.W. France, "Optical Fiber Lasers and Amplifiers", Blackie and Son Ltd, (1991).

27. Alex C. Hannon, Daniela Di Martino, Luis F. Santos, Rui M. Almeida, Journal of Non-Crystalline Solids **353,** 1688, (2007).

28. Zhongmin Yang, Zhonghong Jiang, Journal of Luminescence **121,** 149, (2006).

29. Shiqing Xu, Zhongmin Yang, Shixun Dai, Guonian Wang, Lili Hu, Zhonghong Jiang, Materials Letters **58,** 1026,(2004).

30. Qiuhua Nie, Cheng Jiang, Xunsi Wang, Tiefeng Xu, Haoquan Li, Materials Research Bulletin **41,** 1496, (2006).

31. Jenn-Shing Wang, Kuen-Ming Hon, Ko-Ho Yang, Moo-Chin Wang,
Min-Hsiung Hon, Ceramics International **23,** 153, (1997).

32. Luıs M. Fortes, Luıs F. Santos, M. Clara Goncalves, Rui M. Almeida, M. Mattarelli, M. Montagna, A. Chiasera, M. Ferrari, A. Monteil, S. Chaussedent, G.C. Righini, Optical Materials **29,** 503,(2007).

33. Michel J. F. Digonnet, "Rare Earth Doped Fiber Lasers and Amplifiers", 2nd Edition, Marcel Dekker, Inc., (2002).

34. Fiorenzo Vetrone, John-Christopher Boyer, John A. Capobianco, Adolfo Speghini, Marco Bettinelli, Applied Physics Letters Volume 80, Number **10,** 1752, (2002).

35. G.A. Kumar, E. De La Rosa, H. Desirena, Optics Communications **260,** 601, (2006).

36. R. Balda, A. Oleaga, J. Fernández, J.M. Fdez-Navarro, Optical Materials **24,** 83, (2003).

37. Hongtao Sun, Shiqing Xu, Shixun Dai, Junjie Zhang, Lili Hu, Zhonghong Jiang, Journal of Alloys and Compounds **391,** 198, (2005).

38. Long Zhang, Hefang Hu, Changhong Qi, Fengying Lin, Optical Materials **17,** 371, (2001).

39. Hongtao Sun, Shixun Dai, Shiqing Xu, Junjie Zhang, Lili Hu, Zhonghong Jiang, Physica B **352,** 366, (2004).

40. Yuan Gao, Qiu-Hua Nie, Tie-Feng Xu, Xiang Shen, Spectrochimica Acta Part A **61,** 1259, (2005).

41. Q.Y. Zhang, Z.M. Feng, Z.M. Yang, Z.H. Jiang, Journal of Quantitative Spectroscopy & Radiative Transfer **98,** 167, (2006).

42. Liyan Zhang, Hongtao Sun, Shiqing Xu, Kefeng Li, Lili Hu, Solid State Communications **135,** 449, (2005).

43. Li Feng, Jing Wang, Qiang Tang, Haili Hu, Hongbin Liang, Qiang Su, Journal of Non-Crystalline Solids **352,** 2090, (2006).

EFFECT OF HOST MEDIUM ON THE FLUORESCENCE EMISSION INTENSITY OF RHODAMINE B IN LIQUID AND SOLID PHASE

M. FIKRY, M. M. OMAR*, AND LOTFI Z. ISMAIL

Physics Department, Faculty of Science, Cairo University, Egypt

In this work, we study the effect of concentration, host medium, PH, ions complex and phase states on the fluorescence emission from the laser dye, Rhodamine B, pumping by UV laser as exited source. The polymethylmethacrylate PMMA used as host medium in case of solid phase samples while, ethanol and Tetrahydrofuran (THF) are used in case of liquid one. The Laser Induced Fluorescence (LIF) technique was used to study the fluorescence properties of the both cases liquid and thin film solid-state samples. In addition, the Dual Thermal Lens (DTL) technique was used to study the quantum yield of these samples. The maximum fluorescence emission observed at concentration of Rhodamine B $C=3\times10^{-4}$M. At this concentration of Rhodamine B, the type of solvent and polarity of the medium affect on the fluorescence emission intensity of Rhodamine B with . The measurements revile that, the behavior of both phases state was analogous and Rhodamine B/PMMA thin film sample by ratio of 4:1 and thickness 0.12 mm is the best photostability sample and its quantum yield about ≈ 0.82. Also, the fluorescence emission intensity of Rhodamine B was quenched by complex formation of Co, Al, Cu and iodide ions with Rhodamine B due to the increase of the charge density of the ions.

1. Introduction

Recently, electroluminescent devices based on organic thin layers have attracted much interest due to their potential application as large area light-emitting displays. Organic dyes are also candidates for making LEDs, and this field based on the semiconductor and conductor polymer with the fluorescent dye to enhancement the emission the dye lasers [1-3]. The fluorescent dyes it self can to be using to quantifying fluid mixing in aqueous solution. The general technique involves rationing optically separable fluorescence from two fluorescent dyes. The resulting ratio normalizes for laser intensity distribution and any laser reflections and by properly selecting the fluorescent dyes, fluorescence ratios used to either measure the concentration of a passive scalar or temperature fields [4].

The Rhodamine dyes its self-association in the different liquid crystal (Anisotropic solvent) host materials like the ethanol (isotropic solvent) and this is very important in the new high technology application in the display and electronic technology [5]. Solar cells based on dye-sensitized mesoporous films of TiO2 are low-cost alternatives to conventional solid-state devices. This dye is also used as sensor [6], nonlinear optical material, photosensitizer, malarial protease labels, active laser element, passive Q-switches and optical wave-guide [7,8]. From the early days of development of dye lasers, attempts were made to over came the problems posed by dye solutions, by incorporating dye molecules into solid matrices. A solid-state dye lasers avoids the problem of toxicity, flammability, compact, versatile, easy to operate, maintain and have high photostability properties [9].

In recent years, the synthesis of high performance dyes into the solid matrix use the polymer host present advantages as these materials show much better compatibility with organic laser dyes and its inexpensive fabrication technique, high power, high efficiency [10]. Hence, the thermal and optical properties of dye-doped polymers are important in identifying suitable laser media, where Rhodamine B is the most photostability dye ≈ 9000 GJ/mol. In addition, Polymethylmethacrylate PMMA is one of the most amorphous network stretcher polymers [11].

---

* To whom all correspondences should be addressed. E-mail address: magdyomar@hotmail.com.

C. V. Bindhu et al. using the dual beam thermal lens (DTL) technique as a quantitative method to determine absolute fluorescence quantum yield and effect of concentration of Rhodamine B in different solvent. They investigated the effect of excitation source on the absolute fluorescence quantum yield of Rhodamine B 514 nm radiation from an argon ion laser was used as a cw exited source and 532 nm pulses from a Q-switched Nd:YAG laser was used as a pulsed excitation source [12]. While, A. Kurian, N. A. George, B. Paul, V. P. N. Nampoori and C. P. G. Vallabhan used the dual beam thermal lens technique as a quantitative method to determine absolute fluorescence quantum efficiency and concentration quenching of fluorescence emission from Rhodamine 6G doped Polymethylmethacrylate (PMMA), prepared with different concentrations of the dye. A comparison of the present data with that reported in the literature indicates that the observed variation of fluorescence quantum yield with respect to the dye concentration follows a similar profile as in the earlier reported observations on Rhodamine 6G in solutions. The photodegardation of the dye molecules under cw laser excitation is also studied using the present method [9].

In addition, T. H. Nhung et al. [11] studied the laser dyes Rhodamine B and Perylene Red incorporated at different relative concentrations into hybrid matrices synthesized using the sol–gel process. Energy transfer from Rhodamine B-donor to Perylene Red-acceptor molecules was observed. Using the different co-doped samples, solid-state dye laser systems were achieved with tuning band position control and increased efficiency with respect to the materials using solely one type of dye.

2. Experimental

2.1. Materials

Rhodamine B (Rh B) of M.wt 479.02 gm from Merack used as fluorescent laser dye material. Polymethylmethacrylate (PMMA) of M.wt 100000 gm from Aldrich used as polymer host for the dye to preparing the thin film of Rhodamine B. The tetrahydrofuran (THF) from Aldrich using as common solvent for both (PMMA) and (Rh B), to reduce the problems of non-miscibility and non-homogeneity of the different solvent in each others.

The absolute ethanol (99.9%) from Chema.jet used as common solvent for Rhodamine B.

2.2. Preparation of thin film

At once, time the required concentration of Rhodamine B in THF is prepared and the required concentration of (PMMA) in THF is prepared. Then, the tow solutions are then mixed with continues stirring for at least 5 minutes. This mixture is left in dishes inside oven at constant temperature 30 Co for 24 h to evaporate the solvent with constant rate to produce homogenous thin film [13]. One of the most important notes to avoid all photodegardation processes the energy of nitrogen laser and the exposure time of nitrogen to the sample are reduced and the samples must be clean from any photocatalytic materials [9,10].

2.3. LASER Induced fluorescence (LIF)

The Pulsed N2-laser (337 nm) is delivered to sample by incident angle 45° for thin film of dye sample and 90o for the liquid sample, which focused by quartz cylindrical lens of focal length (156 mm) in a dark chamber. In addition, the emission spectra during the irradiation process have collected by similar lens to monochromater. Which the monochromater has input slit width = 3.5 mm and output slit width = 2.5 mm. The spectrum signals are detected by photomultiplier (face type). The capture information by the data acquisition card are analyzed with pc-computer software as written specially to scanning the wavelength using steeper motor as shown in Fig. (1).

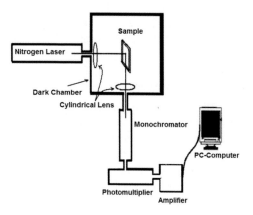

Fig (1): Laser Induced Luminescence (LIF) set-up.

212

2.4. Dual Thermal lens (DTL)

The experimental set-up of the dual beam thermal lens technique shown in Fig. (2).

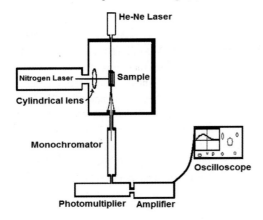

Fig (2): Dual beam thermal lens technique.

In the dual beam technique, separate lasers are used, nitrogen laser (337nm) as exited source and He-Ne laser (635 nm) as probe beam. This technique is more advantageous since only a single wavelength of the probe beam is always detected. The excitation source generate the thermal lens in the samples of dye medium. The beam of low intensity of He-Ne is diverted due to the thermal lens which generated by the exited source. This produced thermal lens signal depends on the type of samples .The probe beam are made collinear and made to pass through the samples. The thermal lens signal is detected by sampling the intensity at the center portion of the probe beam. This signal is transmitted only through the monochromator and detected by photomultiplier which connected to the oscilloscope.

3. Results and Discussion

The fluorescence emission spectra of the Rhodamine B samples are proportional to the different in concentrations as shown in Fig. (3). From this figure, we obtain that the fluorescence emission spectra of Rhodamine B at different concentrations has band tuneability of wavelengths from 520 nm to 765 nm. Where, the band broader of high concentration sample is wider than low concentration sample.

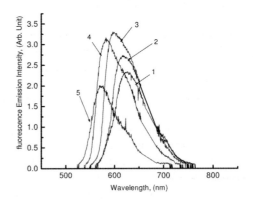

Fig (3): Effect of concentration on the fluorescence emission of Rhodamine B [in curves, (1) C=7×10^{-3}M, (2) C=3×10^{-3}M, (3) C=3×10^{-4}M, (4) C=5×10^{-5} and (5) C=1×10^{-5}M].

This is return to increase of the monomers and aggregates of dye molecules. The fluorescence emission intensity varied corresponding to the concentration of Rhodamine B as shown in Fig. (4).

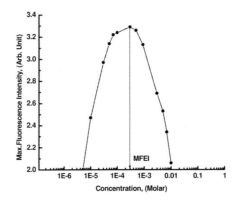

Fig (4): Effect of concentration on the fluorescence emission of Rhodamine B.

In this figure, at low concentration from 10^{-6} M to 10^{-4} M, the fluorescence emission intensity increases with increasing the concentration of Rhodamine B. Because at low concentration C= 5×10^{-6} M, the dye–dye molecules interaction is negligible due to the large average distance between them due to the increase the monomers and aggregates [9].

At high concentration C=3×10⁻⁴ M, the fluorescence emission intensity return to decease with farther increase in the concentration from 10⁻⁴ M to 10⁻² M. This reduces related to different mechanisms. First, this related to the phenomenon of re-absorption and re-emission, with increasing dye concentration the formation of dimers and higher aggregates decreases the fluorescence emission by a combination of monomer dimers and higher aggregates, where the energy transfer and re-absorption of radiation occur by non-fluorescent dimmers and higher aggregates. This transfer of energy between molecules by collisional mechanisms makes the nonradiative part prominent and hence fluorescence decreases. Second, also one of the factors, which reduced the fluorescence intensity, is the relaxation processes between various vibronic levels of dye [9,12]. Fig. (4) shows the maximum fluorescence intensity observed in range of concentration about10⁻⁴ M. This result is agreed with the results of K. Achamma et al. [9], C. V. Bindhu et al. [12], A. Santhi et al. [14] and L. Z. Ismail et al.[13]. In addition, we observed the wavelengths of the fluorescence peak shifted corresponding to the concentration of Rhodamine B as shown in Fig. (5).

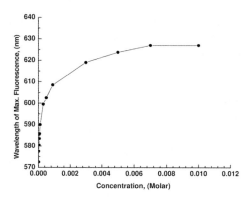

Fig (5): Effect of concentration on the wavelength of the maximum Fluorescence emission of Rhodamine B.

This peak shift with increase the concentration to certain level and after that the shift is very small and can to be negligible. First, this returned to a consequence of re-absorption and re-emission. Since the low frequency tail of the absorption spectrum of the dye molecule overlaps with the high frequency

end of its fluorescence emission spectrum, the fluorescence from the excited state dye molecule reabsorbed by the ground state molecule, which shifts the fluorescence peak to lower energies [9]. Second, absorption saturation would lead to a similar peak shift, where, if the molecules have closely laying levels of different absorption cross section absorption saturation to each level would lead to wavelength shift [12]. However, in general the peak shift in dyes is not due to saturation but they related to the re-absorption and re-emission only [9,12]. At high concentration the aggregation formation, restrict the peak shift [12].

The fluorescence emission spectra of Rhodamine B in liquid phase effected by the type of solvent and polarity of the medium, where the fluorescence emission intensity of Rhodamine B with concentration 3×10⁻⁴ M in ethanol is higher than in THF as shown in Fig. (6).

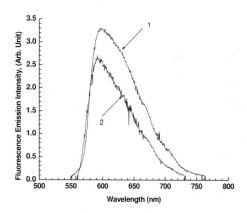

Fig (6): Effect of solvent on the fluorescence emission intensity of Rhodamine B with concentration 3×10⁻⁴ M. Curve (1) in ethanol and curve (2) in THF

Rhodamine B belongs to xanthenes series of dyes with non-esterified phenyl carboxylic group. It shows great sensitivity to solvent polarity and viscosity because of the existence of number of possible molecular forms such as cationic, zwitterionic and lactonic. The zwitterionic form of Rhodamine B appears in the basic ethanolic solution. Cationic form of Rhodamine B absorbs at the longer wavelength compared to the zwitterionic form. The form of Rhodamine B depends on the type of the medium (acidic, basic, polar or non-polar). In addition, the

214

fluorescence emission intensity of Rhodamine B affected by the viscosity of medium which effect on the refractive index of the medium. This suggests that in low viscosity, the chromophore is fully rigid in the ground state and loosens up only after excitation. While in high viscosity solvents or medium, the viscosity is sufficiently high to prevent thermal equilibrium being reaches during the radiative lifetime of a few nanoseconds. Hence, planarity of the ground state is not lost before light emission takes place [10,15].

Hence, Rhodamine B shows high fluorescence emission intensity and broader band of Rhodamine B dissolved in ethanol than THF, due to ethanol is alkaline polar medium according to that the Rhodamine B has zwitterionic form and absorbed more intensity of nitrogen pumped laser (excitation source) and more viscosity and high refractive index than THF samples. In addition, the fluorescence emission intensity of Rhodamine B depends on the dye–dye molecules interaction, which depends on the distance between dye molecules, which the fluorescence emission intensity increases with increasing the distance between molecules [9]. Hence, the ethanol molecules are smaller than THF solvent this development large distance and less dye molecules aggregation in ethanol than THF, where the aggregation reduces the carrier mobility producing decrease in the fluorescence emission intensity [16]. The polymethylmethacrylate (PMMA) has concentration $C=8\times10^{-4}$ M, it used as matrix for the Rhodamine B which has $C=3\times10^{-4}$ M. Hence, PMMA is one of the most amorphous network stretcher polymers [11]. Rhodamine B and PMMA with different in the ratio of Rhodamine B to PMMA are dissolved in THF. The fluorescence emission intensity of the Rhodamine B/PMMA in liquid phase are varying according to the different in the ratios of Rhodamine B to PMMA as shown in Fig. (7). From this figure, the fluorescence emission spectra of Rhodamine B with different ratios of Rhodamine to PMMA have band tuneability of wavelengths from 550 nm to 745 nm. Where the band broader of all ratios may be equal or small and can be negligible.

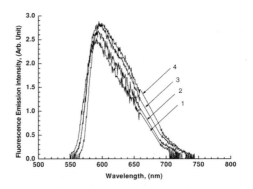

Fig (7): Effect of ratio of Rhodamine B $C=3\times10^{-4}$ M to PMMA $C=8\times10^{-4}$ M on the fluorescence emission of the liquid samples in curves, (1)1:1≈100%, (2) 3:1≈300%, (3) 4:1≈400% and (4) 6:1≈600%.

This return to the aggregation formation restrict the broader of the fluorescence emission band [12]. In Fig. (8), the fluorescence emission intensity increases with increasing the ratio of Rhodamine B. Where at low concentration the dye dissolved as monomers and the dye–dye molecules interaction, the interaction between dye molecules and the macromolecules (aggregated molecules) are negligible. Hence, at low dye ratios the fluorescence intensity increases linearly with Rhodamine B ratios. At ratio about ≈ 4:1 ≈ (400%) the fluorescence emission intensity reach to limit value and any further increase in the dye ratios of Rhodamine B not effect on the fluorescence emission intensity, this related to the monomer dimmers and aggregates formation restrict the fluorescence emission intensity increases [9–12].

215

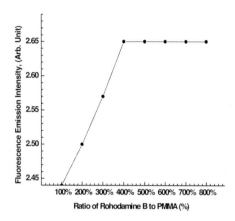

Fig (8): The fluorescence maximum intensity with ratio of Rhodamine B C=3×10⁻⁴ M to PMMA C=8×10⁻⁴M liquid samples.

In addition, in Fig. (9), the maximum of the fluorescence emission peak shifted with increasing the ratios of Rhodamine B to PMMA. After certain ratio about 2:1≈ 200% the shift is very small and can to be negligible, this is related to the phenomenon of re-absorption and re-emission [9].

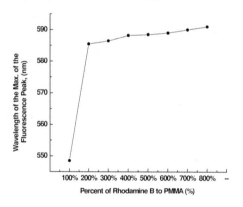

Fig (9): Effect of ratio of Rhodamine B C=3×10⁻⁴ M to PMMA C=8×10⁻⁴M in liquid phase on the wavelength of the maximum fluorescence emission.

The fluorescence emission spectra of the thin film samples of Rhodamine B/PMMA have similar behavior to the liquid phase samples with different ratios as shown in Fig. (10).

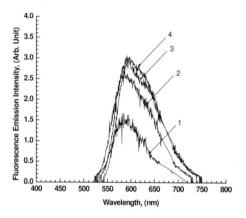

Fig (10): Effect of percent of Rhodamine B C=3×10⁻⁴ M to PMMA = 8×10⁻⁴M on the fluorescence emission of the thin films in curves, (1) 1:1≈ 100%, (2) 3:1≈ 300% (3) 4:1≈ 400% and (4) 5:1≈ 500%.

From this figure, the fluorescence emission spectra of Rhodamine B thin film of thickness=0.12 mm with different ratio of Rhodamine to PMMA has band tuneability of wavelengths from 520 nm to 750 nm. Where the band broader of all ratios may be equal or small and can to be negligible. Fig. (11) illustrates that the comparison between the fluorescence emission intensity of the thin film samples and the liquid samples of the ratio of Rhodamine B to PMMA. The fluorescence emission intensity of thin film of Rhodamine B/PMMA samples and band broader of these samples are grater than liquid samples of Rhodamine B/PMMA dissolved in THF. The main reason of this result is PMMA thin film which is not alkaline medium according to that the Rhodamine B has cationic form and absorbed more intensity of nitrogen pumped laser (excitation source) and thin film of PMMA has more viscosity and high refractive index than liquid samples [9,15].

In addition, Fig. (11) shows the greatest fluorescence emission intensity start observed in range of ratio of Rhodamine B to PMMA about 4:1≈ 400%. This result is agreed with the results of L. Z. Ismail et al. [13] and K. Achamma et al. [9]. Rhodamine B is a kind of xanthenes dye, whose optical properties depends on many factors, such as solvents (polarity and aprotic character) and pH value.

216

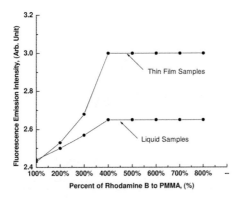

Fig (11): The fluorescence maximum intensity with percent of Rhodamine B C=3×10⁻⁴ M to PMMA C=8×10⁻⁴ M thin films and liquid samples.

Fig. (12) illustrates that the effect of PH on the Rhodamine B dissolving in ethanol (polar solvent), while Fig. (13) for Rhodamine B dissolving in THF (non-polar solvent).

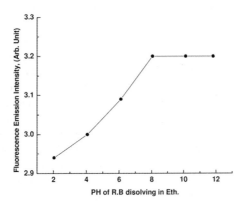

Fig (12): Effect of pH on the fluorescence emission intensity of Rh B C=3×10⁻⁴M dissolving in ethanol.

At low PH, the fluorescence emission intensity is small while the PH increased, the fluorescence emission intensity enhanced and after certain PH value about ≈ 8, the fluorescence emission intensity became constant with increasing PH as shown in Fig. (12). While, in case of Rhodamine B dissolving in THF as shown in Fig. (13), the behavior will be inversed.

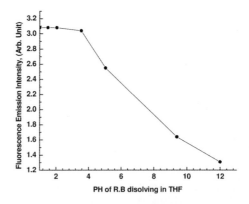

Fig (13): Effect of pH on the fluorescence emission intensity of Rh B C=3×10⁻⁴M dissolving in THF.

Where, at high PH the fluorescence emission intensity is small while the PH decreased, the fluorescence emission intensity enhanced and after certain PH value about ≈ 3.75 the fluorescence emission intensity reach to constant value, where the change in the constant region is very small and can to be negligible. This return to Rhodamine B which is exist with number of possible molecular forms such as cationic, zwitterionic, quinoic and lactonic [10] as shown in Fig. (14).

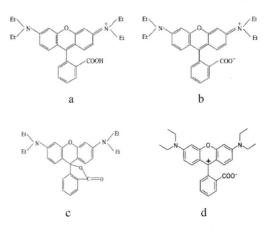

Fig (14): Different formula of Rhodamine a) Cationic formula, b) Quinioic formula, c) Lactonic formula and d) Zwitterionic

The forms of Rhodamine B depend on the polarity and pH of the medium. Where the cationic and quinoic forms are strongly colored and emissive, while lactonic forms is colorless and shows no emission because the π-electron system of the dye chromorphore is interrupted [17]. In ethanol (polar solvent), the zwitterionic form of Rhodamine B appears in the basic ethanolic solution where as a drop of HCl in ethanolic solution produces cationic form. Cationic form of Rhodamine B absorbs at the longer wavelength compared to the zwitterionic form [1]. Also in the ethanolic solvent, the quinoid form present due to the opening of the lactone ring and the formation of the quinoid form is in the ground state as shown in Fig. (15) [18].

Lacton   Quinion

Fig (15): Equilibrium between lacton and quinion form in the polar solvent.

While in the non-polar solvent (THF), at low PH the cation and quinion forms are present. By increasing the PH, the lacton form appears and this leads to the colorless of the dye solution due to the transfer of the cation and quinion to lacton in processes opposite to the last in Fig. (15) [17–19]. In addition, we show with fix the molecular dye form, the change in pH producing small change in the fluorescence emission intensity, this return to the change in the number of hydrogen bonding between molecules [20]. The fluorescence emission intensity of the Rhodamine B quenching due to present ions in the ethanol as solvent as shown in Fig. (16), where fluorescence quenching refers to any process which decreases the intensity of the fluorescence emission of a sample. A variety of interactions can result in quenching. The mechanisms of these interactions include collisional or dynamic quenching, static quenching, excited-state reactions ground-state

complex formation, molecular rearrangements, quenching by energy transfer and charge transfer reactions.

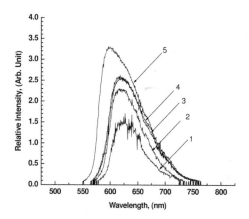

Fig (16): Effect of ions on the fluorescence emission intensity of the Rhodamine B of C=3×10⁻⁴M [Curves: (1) Rh B only, (2) Rh B and Co ions, (3) Rh B and Al ions, (4) Rh B and Cu ions and (5) Rh B and iodine ions.

Dynamic quenching defined as the quenching result due to the collision between a fluorophore in its excited state and the quencher. When quenching occurs by a dynamic or collisional mechanism, the quencher must diffuse to the fluorophore during the lifetime of the excited state. Upon physical contact, the fluorophore returns to the ground state, without emission of a photon. Quenching occurs without any permanent change in the molecules, i.e. without a photochemical reaction. Static quenching occurs if a non-fluorescent ground state complex is form between the fluorophore F and the quencher Q as shown in the equation. When this complex absorbs light, it immediately returns to the ground state without emission of a photon.

$$F + Q \rightleftharpoons (F-Q)$$

Both dynamic and static quenching requires molecular interaction between the fluorophore and the quencher. The measurement of fluorescence intensity cannot differentiate between dynamic and static quenching. The most effective way to distinguish between dynamic and static quenching is the measurement of luminescence decay times since in case of static quenching a fraction of fluorophores is removed by complex formation. While in the

Rhodamine B must the photophysical change in the fluorescence emission intensity related to the complexation with an ions due to the presence of active carboxylic group on the cationic form of Rhodamine B in the ethanol. Then the photophysical changes upon complexation with ions described in terms of enhancement of the intermolecular charge transfer in the Rhodamine B owing to direct interaction between the bound ions and the carboxylic group [21]. The measurement of the quantum yield of the Rhodamine B in different host medium [10,15] according to the dual thermal lens technique required to measure the thermal lens (TL) signal η and thermal lens (TL) signal η_α corresponding to the concentration at which the fluorescence intensity quenched completely. we can calculate the ratios of η to η_α and the quantum yield (Q) [9,22]. The quantum yield and the ratios of η to η_α of Rhodamine B of different sample of the Rhodamine B in different host medium (Ethanol, THF, PMMA in THF and PMMA thin film) are shown in table (1).

Table (1): The variation of quantum yield and the Ratios of η to η_α of Rhodamine B as function of the type of the medium.

Sample type	η/η_α	Q
1) Rhodamine B in ethanol	0.792	0.36
2) Rhodamine B in THF	81	0.31
3) Rhodamine B and PMA in THF	0.834	0.27
4) Rhodamine B and PMMA thin film	0.529	0.82

The quantum yield of Rhodamine B in ethanol is less than in the PMMA thin film. This mean the best lasing samples is the Rhodamine B in PMMA thin film, where they have the highest quantum yield. From the results, the Rhodamine B/PMMA with ratio about 4:1 has highest quantum yield $\Phi \approx 0.82$ and good lasing properties.

Conclusion

By using the Laser Induced Fluorescence (LIF) technique, the effect of concentrations on the laser dye Rhodamine B dissolved in ethanol were studied. It was found that, the fluorescence emission intensity varied corresponding to the concentration of Rhodamine B. Where at low concentration from 10^{-6} M to 10^{-4} M, the fluorescence emission intensity increases with increasing the concentration of Rhodamine B, this return to the dye–dye molecules interaction is negligible with increase the monomers and aggregates. At high concentration C=3×10^{-4} M the fluorescence emission intensity return to decease with farther increase in the concentration from 10^{-4} M to 10^{-2} M. This reduces related to re-absorption and re-emission by a combination of monomer dimers and higher aggregates. The greatest maximum fluorescence intensity was observed at concentration about 3×10^{-4} M.

In addition, the wavelengths of the fluorescence peak shifted corresponding to the increase concentration of Rhodamine B to certain level and after that the shift is very small and can to be negligible this returned to a consequence of re-absorption and re-emission and absorption saturation. The type of solvent and polarity of the medium also affect on the fluorescence emission intensity of Rhodamine B with concentration 3×10^{-4} M. Where, the fluorescence emission intensity of Rhodamine B in ethanol is higher than in THF, because the Rhodamine B shows a great sensitivity to solvent polarity and viscosity due to the existence of number of possible molecular forms Rhodamine B. The present of PMMA with Rhodamine B on THF will decrease the fluorescence emission intensity of Rhodamine B, while this fluorescence emission intensity of Rhodamine B can be increased by increasing the ratios of Rhodamine B to PMMA to certain ratios about 4:1 after that any increase in the ratios not affect.

Because at low Rhodamine B/PMMA ratios, the increase in ratios tend to increase the concentration of Rhodamine B and dye–dye molecules interaction are negligible. At ratio about $\approx$ 4:1 $\approx$ (400%) the monomer dimmers and aggregates formation restrict the fluorescence emission intensity increases. The fluorescence emission spectra of the thin film samples of Rhodamine B/PMMA have similar with the behavior of the liquid phase samples. Where, the fluorescence emission spectrum of thin film of Rhodamine B/PMMA samples are grater than liquid samples of Rhodamine B/PMMA dissolved in THF due to the higher viscosity and refractive index of PMMA thin film. In addition, the effect of PH on the different solvent (ethanol and THF) varied the fluorescence emission intensity. This variation depends on the type, polarity, viscosity and PH of the samples.

Finally, the fluorescence emission intensity of Rhodamine B was quenched by complex formation of Co, Al, Cu and iodide ions with Rhodamine B due to the increase of the charge density of the ions. From the results, the Rhodamine B/PMMA with ratio about 4:1 has highest quantum yield $\Phi \approx 0.82$ and good lasing properties.

References

[1] David Braun, Elsevier Science Ltd, (2002).

[2] J. Huang, V. Bekiari, P. Lianos, S. Couris, (1999), Journal of Luminescence 81, 285-291.

[3] Oscar G. Calder´on, I. Leyva, and J. Guerra, (1999), IEEE Journal of Quantum Electronic, Vol. 35, NO.1.

[4] J. R. Coppeta and C. B. Rogers, (1995), American Institute of Aeronautics and Astronautics.

[5] A. Ghanadzadeh, M.A. Zanjanchi, (2001), Spectrochimica Acta Part A 57, 1865–1871.

[6] M. Ahmad, K. Chang, T.A. King and L. L. Hench, (2005), Sensors and Actuators A 119, 84–89.

[7] N. A. George, B Aneeshkumar, P Radhakrishnan and C.P.G. Vallabhan, (1999), J. Phys. D; APPL. Phys. 32, 1745-1749.

[8] G.S.S. Saini, S. Kaur, S.K. Tripathi, C.G. Mahajan, H.H. Thanga, A.L. Verma, (2005), Spectrochimica Acta Part A 61, 653–658. of Luminescence 91, 25-31.

[9] A. Kurian, N. A. Georgy, B. Paul, V. P. N. Nampoori and C. P. G. Vallbhan, (2002), Laser Chemistry, Vol. 20, (2-4), 99-110.

[10] A. V. Deshpande, E. B. Namdas, (2000), Journal

[11] T. H. Nhung, M. Canva, F. Chaput, H. Goudket, G. Roger, A. Brun, D. D. Manh, N. D. Hung and J. Boilot, (2004), Optics Communications 232, 343–351.

[12] C. V. Bindhu and S. S. Hariial, (2001), Analytical sciences, Vol. 17

[13] L.Z. Ismail, H. M. abdel Moneim, G. Abdel Fatah and Z. A. Zohdy, (2001) , Polymer Testing 20, 135-139

[14] A. Santhi, M. Umadevi, V. Ramakrishnan, P. Radhakrishnan and V.P.N. Nampoori, (2004), Spectrochimica Acta Part A 60, 1077–1083.

[15] A. G. Knospe, A. S. Kwok, (2004), Chemical Physics Letters 390, 130–135.

[16] S. Sinha and A. P. Monkman, (2003), Journal of Appl. Phys. Vol. 93, No. 9 1.

[17] X.M. Hana, J. Lina, R.B. Xingb, J. Fub and S.B. Wanga, (2003), Materials Letters 57, 1355–1360.

[18] A. A. El-Rayyes, A. Al-Betar, T. Htun, U. K.A. Klein, (2005), Chemical Physics Letters 414, 287–291.

[19] X. Wang, M. Song, and Y. Long, (2001), Journal of Solid State Chemistry 156, 325-330.

[20] R. R. Amin, I. S. Al Namimi and L. Z. Ismail, (1998), Asin J. of Chem. Vol 10, No. 2, 347-355.

[21] I. Leray, J. L. Habib-Jiwan, C. Branger, J.-Ph. Soumillion and B. Valeur, (2000), Journal of Photochemistry and Photobiology A: Chemistry, 135, 163–169.

[22] A. Kurian, S. D. George, V.P.N. Nampoori and C.P.G. Vallabhan, (2005), Spectrochimica Acta, Part A, 61, 2799–2802.

II-3 CONTRIBUTING PAPERS

III. CONTRIBUTING PAPERS

LOW TEMPERATURE PHOTOLUMINESCENCE AND PHOTOCONDUCTIVITY OF ZnSe$_x$Te$_{1-x}$ TERNARY ALLOYS

A. SALAH[a], G. ABDEL FATTAH[a], I.K. ELZAWAWY[b] AND Y. BADR[a]

[a] *National Institute of Laser Enhanced Science (NILES), Cairo University, Egypt*
[b] *Solid State Physics Department, National Research Center (NRC), Cairo, Egypt*

Abstract. We investigated Low-Temperature Photoluminescence (PL) spectra of ZnSe$_x$Te$_{1-x}$ were grown from the melt where $0 \leq x \leq 0.202$, the spectra of ZnSe$_x$Te$_{1-x}$ showing a broad band which may be attributed to self activated emission, The broad self activated (SA) emission band have been assigned to various crystalline defects, such as dislocations and vacancies or their combination with impurities . The room temperature photoelectric response spectra of ZnSe$_x$Te$_{1-x}$ samples ($0 \leq x \leq 1$) were measured, a single band was observed in the band edge region which attributed to the generation of more number of free charge carriers in the band gap region. The relaxation time was determined from studying the kinetics of photoconductivity
Keywords: Photoluminescence -Photoconductivity - ZnSe$_x$Te$_{1-x}$ –Ternary alloys.

PACS: 71.20Nr, 71.55Gs,72.40+w

1. Introduction

The mixed crystals of II-VI compound semiconductors have attracted much attention for applications to optical devices. Zinc tellurides have a direct band gap corresponding to a wavelength of the green-light region at room temperatures, and it is one of the promising materials for green light emitting devices. Graded alloys of ZnSe$_x$Te$_{1-x}$ and digital alloys utilizing thin ZnSe and ZnTe layers are used as contact layers for Zinc selenide based optoelectronic devices in order to increase p-type doping [1, 2].

The study of the photoconductivity (PC) provides an understanding of the photo generations and the transport of the free carriers. The PC represents a valuable tool in understanding the recombination kinetics, which, in turn, provides information about the localized states in the amorphous materials [3]. Photoluminescence spectroscopy is a contactless, nondestructive method of probing the electronic structure of materials. Results of photoconductivity and photoluminescence emission spectra are presented in this paper.

2. Experiment

ZnSe$_x$Te$_{1-x}$ ternary alloys were grown from the melt. The compositions of the ZnSe$_x$Te$_{1-x}$ were analyzed by the (Energy Dispersive X-ray Analysis)EDX micro analytic unit attached to the SEM. with (Li/Si) detector at accelerated voltage 25 KeV.
The Photoluminescence Spectra were measured using Ar$^+$ laser 488 nm with a power of 90 mW , a monochromator of a length of 750 mm [SPEX 750M] with a grating (1200 gr/mm), the resolution of the monochromator is 1 A^0 and photomultiplier detector (185-850nm) uv-glass; the output signal was amplified using the lock-in technique [SR510], the sample was placed in the pumped liquid helium path cryostat (CTI-CRYOGENICS), with an electrical heater (SCIENTIFIC INSTRUMENT INC. 9620-1) and control equipment, to reach and hold any temperature from 8 K to 300 K.

224

White light from Halogen lamp (150 W) into a monochromator (Model 285 GCA McPHERSON instrument double beam monochromator), goes through a chopper and collimating lens, into the surface of the sample. Two silver electrodes are displaced on the surface of the samples. External source (0 – 9 V) of applied voltage is used and the photocurrent response is detected by a double channel Lock-In amplifier (Model SR-530)through load resistance R_L=32 K Ω, In order to determine the life time of the carriers, we plot a relation between the frequency of the chopper and the alternating component of the photoconductivity.

3. Results and Discussion

It was observed that the emission lines of Zn, Se and Te were present in the energy range investigated (0-20 KeV) which is represented in Fig. (1).and the chemical composition results of $ZnSe_xTe_{1-x}$ are shown in Table (1).

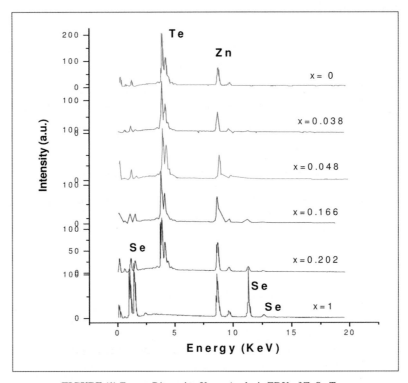

FIGURE (1) Energy Dispersive X-ray Analysis EDX of $ZnSe_xTe_{1-x}$

TABLE (1) Chemical composition of $ZnSe_xTe_{1-x}$ by EDX

	Se	Te	Zn
0	0	50.18	49.82
0.038	1.85	46.31	51.84
0.048	2.33	46.15	51.52
0.166	7.54	37.98	54.48
0.202	8.93	35.22	55.85
1	34.94	0	65.06

3-1 Photoluminescence measurements

The PL spectra of ZnSe$_x$Te$_{1-x}$ Crystals were measured in the 500-950 nm wavelengths using Ar$^+$ laser of wavelength 488 nm. At a constant excitation laser intensity 90 Mw, The low temperature PL measurements of our ZnSe$_x$Te$_{1-x}$ at 25 K are shown in Fig (2).

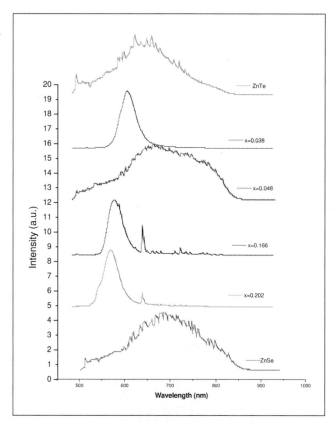

FIGURE (2) The low temperature PL measurements of ZnSe$_x$Te$_{1-x}$ at 25 K

TABLE (2) the broad bands of the ZnSe$_x$Te$_{1-x}$ at 25 K

x	λ (nm)	Eg(eV)	The assignments	Ref.
	The main broad band			
0	663.2	1.868		
0.038	623	1.989		
0.048	680	1.822	self activated (SA)	[4]
0.166	595.25	2.081		[5]
0.202	587	2.111		[5]
1	695	1.783		[5]

226

| x | The band | | The assignments | Ref. |
	λ (nm)	Eg(eV)		
0	539	2.299		
0.048	535	2.316	(DAP) about 2.3 eV	[4]
0.202	542	2.286		
1	539	2.299		

TABLE (3) DAP is observed about 2.3 eV at 25 K

Deep level emission of Zn (SA) observed in both ZnSe and ZnTe, But the y band is observed in ZnSe, ZnTe and at x=0.048. The appearance of this band might be attributed to the lattice imperfection; this y band was observed in other II-VI semiconductors [6]. The broad self activated (SA) emission band has been assigned to various crystalline defects, such as dislocations and vacancies or their combination with impurities [5]. The band of a self activated (SA) photoluminescence is properly studied in A2B6 compounds. It is a donor –acceptor recombination nature and is determined to DA associate $\{V_{Zn} + D^+)^o$ as acceptor, with the components in the nearest points in a unit cell.

3-1-1 PL spectra of ZnSe$_{0.048}$Te$_{0.952}$

Fig. (3) shows the main bands of ZnSe$_{0.048}$Te$_{0.0952}$ appearing at 680.3 nm at 25 K, the intensity decreases with increasing the temperature, the band position was shifted to higher wavelength in the range (25 K-55 K). The band 680.3 nm is attributed to SA emission [4]; it seems that our spectrum consists of two bands. A similar feature has been observed on the Bridgman grown ZnSe $_{0.69}$Te $_{0.31}$ [7]. The origin of it is not yet clear.

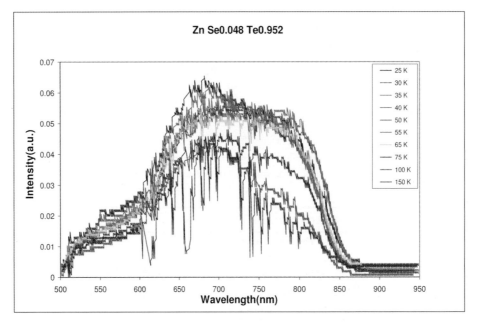

FIGURE (3) PL Spectra of ZnSe$_{0.048}$Te$_{0.952}$ in the 25-150 K temperature range.

3-1-2 Temperature dependence of the PL intensity

Temperature dependence of the PL intensities of the main broad bands for Zn Se_x Te_{1-x} where x=0, 0.038, 0.048, 0.166, 0.202, 1 are shown in the figures (4-a-f), The intensity of the main broad band of Zn $Se_x Te_{1-x}$ decreased as the temperature increased and this can be due the ionization of shallow acceptors [8].

The thermal quenching is expressed by [8]

$$I = I_0 e^{E_A/KT}$$

where E_A is the activation energy of a certain thermal quenching process. I_0 is a constant, K is Boltzmann constant. This equation can be written as [6, 9]:

$$I(T) = \frac{I_0}{1 + C \exp(-E_A / K_B T)}$$

where C is a fitting parameter to fit the calculated value to the experimental data.

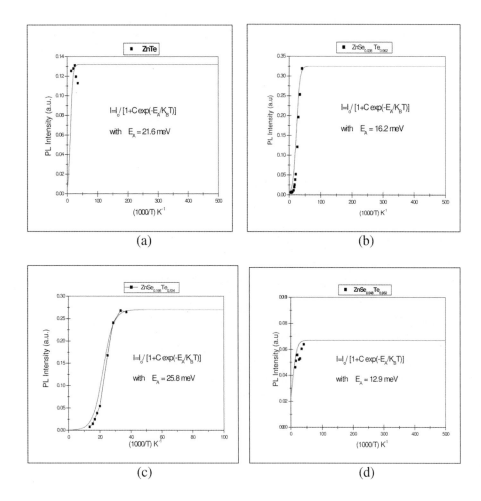

(a)   (b)

(c)   (d)

228

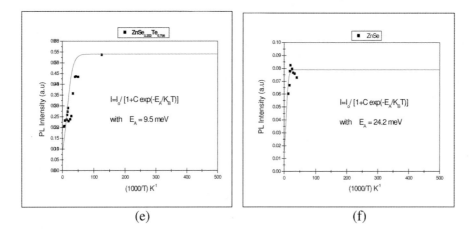

(e)   (f)

FIGURE (4 a-f) Temperature dependence of ZnSe$_x$Te$_{1-x}$ PL intensity at the emission band maximum.

TABLE (4) The calculated activation energy
for samples of different Se concentration

ZnSe$_x$Te$_{1-x}$	E$_A$(activation energy of the transition)
x	meV
0	21.6
0.038	16.2
0.048	12.9
0.166	25.8
0.202	9.5
1	24.2

3-2 Photoconductivity of ZnSe$_x$Te$_{1-x}$ crystals

The ability of the contacts to replenish carriers to maintain charge neutrality in the material (properties of an ohmic contact) if carriers are drawn out of the opposite contact by an electric field plays a key role in determining the kind of the phenomena observed [10].

Current-voltage (I_V) characteristics of ZnSe$_x$Te$_{1-x}$ samples sandwiched between Ag layers of 99.999 % purity were measured using I-V circuit. The Ag layer was deposited as a paste. The measurements were carried out for ZnSe$_x$Te$_{1-x}$ samples of different x content in ambient atmosphere at 27 C^0.

 The obtained data showed a straight line relationship for different samples as shown in Fig. (5) where the characteristic dependence of current on applied voltage for ohmic contact is linear. This behavior proves that Ag makes ohmic contacts with ZnSe$_x$Te$_{1-x}$ bulk samples where x ranged from 0 to 1.

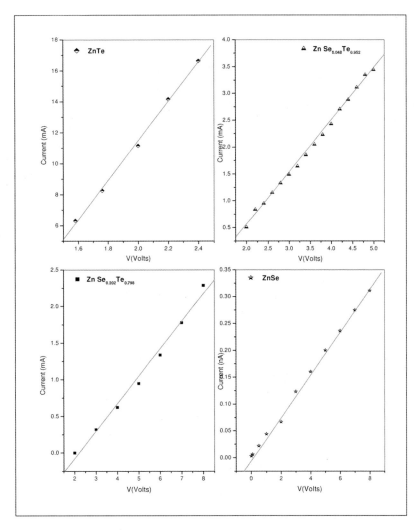

FIGURE (5) I-V characteristics of ZnSe$_x$Se$_{1-x}$ crystals

3-2-1 Photoresponse spectra of ZnSe$_x$Te$_{1-x}$ crystals

The room temperature photoelectric response spectra of ZnSe$_x$Te$_{1-x}$ samples ($0 \leq x \leq 1$), in the wavelength range (0.2-2.5) μm is shown in Fig (6). The spectra are characterized by a single peak situated in the band edge region. This can be attributed to the generation of more number of free charge carriers in the band gap region. The longer wavelength side of the peak extends over a wider range of wavelength as x increases. This is possibly associated with the formation density –of – states tail at the band edges due to alloying effects. For a semiconductor in which the limits of the energy bands occur at the same value of the wave vector K, the onset of absorption will occur at hν=E$_g$ as the result of direct transitions. The optical energy band gap E$_g$ was determined from the position of the photoresponse maximum.

The composition dependence of E$_g$ is shown in Fig. (7). the experimental results of the band gap determined from the photoconductivity spectra were compared the theoretical results obtained from Vegard's law [11]. Vegard's law states that Eg(x)=a+ b x+ c x^2 where Eg(x) is the band gap of ZnSe$_x$Te$_{1-x}$ using Eg(0) the optical energy gap of ZnTe which equal to 2.26 eV[12] , Eg(1) the optical energy gap of ZnSe which equal to 2.67 [13] at room temperature and Eg(Se=0.68)=2.18 eV [14] , by Substituting in Vegard's law for determining the constants a, b and c from the three equations,

230

We get the following equation:

$$Eg(x) = 2.26 - 1.238x + 1.648x^2 \qquad (1)$$

Theoretical results of $E_g(x)$ given by equation (1) were calculated and shown in Fig. (7), in addition to the fitting of the experimental results shown as curved line, the experimental and theoretical data are in good agreement within the range of experimental error arising from devices and inhomogenity of the samples.

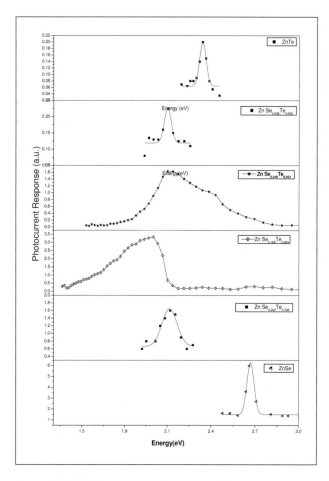

FIGURE (6) Photoconductivity spectra of Zn Se_x Te_{1-x}

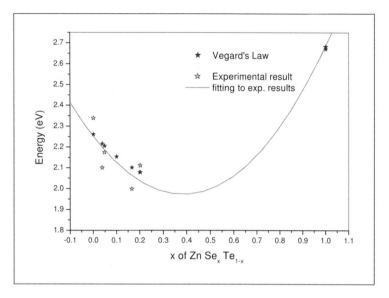

FIGURE (7) Photoconductivity measurement as a function of x

3-2-2 The kinetics of photoconductivity

The method used for determining the relaxation time is the frequency dependence of the alternating component of photoconductivity signal $\Delta\sigma$ [15] is given by the following relation

$$\frac{\Delta\sigma}{\Delta\sigma_{st}} = \tanh(\frac{1}{4\tau_{eff}f}) \tag{2}$$

Where $\Delta\sigma_{st}$ is the photoconductivity when $f \to 0$, τ_{eff} is the effective relaxation time. Fig. (8) shows the results of the frequency dependence of photoconductivity at room temperature.

In order to determine the life time of the carrier, by using equation (2) a straight line was plotted parallel to the frequency axis at height 0.76 of the $\frac{\Delta\sigma}{\Delta\sigma_{sr}}$ axis (where tanh1=0.76) to intersect the curves at certain point. Then if straight lines were drawn from these points perpendicular to the frequency axis, they cut off segments of the axis equal to $\frac{1}{4\tau}$ for each curve [16.] And we get the life time of the carriers in ZnSe$_x$Te$_{1-x}$ crystals in the range of a few milliseconds and the calculated life time in table (5).

The values of τ_{eff} for different samples of Zn Se$_x$ Te$_{1-x}$ were indicated in table (5). Such large values of τ_{eff} usually characterize processes of carrier release from trap levels by thermal excitation, the behavior of the relation $\frac{\Delta\sigma}{\Delta\sigma_{st}}$ (f)

for all samples of different compositions are the same, but the relaxation time decreases with increasing intensity of light. The magnitude of this parameter depends on the density of trapping centers available for the carrier density induced by the illumination. This behavior of τ with different values of x cannot be explained by alloying effect only. This suggests that, the contribution of residual impurities and native defects in the sample is significant.

232

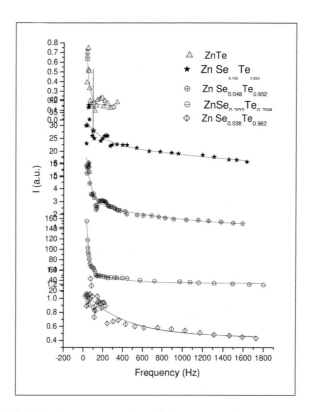

FIGURE (7) The frequency dependence of photoconductivity at room temperature

TABLE (5) The life time of the carriers

x	τ (msec)
0	7.78
0.038	1.14
0.048	3.75
0.166	2.5
0.202	6.25

4. Conclusion

Low-Temperature Photoluminescence (PL) spectra were measured on six samples, the spectra of $ZnSe_xTe_{1-x}$ showed a broad band which may be attributed to self activated emission, The broad self activated(SA) emission band have been assigned to various crystalline defects, such as dislocations and vacancies or their combination with impurities. DAP is observed about 2.3 eV at x=0, 0.048, 0.202 and 1.

Experimental results of the band gap determined from the photoconductivity spectra were compared with the theoretical results obtained from Vegard's law.

$Eg(x) = 2.26 - 1.238x + 1.648x^2$

The relaxation time was determined from the frequency dependence of the alternating component of photoconductivity signal $\Delta\sigma$

The behavior of the relation $\dfrac{\Delta\sigma}{\Delta\sigma_{st}}$ (f) for all samples of different composition are the same, but the relaxation time decreases with increasing intensity of light, The magnitude of this parameter depends on the density of trapping centers available for the carrier density induced by the illumination. This behavior of τ with different values of x cannot be explained by alloying effect only. This suggests that, the contribution of residual impurities and native defects in the sample is significant.

REFERENCES

1. W. Lin, X. Yang, S. P. Guo, A. Elmoumni, F. Fernandez, and M. C. Tamargo, Appl. Phys. Lett. 75, 2608 (1999).
2. D. Albert, J. Nürnburger, V. Hock, M. Ehinger, W. Faschinger, and G. Landwehr, Appl. Phys. Lett. 74, 1957 (1999).
3. V. Sharma, A. Thakur, P.S. Chandel, N. Goyal, G.S.S. Saini, S.K. Tripathi, J. Optoelectron. Adv. Mater. 5 (2003) 1243.
4. Choon-Ho Lee, Gyoung-Nam Jeon, Seung-Cheol Yu and Seok-Yong Ko, J. Phys. D: Appl. Phys. 28 (1995) 1951-1957.
5. Q. Liu, H. Lakner, C. Mendorf, W. Taudt, M. Heuken, K. Heime, J. Phys. D: Appl. Phys. 31(1998)2421-2425.
6. Young-Moon Yu, Sungun Nam, and Ki-Seon Lee, J. Appl. Phys., Vol. 90, No. 2, 15July 2001
7. A. Yu. Naumov, S. A. Permogorov, T. B. Popova, A. N. Reznitsky, V. Ya. Zhulai, V. A. Novozhilov, and N. N. Spendiarov, Sov. Phys. Semicond. 21, 213 (1987).
8. İbrahim İnanç, M.Sc. thesis in temperature and laser power dependence of photoluminescence in nanocrystalline silicon, BoğaZiçi University 2004.
9. A. P. Jacob, Q. X. Zhao, M. Willander, C. S. Yang and W. C. Chou, J. Appl. Phys., Vol. 94, No. 4, 2337 (2003).
10. Richard H. Bube, *Photoelectronic Properties of Semiconductors*, Cambridge University Press 1992.
11. D. Zhou and B. F. Usher, J. Phys. D: Appl. Phys. 34 (2001) 1461-1465.
12. T. Mahalingam, V. S. John and P. J. Sebastian, J. Phys.: Condens. Matter 14(2002) 5367-5375.
13. S. Darwish, A. S. Riad and H. S. Soliman. Semicond. Sci.Technol. 1995 (10) 1-7.
14. Ching-Hua Su, S. Feth, Shen Zhu and S. L. Lehoczky, J. Appl. Phys. Vol. 88, No. 9 (Nov. 2000) 5148-5152.
15. G. Revel, J. Pastol, J. Rouchard and R. Fromageru, Metal Trans., (1978), 665.
16. I. M. Ashraf, H. A. Elshaikh, A. M. Badr, Crys. Res. Technol. 39, 63-70 (2004).

OPTICAL PROPERTIES OF Ga OXIDE NANOSTRUCTURES

LOTFIA EL NADI*, GALILA ABDELATIF, HUSSEIN M. MONIEM AND MAGDY M. OMAR

Physics Dept., Laser Communication Lab., Faculty of Science,
Cairo University, Giza, EGYPT
lotfianadi@gmail.com

Abstract. Gallium oxide nanodots of diameter 300 to 220 nm have been grown by SiO_2 assisted thermal evaporation. The formed grayish white crust deposited on the crucible walls was proved to be crystalline Ga_2O_3 through X-ray diffraction and TEM electron diffraction. The morphology of the samples examined by SEM, confirmed the nanodot structure formation of average diameter 220 ± 30 nm and average density of 1.77×10^8 cm^{-2}. The absorption spectra of the Ga_2O_3 suspension in DMF solution revealed two absorption peaks at 329.99nm (3.76 eV) and 338.60 nm (3.67 eV). Photoemission of blue light at room temperature was observed at 410.3 nm with FWHM 56.7 nm under excitation by 330.00 nm. The growth mechanism of the nanodots is explained in terms of liquid-vapor-solid mechanism (LVS). The remaining hard ingot was found to have 16.6 Vicker's hardness. The silicon glass having the value of Vicker's hardness 15 indicates that the ingot material is 10 % harder than that of glass.

Keywords: nanodots, Gallium oxide, SiO_2, X-ray diffraction, TEM electron diffraction, absorption spectra, Photoluminescence.

PACS : 78.20.e

1. Introduction. Material science is introducing revolutionary new field of scientific discipline of nanotechnology. Nano meter size develops quantum one dimensional domain that provides special physical properties different from those of bulk materials. Consequently a new era is opening in science that needs deep investigation of such domains both theoretically and experimentally. Data is still needed about the optical and electronic properties of systems in nanometer sizes. Also the structural type of such domain being dots, wires, tubes etc.... resulting in different properties for one and the same material, call for new methods to grow and identify nanostructures [1-4]. Important mechanisms encountered during growth are vapor-liquid-solid for thermal treatment at high temperatures or solution-liquid-solid growth at lower temperatures. These were used for growing oxide, carbide and nitride nanostructures [5-8]. Other recent growth method known as Si oxide-assisted nanostructure growth developed Si and Ge and Zn oxides in nanostructures [9,10]. Silica assisted catalytic growth of tungsten nanowires has been reported by Zhu et al. [11].

Since Ga_2O_3 has interesting optical and electrical properties, it has been prepared by X.C.Wu et al. [12] by carbothermal reduction from gallium oxide powder on silicon substrate.

Our group succeeded to grow Ga_2O_3 nanowires by laser ablation technique [13]. We decided to study further the physical properties by growing Ga_2O_3 nanostructures using the silica assisted thermal process to investigate the effect of growth mechanism on the properties of the obtained nanostructures.

CP 998, Modern Trends in Physics Research
Third International Conference MTPR-08
edited by L. El Nadi

2. Experimental. The silica assisted catalytic growth of Ga oxide nanostructures was carried out using equal weight amounts of the gallium metal and fine graphite powder of high purity 99.99% which played the role of a catalyst. The gallium metal and carbon powder were mixed thoroughly and homogenized in a porcelain morter for 1 hour. The mixture was then placed in a small porcelain crucible, topped by SiO_2 plate and covered by the crucible porcelain cover. It was then mounted into high temperature small compartment furnace {Napertherm}. The temperature was raised to 950 $^{\circ}$C during 45 minutes in low flow of atmospheric air. When the temperature reached 750 $^{\circ}$C, the SiO_2 plates melted, at lower temperature than expected (SiO_2 Mp=1610°C). The melt mixture of gallium metals + carbon powder in presence of silica melt was formed. The melts were allowed to cool down slowly reaching room temperature after approximately 4 hours. The grayish white crust on the crucible cover and walls, were then collected as fine powder and prepared for imaging by transmission electron microscopy and for spectral measurements. The samples of the crust powder, dispersed in acetone were pipetted onto the TEM Cu grids covered with carbon thin films. The grids were examined by "TEM 10 Zeiss WW EM", and constituents were identified by EDX. The samples of the crust powder to be spectroscopically examined were mixed in a test tube with micelle in DMF solution and homogenized in ultrasound basin. They were then placed in the spectrophotometer (Perkin Elmer λ - 40) with the reference micelle sample for the absorption measuring the spectrum. The emission spectrum was measured by Perkin Elmer LS55 applying excitation line at 330 nm. The products left overnight in the crucibles developed an extraordinary hard ingot, sticking to the crucible bottom. The ingots were crushed in a marble mortar and used for measuring the X-ray diffraction applying a Phillips X-ray diffractometer with CoKα line.

3. Results and discussion. The X-ray diffraction pattern as shown in Fig. 1 can be indexed in peak position to Ga_2O_3 (Phil. CoKa 1 :1.79 Å) although the relative intensities of the peaks are not consistent with that of bulk Ga_2O_3. Amorphous structure is proved to exist in addition to crystalline structure formation. Absence of metal gallium structure is noticed.

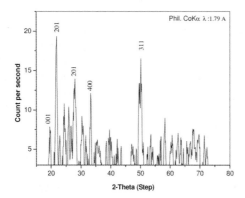

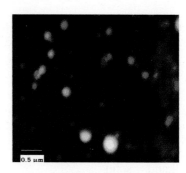

Figure 1. X-ray diffreaction patterns of Ga_2O_3 nanodots.

Figure 2. TEM micrograph of Ga_2O_3 nanodots.

Figure 2 (a,b) represent the TEM images revealing the growth of only nanodots Ga_2O_3 of average diameter 200 ±2 nm and average density of 1.77×10^8 cm^{-2}. The absorption spectrum in figure 3 indicates that gallium oxide nanodots can absorb at 329.9 nm (3.76 eV) and 338.6 nm (3.67 eV).

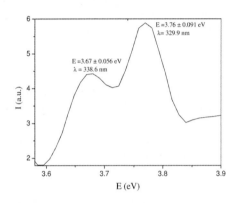

Figure 3. The absorption spectrum of Ga_2O_3 suspension in DMF solution.

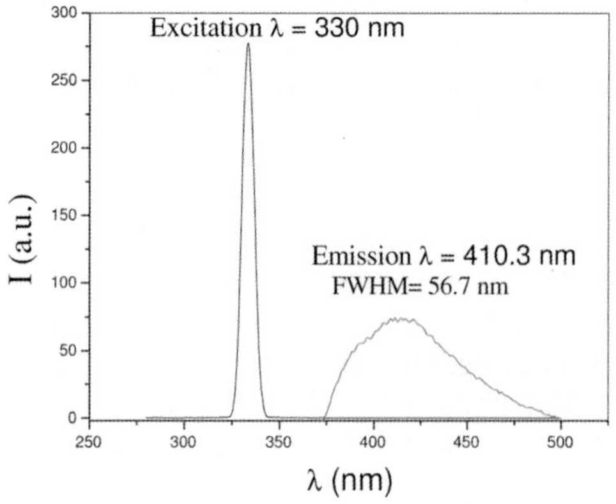

Figure 4. Emission spectrum of Ga_2O_3 under excitation with λ= 330 nm.

The emission spectrum in figure 4 proves that gallium oxide nanodots luminesces at 410.3 ± 28.3 nm (3.02 ± 0.21 eV) in the blue region. This is very near to the PL peak position 2.85 eV of β- Ga_2O_3 single crystal.

In order to investigate the growth mechanism one might easily consider that the original reactants to be:

1- Ga metal liquid at room temperature since it has a low melting point,
2- carbon solid in the graphite powder,
3- SiO_2 solid as the reaction started,
4- O_2 as a gas in the air flowing in the furnace.

Then one may suggest that, during the process of heating in the oven the following reactions most probably occur according to the following processes:

REACTANTS		*PRODUCTS*
$4 Ga + C + 2 O_2$ liquid solid gas	$\rightarrow$	$2 Ga_2O + CO_2$ gas gas
$4 Ga + SiO_2$ liquid solid	$\rightarrow$	$2Ga_2O + Si$ gas solid

The expected reactions between the products in 1 and 2

$2 Ga_2O + 2CO_2$ products (1)		$2Ga_2O_3 + 2C$ solid solid
$C + Si$ products of (2)& (3)	$\rightarrow$	SiC tiny droplets

The products of the reaction (3) super saturates when cooling takes place and solidify in the tiny droplets of SiC products of reaction (4). Such tiny droplets of SiC enhance the further formation of the nanodots of Ga_2O_3.

Figure 5 (a, b) shows the image of two samples from the solid ingot which were removed from the crucible bottom when illuminated by UV light and blue light respectively. It is clear that they are luminescent and emit visible light in agreement with the results of the emission spectra of Ga_2O_3.

a- 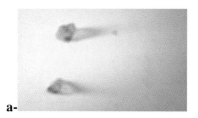   b-

Figure 5. Ingots emission images under perpendicular illumination with a) UV light, b) blue light.

The result for the Vicker's hardness number of Ga_2O_3 ingot sample was found to be 16.6. This value compared to that of the silicon glass having the value of 15 indicates that the ingot material is 10 % harder than that of glass.

4. Conclusions. Gallium oxide nanodots can be prepared by silica assisted thermal vaporization, of the mixture of metal, solid carbon and SiO_2 as clear from the above reactions. The SiO_2 has an important role in the catalytic growth of the metal oxide Nanoparticle. The SiC tiny droplets formed in reaction (4) enhances the formation of Ga_2O_3 during the super saturation stage. The final products in the gas phases Ga_2O and Co_2 gases supersaturate in the SiC tinydots when cooling takes place providing Nanodots of the higher stage metal oxide Ga_2O_3. Accordingly the absence of nanowires could easily be explained.

References

[1] Seung Bum Suh, Jong Chan Kim, Young Cheol Choi, Sunggoo Yun, Kwang S. Kim 2004 J.Am.Chem.Soc.**126** 2198.
[2] Seung Bum Suh, Byung Hee Hong, P. Tarakeshwar, Suk Joo Youn, Sukmin Jeong, Kwang S. Kim 2003 Phys. Rev. **B67** 241402.
[3] Gilles Renaud 2005 Aip conf. proc. MTTPR-04, **748** 63.
[4] Poul L. Hansen, Jakob B. Wagner, Stig Helveg, Jens R. Rostrup-Nielsen, Bjerne S. Clausen, Henrik Topsøe 2002 Science **295** 2053.
[5] X.F. Duang, C.M. Lieber 2000 J. Am. Chem. Soc.**122** 188.
[6] D.P.Yu, Q.L.Hang, Y.Ding, H.Z.Zhang, Z.G.Bai, J.J. Wang, Y.H.Zou, W.Qian, G.C. Xiong, S.Q.Feng 1998 Appl.Phys.Lett.**73** 3076.
[7] N.Wang, Y.H.Tang, Y, F, Zhang, D.P.Yu, C.S.Lee, I.Bello, S.T.Lee 1998 Chem. Phys. Lett. **283** 368.
[8] K.W.Wong, X.T.Zhou, F.C.K.Au, H.L.Lai, C.S.Lee, S.T.Lee 1999 App. Phys. Lett. **75** 1925.
[9] N.Wang, Y.F.Zhang, Y.H.Tang, C.S.Lee, S.T. Lee 1998 Phys. Rev. **B58** 1604.

238

[10] N.Wang, Y.H. Tang, Y.F. Zhang, C.S. Lee, I. Bello, S.T. Lee 1999 Chem. Phys. Lett. **299** 237.

[11] Y.Q.Zhu, W.B.Hu, W.K.Hsu, M.Terrones, N.Grobert, J.P. Hare, H.W. Kroto, D.R.M.Walton, H. Terrones 1999 Chem. Phys. Lett. **309** 327.

[12] X.C.Wu, W.H.Song, W.D. Huang, M.H.Pu, B. Zhao, Y.P. Sun, J.J. Du 2000 Chem. Phys. Lett. **328** 5.

A COMPARATIVE STUDY OF LASER CLEANING OF ARCHAEOLOGICAL INORGANIC MATERIALS WITH TRADITIONAL METHODS

HISHAM IMAM[1], KHALED ELSAYED[2], FATMA MADKOUR[3]

[1] *NILES, Cairo University, Egypt*
Hishamimam@mail.niles.edu.eg

[2] *Physics Dept, Faculty of Science, Cairo University, Egypt*

[3] *Conservation Dept, Faculty of Fine Arts, Minia University, Egypt*

Abstract:

Ancient artifacts excavated from archaeological site were covered with different soil contaminates and stains which changed their chemical composition and aesthetic appearance. Ancient inorganic materials such as bronze, glass and pottery covered with different contaminates such as corrosion products, soil deposits, organic stains and gray white encrustations. Lasers are currently being tested for a wide range of conservation applications. Since they are highly controllable and can be selectively applied, lasers can be used to achieve more effective and safer cleaning of archaeological artifacts and protect their surface details. In the present work we investigated in a general way the laser cleaning of bronze corrosion products, glass, and pottery by Q-switched Nd:YAG Lasers. The results were compared with conventional methods. The artifact samples were examined by Light Optical Microscope (LOM) and showed no noticeable damage.

Keywords: Archaeological artifacts, Laser cleaning, Traditional artifact cleaning

CP 998, Modern Trends in Physics Research
Third International Conference MTPR-08
edited by L. El Nadi
Copyright @ 2011 by World Scientific Publishing Co. 978-981-4317-50-4 / 981-4317-50-0

240

1. Introduction

Cleaning is the most common conservation procedure for art objects without damaging or altering the mechanical, chemical or optical properties of the original surface [1]. Nowadays a number of laser material processing and spectroscopic techniques are being used widely for variety of conservation applications [2, 3, 4, 5] Laser cleaning and diagnosis of artworks and antiques have proven successful in many cases [6, 7]. Laser based techniques are largely non-intrusive and appropriate for the in-situ analysis of composition and structure diagnosis of the object. Laser cleaning has been applied to in-organic archeological objects such as glass, pottery and Bronze, [8, 9, 10]. The fluence of the laser should be adjusted and kept below a certain threshold, otherwise morphological as well as chromatic changes of the material will be observed. The samples were excavated from El-Fustat excavation in Old Cairo and may be belong to Abbasid Period (132 AH/AD 750). These samples of glass, pottery and bronze were contaminated with different deposits. The glass sample was covered with different corrosion layers resulted of contact with water in the soil during the burial for a long time which resulted in migration of the alkali ions (sodium or potassium) together with hydroxyl ions out of the glass and this has the effect of creating an alkali deficient. [11,12]. Encrustations and different stains covered the surfaces of pottery samples due to the burial for prolonged periods. [13,14]. The bronze samples were covered with porous thick layers of green corrosion. Cleaning of archaeological objects often forms an important part of the stabilizing process because of the exist of dirt on an object can be a potent source of deterioration [15].

2. Experimental Section

2.1. Samples description

All samples were excavated from El-Fustat excavation in Old Cairo which founded as a first capital of Islamic Egypt (21AH / 641 AD) [16]. The samples may be belong to Abbasid period (132 AH / 750 AD).

2.1.1. Glass sample

This sample has transparent green color but during the burial in the soil it has been covered with hard crust of corrosion products.Figs.(1,2).

Fig. (1) Shows glass sample before cleaning

The weathering of such glass proceeding by chemical and physical changes in the glass surface resulting of reaction with water in the soil during the burial, produces a range of decay symptoms including loss of transparency, the formation of cracks and corrosion products in the surface layers, and, in some cases, the flaking and the spalling away of areas of glass from the surface [17,18].

241

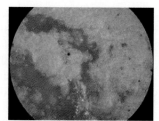

Fig. (2) Shows investigation corrosion
of glass by LOM with 25X

The glass objects showing this type of deterioration have compositions that are high in alkali (soda and / or potash) and low in alkaline earth stabilizers (lime and magnesia) [11,18,19]. The worst conditions for the preservation of glass are found in alkaline soils because the silica skeleton of the glass is attacked and soluble salts (sodium and potassium) are gradually removed from the surface allowing fresh potassium and sodium ions from the glass to react with water whilst calcium salts dissolve were slowly and will delay the decay process [20]. By these processes of ion lacking and silica bond breaking, the surface of the glass begins to become opaque as in this sample [21].

2.1.2. Pottery samples

Fig.(3) Shows pottery sample no.1
before cleaning

Three samples of archaeological pottery were taken after excavation. The first sample was covered with hard gray-white

encrustation as shown in Fig. (3). These deposits of insoluble salts consist mainly of calcium readily from on pottery during burial [13]. The result of XRD analysis of this layer indicated that it contains calcite as a main compound with ratio about 91%, halite with ratio 5% and quartz with ratio 4% as shown in table (1).

Fig.(4) Shows pottery sample no.2
before cleaning

Fig.(5) Shows pottery sample no.3
before cleaning

The other two samples were covered with soot layers and black spots which mixed with the soil residues. Figs (4, 5)

2.1.3. Bronze samples

The bronze samples were covered with thick porous layers of green corrosion products. Figures (6, 7) Corrosion of copper and copper alloys buried in the ground is governed by oxygen, water and the salts which circulate in the soil and the bronzes from sandy soils may be

242

Fig.(6) Shows bronze sample before cleaning

Fig.(7) Shows cross section in bronze sample

strongly degraded [22]. The green crusts on these samples and which common on excavated objects often contain under different conditions that create rough uneven encrustations. These crusts would seem to be due mainly to malachite $CuCO_3$ $Cu(OH)_2$ and may contain chloride minerals such as atacamite or paratacamite $CuCl_2$ $3Cu(OH)_2$ forming dark-green crystals beside copper oxides and tin, lead corrosion products. [13]. XRD analysis of corroded sample indicated that it contains both of paratacamite and quartz as minor compounds with ratio 38%, 32% and calcite, cuprite, tenorite and prochantite as traces as shown in table (1).

Table (1) shows the results of XRD analysis of some deposits layers

Kind of sample	Compounds		
	Major	Minor	Traces
Gray-white layer (sample no.1)	Calcite CaCO3	-	Halite NaCL Quartz SiO_2
Corrosion sample	-	Paratacamite $CuCl_2$.$3Cu(OH)_2$ Quartz SiO_2	Calcite $CaCO_3$ Cuprite Cu_2O Tenorite CuO Prochantite $CuSO_4$. $3Cu(OH)_2$

2.2. Methodology

2.2.1. Traditional cleaning methods Glass sample

Different solvents were tested to remove the corrosion layer which was covered glass samples such as acetone and IMS but they didn't achieve sufficient results. The chelating agent EDTA solution [Ethylene-diamintetraacetic acid, $(HOOCCH_2)_2NCH_2$] was carried out to remove this layer. The cleaning efficiency of this solution depending on its concentration and pH value. It is more effective at neutral pH with low concentrations between 5-10 %.[23]. EDTA solution was used with concentration 5 % to remove this layer where cotton poultices emerged in this solution and were applied on the corroded surface. [24]

Pottery samples

The hard gray white encrustation (insoluble salts) which was covered sample no.1 was cleaned with mechanical methods to reduce this crust and then applying EDTA solution

(Ethylene- diamintetraacetic acid) with concentration 5 % to remove the thin hard crust adhered strongly with the surface[24,25]. The dark stains which were mixed with soil residues that covered the surface of sample no. 2 and the black soot which covered sample no.3 were cleaned with a mixture of hydrogen peroxide H_2O_2 and Industrial Methylated Sprit (IMS) 50/50. EDTA solution 5% was used also to remove the dark stain and black spots [26]

All previous treatments were followed with good rinsing with clean water [21].

Bronze samples

Mechanical cleaning was used to remove the superficial layers of corrosion products from the sample [27]. After that chemical cleaning by soaking the sample in Rochelle Salt solution was applied. The solution was made by mixing 150 grams of Rochelle Salt (sodium potassium tartrate) and 50 grams of sodium hydroxide in 1 liter of water. [28]. This solution succeeded in dissolving most of the green corrosion products.

2.2.2. Laser Cleaning

Laser cleaning were performed with different irradiances according to the kind of sample. Various conditions of energy per pulse, no of shots and the distance between the lens and sample were chosen to irradiate rectangle zone.

Experimental Setup

The cleaning of the sample was done by focusing of the laser beam of the Nd:YAG laser at 1064-nm via a quartz cylindrical lens of 30-cm focal length on the sample. The sample was positioned on the x-y-z micro-translation stage.

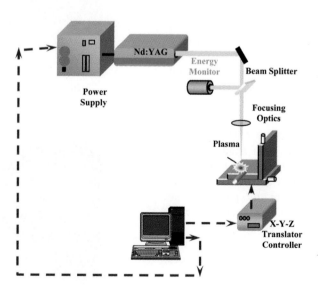

Figure (8): LIBS experimental setup

244

The cleaning of the sample was done by focusing of the laser beam of the Nd:YAG laser at 1064-nm via a quartz cylindrical lens of 30-cm focal length on the sample. The sample was positioned on the x-y-z micro-translation stage.

Glass sample

The glass sample was cleaned with the Q-switched Nd:YAG laser. The pulse energy was set to 250 mJ/pulse. The focused beam shape had a cylindrical shape with dimensions of 0.3-mm width by 10-mm long. The micro-translation stage was set to 0.25-mm step to obtain a good resolution of cleaning. The sample irradiated with 8.3 J cm^{-2} for 10 shots.

Bronze sample

The same experimental condition used to clean glass sample was used to clean the Bronze sample but with 50 number of shots.

Pottery samples

There were three pottery samples with different specific layers of impurities. The distance between the sample and the lens was set to 42-cm and the dimension of focused beam was 3.22×7.6-mm^2 long. The sample number 1 was cleaned at different irradiance (9, 6, 2.3, 1.5 J cm^{-2}) according to thickness of layer need to be removed. The high irradiance was used to remove thick crust layer while the low irradiance used to remove the thin layer cement to pottery. The sample number 2 was irradiated with different irradiances (0.6, 0.49, 0.3 J.cm^{-2}). The sample number 3 was cleaned by two different irradiances 0.3 J.cm^{-2} and 0.6 J.cm^{-2}.

3. Evaluation of cleaning methods

All samples that have been cleaned with both of conventional cleaning and laser cleaning methods were investigated with light optical microscope to evaluate the quality of both mentioned cleaning methods.

3.1. Glass sample

Investigation of first area of this sample which was cleaned with laser showed that laser beams could remove all the corrosion layer which was coated the surface without any damage of the original glass surface as shown in Figs.(9,10).

The chemical cleaning was applied to another area of the same glass sample as shown in Figs (11, 12). The results were similar to that obtained by laser cleaning but there was a risk of damaging the glass surface due to the chemicals used in the cleaning processes. [29].

Fig.(9) Shows different areas of glass cleaned with laser

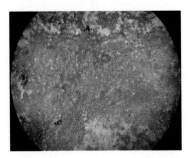

Fig.(10) Shows investigation the area was cleaned by laser by LOM with 16 X

245

Fig.(11) Shows the areas of glass was cleaned with chemical method

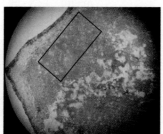

Fig.(12) Shows investigation the area was cleaned by chemical method by LOM with 12 X

3.2. Pottery samples

Sample no. 1 which was coated with hard gray–white encrustation, was cleaned at different irradiance (9, 6, 2.3, 1.5 J cm^{-2}) according to thickness of layer need to be removed. The high irradiance was used to remove thick crust layer while the low irradiance used to remove the thin layer cement to pottery as illustrated in figs (13, 14). The laser cleaning in this sample showed a good result. No evidence of damage or color change on the original surface of the sample has been noticed. However chemical cleaning caused of staining, erosion and pitting of the original surface due to the penetration of chemical materials within the porous object and softening the surface as it was obvious in figs. (15, 16). Better results were achieved by applying conventional technique to reduce the hard crust deposited on the

samples. Reduction of the thickness of this crust layer of the deposited dirt ensured an efficient and more controllable laser cleaning action using different fluencies per pulse and number of shots confined to narrow

Fig.(13) Shows the area of pottery sample no.1 after cleaning

(a) (b)

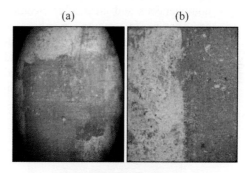

Figs. (14a-b) Shows the area of pottery sample no.1 after cleaning by laser with magni.12 X- 25 X

Fig.(15) Shows the area of pottery sample no. 1 after cleaning by chemical methods

246

(a)   (b)

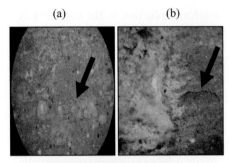

Figs.(16 a-b) Shows staining and
erosion in surface of pottery sample
no.1 after cleaninbg with chemicals

The sample no. 2 coated with soot
mixed with soil residues was irradiated
with different irradiances (0.6, 0.49,
0.3 J.cm^{-2}) achieved a good result at
irradiance 0.49 J.cm^{-2} where the
pottery surface appeared very
homogeneous and clear after cleaning
as shown in region number 3,
Fig(17,18). Though the chemical
cleaning of this sample was very bad
where many hairline cracks appeared
on surface after cleaning as shown in
Fig.(19, 20).

(a)   (b)

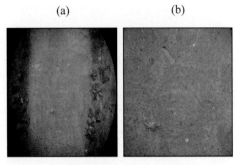

Figs.(18 a-b) Shows the surface of pottery
sample no.2 after cleaning by laser with
magni. 12X- 50X

Fig.(19) Shows the area of
after pottery sample no 2
cleaning by laser

Fig.(17) Shows the area of
pottery sample no 2 after
cleaning by laser

(a)   (b)

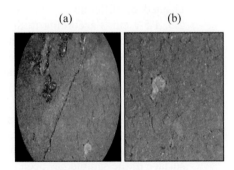

Figs.(20a-b) Shows the cracks in the
surface of sample no.2 after chemical
cleaning with magni.12 X- 50 X

(a)   (b)

 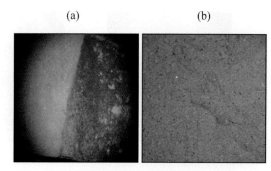

Fig.(21) Shows the area of pottery sample no 3 after cleaning by chemical method

Figs.(22 a-b) Shows the the surface of pottery sample no.3 after cleaning by chemicals with magni. 12 X- 32 X

(a)   (b)

 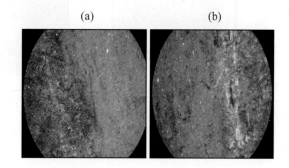

Fig.(23) Shows the area of pottery sample no 3 after cleaning by laser

Figs.(24 a-b) Shows the surface of pottery sample no.3 after cleaning by laser with magni. 40 X- 40 X

Sample no.3 which coated with a layer of black soot was cleaned by two different irradiances 0.3 J.cm^{-2} and 0.6 J.cm^{-2}. Better results were obtained at the higher irradiance. In this sample chemical cleaning showed a good result better than laser cleaning see Figs (21,22).

In the chemical cleaning, the chemical aqueous penetrate within the porous of pottery surface and remove the soot stains while in the laser cleaning, a thin layer was ablated without penetrating through the porous of the original surface. Figs (23, 24).

248

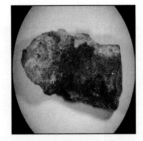

Fig.(25) Shows the area of
bronze sample after cleaning
by laser

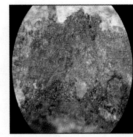

Figs.(26) Shows investigation of the
area of bronze sample after cleaning
by laser by LOM with 40X

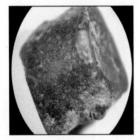

Fig.(27) Shows the area of bronze
sample after cleaning by chemical
method

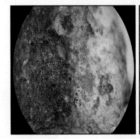

Figs.(28) Shows investigation of the area
of bronze sample after cleaning by
chemical method by LOM with 50X

3.3. Bronze sample

Fig. (25,26) show the bronze sample
cleaned by laser while Figs (27,28)
show the same bronze sample cleaned
chemical cleaning. It was observed that
there no much difference between the
two methods of cleaning. The chemical
method has disadvantage than the laser
method since the chemical aqueous can
penetrate through the porous corrosion
layers where their residues could be
difficult to remove and can lead to
more deterioration [30].

4. Conclusion

Archaeological samples of pottery,
glass, and Bronze can be efficiently
cleaned by using Q-switched laser at
1064nm. Better results were achieved
by applying conventional technique to
reduce the hard crust deposited on the
samples. Reduction of the thickness of
this crust layer of the deposited dirt
ensured an efficient and more
controllable laser cleaning action using
different fluencies per pulse and
number of shots confined to narrow
range.

Chemical-based cleaning techniques have associated problems where often leave residues within the material (object) which can cause problems later and once they have been applied their reaction can not be suitably controlled

Using of chemical materials such as organic solvents and other aqueous materials such as EDTA for removing disfiguring surface layers include corrosion can not applied without any risk of damaging the glass object.

The use of water and solvent cleaning based methods was discarded due to their deleterious effects on fragile and porous materials such as pottery which can absorbed these materials within their pores.

5. References

1- D. Stulik, H. Khanjian, V. Dorge and A. de Tagle, (2002) Scientific investigation of surface cleaning processes : quantitative study of gel residue on porous and topographically complex surfaces, in 13 th Triennial Meeting Rio de Janeiro 22-27 Sept., pp. 245-251.

2- C. Fotakis, D. Anglos, C. Balas, S. Georgiou, N.A. Vainos, I. Zergioti, V. Zafiropulos, (1997) Laser technology in art conservation, in: A.C. Tam(Ed), OSA TOPS on Laser and Optics for Manufacturing, Optical Society of America, p.99 (Vol.9)

3- M. Cooper (1998) Laser cleaning in conservation: An Introduction, Butterworth Heinemann, Oxford.

4- R. Salimbeni, R. Pini, S. Siani (2001) Achievement of optimum laser cleaning in the restoration of artwork: expected improvement by on-line diagnostics, Spectrochimica Acta B 56 pp.877-885.

5- K. Dickmann, C.Fotakis, J.F. Asmus (Eds) (2005) Laser in conservation of artworks, LACONA V Proceedings, Osnabrueck, Springer Proceedings in Physics100, Germany, Sept.15-18.

6- R. Pini, S. Siano, R. Salimbeni, Laser cleaning of stones: Optimizing the laser parameters and treatment methodology, SPIE Vol.4402, pp.32-37.

7- C. Fotakis (2002-2003) Laser technology for the preservation of cultural heritage: Diagnostic and restoration applications, In Colloquia . 2

8- R. Hannelore & W. Arno (2000) Laser cleaning of stained glass windows, Overview on an interdisciplinary project. J. Cult. Heritage 1, pp. 5151-5154.

9- R.Pini, S. Siano, R. Salimbeni, R. Pasquinuccim, and M. Micco (2000) Test of laser cleaning on archaeological metal artifacts, J. of Cultural Hertiage 1, 129-138.

10- O. Mohamed, R. Esther, C. Marta, D. Concepcion, C. Concepcion, G.L. Fernando (2005) Laser cleaning of terracotta decorations of the portal of Palos of the Cathedral of Seville, J. Cult. Heritage 6 pp. 321-327.

11- V. Oakley (1989) The deterioration of vessel glass, In Glass and Enamel Conservation, GW Belton Ltd., England, pp.18-22.

12- S. Davison (2003) Conservation and restoration of glass, Butterworth Heinemann, UK.

13- J. Cronyn (1996) The elements of archaeological conservation, TJ Press Ltd., Great Britain .

250

14- V. Oakley and K. Jain (2002) Essentials in the care and conservation of historical ceramic objects, Archetype Publication, Great Britain

15- A. Moncrieff and G. Weaver (1983) Cleaning, Crafts Council Conservation Science Teaching Series Book 2, Pindar Print Ltd., England .

16- UNESCO (2006) Presents project for National Museum of Egyptian Civilization at Exhibition of Egyptian Antiquities in Paris, UNESCO Press.

17- S.Fitz (1989) Glass objects: Causes, Mechanisms and Measurement of Damage, In, Science, Technology and European Cultural Heritage, Bologna, Italy, 13-16 June, pp.180-189.

18- J. L. Ryan, D. S McPhail. and P. S. Rogers (1996) Glass deterioration in the Museum Environment: A Study of the Mechanisms of Decay Using Secondary Ion Mass Spectrometry, In, 11th Triennial Meeting Edinburgh, Scotland 1-6 Sept. pp. 839-844.

19- P.J. Sirois (1999) The deterioration of glass trade bead from Canadian and textile collections, In, The Conservation and Restoration of Glass and Ceramics, Magnum International Printing Co.Ltd. Hong Kong, pp. 84-95.

20- F. H Goodyear (1971) Archaeological Site Science, Heinemann, London.

21- C. Pearson (1987) Deterioration of ceramic, glass and stone, In, Conservation of Marin Archaeological Objects, Butterworth, London, pp. 99-104.

22- H. B. Madsen (1989) Soils as a medium of preservation for artifacts, In, Science, Technology and European Cultural Heritage, Bologna, Italy, 1316 June, pp. 284-291.

23- R. Abd-Alla (2007) Stabilization and treatment of corroded glass objects displayed in the Museum of Jordanian Heritage, In, The Third International Conference and Workshop on Conservation and restoration, Faculty of Fine Arts, Minia University 2-4 April, Egypt.

24- D. Hamilton (2000) Conservation of Cultural Materials from Underwater Sites, In, Science and Technology in Historic Preservation, KLUWER ACADEMIC/ PLENUM PUBLISHERS, New York, pp. 193-227.

25- F. Hayward (1999) Chelating agents for the removal of iron stains from ceramics, Ceramic and Glass section, In, Conservation News, No.68, March, pp. 28-31.

26- L. Hogan (1998) Islamic Pottery; Methods of Old restoration, Staining and Its removal, In, Interim Meeting of the ICOM-CC Working Group, Sept. 13-16, Vantaa, Finland, pp. 123-133.

27- D. Hamilton (2000) Conservation of Cupreous Metals (Copper, Bronze, Brass) Conservation research Laboratory, Texas A&M University, 2000, pp.1-7.

28- D. Lynch (1991) The Restoration and Conservation of Ancient Copper Coins, Baylor University, Waco, Texas, pp.1-10.

29- H. Romich and A. Weinmann (2000) Laser cleaning of stained glass windows. Overview on an interdisciplinary project, J.Cult. Heritage1, pp.5151-5154.

30- R. Salimbeni (2000) Laser Techniques for Conservation of Artifacts, In, archeometriai M hely.

ASTHMA EARLY WARNING SYSTEM IN NEW YORK CITY (AEWSNYC) USING REMOTE SENSING APPROACHES

YASSER HASSEBO*, ZAHIDUR RAHMAN

Mathematics Engineering Program Department, LaGuardia Community College of the City University of New York, USA
*yhassebo@lagcc.cuny.edu, zrahman@lagcc.cuny.edu

Abstract

Asthma is estimated to affect approximately 17.3 million Americans, including 5 million children less than 18 years of age. Of these 5 million children, 1.3 million are less than 5 years of age. Asthma is a major public health problem in NYC particularly in Bronx. 12.5% of new Yorkers have been diagnosed with asthma. 300,000 children in NYC have been diagnosed with asthma up to year of 2000. NYC children were almost twice as likely to be hospitalized due to asthma attacks as the average of US child in 2000. Queens county's diesel pollution risk ranks as the 10th unhealthiest in the US compared to over than 3000 counties. Asthma symptoms are consistent with exposure to a high level of a respiratory irritant gas, smoke fume, vapor, aerosol, particulate matter (PM_{10} and $PM_{2.5}$), and dust. Some types these environmental gaseous such as sulfur dioxide (SO_2), nitrogen dioxide (NO_2), and ozone (O_3) can exacerbate preexisting respiratory symptoms in the short-term. Control of air pollution related diseases such as asthma, cancer, and bronchitis is difficult and inefficient due to the uncertainty in the air pollution transportation. Asthma control relies on air pollution detection and reduction. Asthma control can be improved by applying spatial tools such as Remote Sensing (RS), Geographical Information Systems (GIS). The project long-term goal is to develop a model to predict an Asthma Early Warning System for NYC (AEWSNYC), using two approaches: (1) satellite data error correction collaboratively with (2) Ground-based multiwavelength lidar measurements and NASA back trajectory tools. The proposed method can be used to create an efficient asthma control model globally.

1 Introduction

In the United States the number of people with asthma is increasing. Asthma is estimated to affect approximately 17.3 million Americans, including 5 million children less than 18 years of age. Of these 5 million children, 1.3 million are less than 5 years of age [1]. Asthma is a major public health problem in New York City (NYC) particularly in Bronx. The World Health Organization (WHO) has estimated that over one million cases of Asthma are reported each year in NYC [2]. More than two million people in New York City live within 500 feet of major roadways [3-6]. In Manhattan, over than 75% of the total population live within 500 feet of a congested road. For example, in Brooklyn, over 35% of both health facilities and standalone playgrounds are within this 500 feet risk zone. 12.5% of new Yorkers have been diagnosed with asthma. 300,000 children in NYC have been diagnosed with asthma up to year of 2000. NYC children were almost twice as likely to be hospitalized due to asthma attacks as the average of US child in 2000. In central Harlem, central Brooklyn, and South Bronx Queens county Children have been diagnosed with asthma with rate of 20-30 % (this rate is more than double the national average). Queens county's diesel pollution risk ranks as the 10th unhealthiest in the US compared to over than 3000 counties [3-6].

Asthma symptoms are consistent with exposure to a high level of a respiratory irritant gas, smoke fume, vapor, aerosol, and dust. Some types these environmental gaseous such as sulfur dioxide (SO_2), nitrogen dioxide (NO_2), and ozone (O_3) can exacerbate preexisting respiratory symptoms in the short-term. The National Ambient Air

252

Quality Standards (NAAQS), required by the Clean Air Act (CAA), are set for six criteria pollutants (O_3, SO_2, NO_2, CO, lead toxicity, and PM_{10} (particulate matter ≤ 10 μm in aerodynamic diameter, respectively) and $PM_{2.5}$ (particulate matter ≤ 2.5 μm) [7].

Control of air pollution related diseases such as asthma, cancer, and bronchitis is difficult and inefficient due to the uncertainty in the air pollution transportation and the traffic congestions. Asthma control relies on air pollution detection and reduction. Earth-observing instruments on environmental satellite platforms provide information on landscape features and climatic factors that are associated with the risk of asthma. The use of RS and GIS for investigating spatial relations between asthma risk and risk factors is rare in NYC. Asthma Early Warning System for NYC (AEWSNYC) can be predicted using two approaches:
1)   Error correction of satellite data
2)   Ground-based multiwavelength lidar measurements

1) The first part of this proposal investigates the Normalized Difference Vegetation Index (NDVI) NOAA/NESDIS Global Vegetation Index (GVI) environmental data stability during 1982-2003 [8]. This investigation is to study and monitor land surface atmosphere, analyze, climate and environmental changes, air pollution, and asthma control. This NDVI data collected from five NOAA series satellites. The satellites data for the years 1988, 1992, 1993, 1994, 1995 and 2000 are not stable enough compared to other years because of satellite orbit drift, and Advanced Very High Resolution Radiometer (AVHRR) sensor degradation. . For our research the data for NOAA-7(1982, 1983), NOAA-9 (1985, 1986), NOAA-11(1989, 1990), NOAA-14(1996, 1997), and NOAA-16 (2001, 2002) are assumed to be standard due to the fact,

that equator crossing time of satellite between 1330 and 1500, which maximized the value of coefficients. For the purpose of this study this years are called standard. Data from this particular period of the day maximized the value of coefficients. The crux of the proposed correction procedure consists of dividing standard year's data sets into two subsets. The subset 1 (1982, 1985, 1989, 1996, 2001) called standard data correction sets is used for correcting unstable years and then corrected or normalized data for this years compared with the standard data in the subset 2 (1983, 1986, 1990, 1997, 2002). The subset 2 also called standard data validation sets that are used for data validation. In this paper, we apply Empirical Distribution Function (EDF) to correct this deficiency of data for the affected years. We normalize or correct data by the method of empirical distribution functions compared with the standard. Using these normalized values, we estimate new NDVI data time series which provides data for these years that match in subset 2 that is used for data validation [9] .The corrected datasets can be used as proxy to study for asthma control in NYC

2) Satellite measurements do not provide the vertical information needed to assess Atmospheric aerosol (smoke plumes, smog, haze, and dust) transport mechanisms. To support this information, Remote Sensing ground-based multiwavelength lidar measurements are needed to study the vertical structure of aerosol events. For example, Fig 1 presents MODIS Satellite image, which identify multiple stratified particulate smoke plumes occurring over New York City on July 21, 2004. In support of NASA-INTEX and NOAA-NEAQS transport experiments; multiwavelength lidar measurements can be used to determine the Angstrom coefficient of the smoke plumes that unambiguously identifies them as fine mode particulates and not cloud

particles. Processing algorithms will be developed to process simultaneous backscatter lidar measurements with satellite data. NASA back trajectory tools can be used to predict an Asthma Early Warning System for NYC (AEWSNYC) based on the air pollutants and smoke plume sources. This project also can be used to create a model for efficient asthma control globally [10-11].

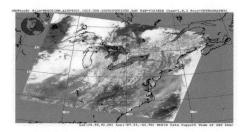

Fig 1. Smoke Plumes: Seen by MODIS Satellite

2   Methodology

Empirical distribution function (EDF) approach is based on the physical reality, that each ecosystem may be characterized by very specific statistical distribution, independent of the time of observation. It has proven to be an efficient technique to normalize satellite data [9]. It allows to represent a full spectrum of global ecosystem from desert to tropical forest and to correct distortions in satellite data related to a variety of satellite technical problems. To generate the normalization data, samples of raw earth-scene data for the wide range intensities are selected. For NOAA satellites, the area will be rectangular, extending several thousand pixels from desert to tropical forest (both east to west and north to south). Corresponding to the incoming radiance from any pixel, the instrument will respond with an output x, in digital counts. In the next step a histogram is compiled. That discrete density function, describing the relative frequency of occurrence of each possible count value, for

each year. For year **i**, which is the year to be normalized, let the histogram be $P_i(x)$. An empirical distribution function (EDF) $P_i(x)$ can then be generated; viz [8-9]

$$P_i(x) = \sum_{t=0}^{x} p_i(t). \qquad (1)$$

The EDF is also known as a cumulative histogram of relative frequency. It is a non-decreasing function of x, and its maximum value is unity. For example, Figure 2 shows the EDF of 1988, which computed from the histogram. The abscissas, labeled "NDVI value (x=counts)" is the output levels and the ordinates, labeled "EDF (0 to 1)", is the percentage of the data with the outputs at or below that level.

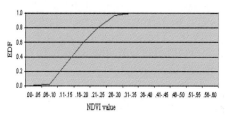

Fig 2. Empirical Distribution Function of un-normalized data, 26th week of 1988

Figure 3 shows how the procedure is applied in actual practice to generate normalization NDVI value [9]. The figure shows idealized EDF's for the standard and the year of 1988. As EDF are based on cumulative histogram, they are discrete. But in Figure 3, they are shown as continuous function

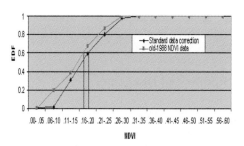

Fig 3. Illustration of procedure to generate normalization NDVI data of year 1988 compared with standard data correction sets (subset 1)

254

For example, for the NDVI value 0.165in year 1988 (Figure 3) find the value from the EDF of year 1988. In the illustration it the EDF_{88} is 0.6. Then find the point on the standard data correction sets EDF with the same EDF value. EDF value can also be expressed as the $EDF_{standard}$ is 0.6. Finally, use the EDF of the standard data correction to find the normalized count value 0.18. Since the data are actually discrete, we will need to interpolate within the EDF of the standard data correction sets to find the value of 0.18. Therefore,

New NDVI value for 1988 = $NDVI_{1988}$ + ($NDVI_{standard}$-$NDVI_{1988}$) or

New NDVI value for 1988 = 0.165+ (0.18-0.165 = 0.18

The proposed technique utilizes EDF's of subset 1 to normalize/correct data for the year 1988. Then corrected or normalized data is validated with the subset 2. This is illustrated in Figure 4.

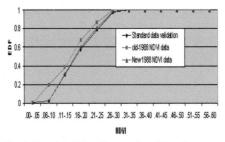

Fig 4. Empirical distribution functions for normalized BT data of year 1988 is compared with standard data validation set (subset 2).

The same technique is applied for normalization of data for years1992, 1993, 1994, 1995, and 2000.

3   Results and Discussion

Original NDVI time series for five NOAA satellites is shown in Fig 5. Data for the years of 1988 (NOAA-9), 1992, 1993, 1994 (NOAA-11), 1995, and 2000 (NOAA-14)

are not sufficiently stable due to satellite orbit drift, and sensor degradation [8-9].

Hence there is need to correct the data.

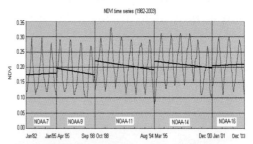

Fig 5. NDVI time series (weekly) for study area

To correction, the method of empirical distribution functions is applied for unstable years to normalize data as described in the previous section.   A corrected NDVI time series is presented in Figure 6 which shows improvement of NDVI data (pink line) for the year 1988, 1992, 1993, 1994, 1995, and 2000 which can be used for predict an asthma control in NYC.

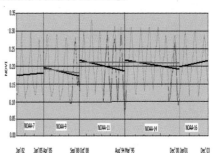

Fig 6. Corrected NDVI time series (weekly) for study area (old NDVI data —— , and new NDVI data ——).

4   Conclusions

Empirical distribution function approach proposed here can be used to correct   GVI and similar satellite date sets. It should also be noted, that resolution of the EDF technique is limited by the available representative sample. EDF improved the time related stability of NDVI for all satellites, especially NOAA-9, -11, and -14 environmental satellites. This is strong evidence that normalization by EDF is an effective method for improving stability of NDVI time series which can be used for prediction in Asthma control in NYC. NASA back trajectory tools can be used to predict an Asthma Early Warning System for NYC (AEWSNYC) based on the air pollutants and smoke plume sources. This method also can be used to create a model for efficient asthma control globally.

References

[1] USA National Institute of Medicine 2000

[2] WHO/UNICEF. 2005. *World Asthma Report*. Roll Back Asthma partnership. World Health Organization/United Nations Children's Fund. http://rbm.who.int/wmr2005/

[3] http://www.scorecard.org/env-releases/cap/index.tcl; last viewed on 3-20-07

[4] NYC Dept of Health and Mental Hygiene: Vital Signs. Vol 2., Number 4, April 2003

[5] http://www.nyc.gov/html/doh/downloads/pdf/survey/survey- 2003asthma.pdf, last viewed March 25, 2007

[6] New York State Department of Health. Data from 2004. http://www.health.state.ny.us/statistics/sparcs/, last viewed March 25, 2007

[7] Department of Health and Human Services Volume 10, No. 2, Summer/Fall 2000, Special Issue on Geographic Information Systems

[8] Mohammed Z. Rahman, et al, "Improving stability for NOAA environmental satellite". Proceedings of SPIE 6412, 641217 (2006)].

[9] Mohammed Z. Rahman, Ph.D Thesis, February 2008, The Graduate Center of the City University of New York, "Error Correction of the NDVI and BT Calculated from AVHRR Observations"

[10] Y. Hassebo, et al. "Multi-wavelength Lidar Measurements at the City College of New York in Support of the NOAA-NEAQS and NASA-INTEX-NA Experiments, Remote Sensing of Atmospheric Aerosols Workshop, pp 16 – 23, IEEE, 2005

[11] Yasser Hassebo, Ph.D. thesis, The Graduate Center of the City University of New York, "Lidar Signal to Noise Ratio Improvements: Considerations and Techniques", May 2007

III. NUCLEAR, PARTICLE PHYSICS & ASTROPHYSICS
III-1 KEYNOTE AND PLENARY PAPERS

DO WE EXPECT A MULTIPLE DIP STRUCTURE AT LHC ENERGIES?

FAZAL-E-ALEEM, HARIS RASHID AND SOHAIL AFZAL TAHIR

High Energy Physics Group, CHEP, University of the Punjab
Lahore-54590, Pakistan

Besides other parameters, measurements are also planned for differential cross section at Large Hadron Collider (LHC). Shrinkage of the diffraction peak and dip structure in the differential cross section are amongst the agenda of measurements at TOTEM Experiment. Many theoretical models predict multiple dip structure at LHC energy. We briefly review the status of shrinkage phenomena and possibility or otherwise of a multiple dip structure in the light of Geometrical models. A comparison has been made with the predictions of other models. We have also undertaken the role of rho (ρ) in the appearance or otherwise of multiple structure. In order to have a better understanding of the evolution of dip structure, we will also include the measurements from PP2PP at RHIC.

1. Introduction

LHC, the highest energy accelerator ever built by the mankind is now ready to take up the exciting challenge of understanding many unknown and unseen physics phenomena. Large hadron Collider (LHC http://public.web.cern.ch/public/en/LHC/LHC-en.html) [1] is a state of the art technology with 27 km long tunnel which is 50-150 m below the ground. Two beams of protons bunches circulating in opposite directions will collide head on. Each bunch contains 200 billion protons while 600 million proton-proton collisions per second are expected to be produced. (www.lhc.ac.uk etc). It is now set to explore physics at TeV scale giving a final word about the Higgs mechanism in Standard Model along with many "New Physics predictions" in this energy scale. Through these measurements, we will also have a better understanding of B-physics, Heavy Ions, Electroweak physics, and Quantum Chromodynamics (QCD). LHC will also shed more light on the supersymmetry, Black Holes, Extra Dimensions and Quark Gluon Plasma (QGP) [1].

One of the difficult to understand area of Particle Physics, known as "Diffractive Physics" is also in the scope of LHC's measurements through TOTEM (**Tot**al and **El**astic **M**easurements) experiment (http://totem.web.cern.ch/Totem/) [2]. "TOTEM is an experiment dedicated to the measurement of total cross section, elastic scattering and diffractive processes at the LHC. The total cross section will be measured using the luminosity independent method which is based on the simultaneous detection of elastic scattering at low momentum transfer and of the inelastic interactions" [2] .

Another experiment PP2PP at Relativistic Heavy Ion Collider (RHIC) (http://www.rhic.bnl.gov/pp2pp/) [3] is also undertaking measurements in the lower energy range. Goals of the PP2PP experiment at RHIC include the measurements of total and elastic scattering in c.m energy range from 60 GeV to 500 GeV using both polarized and unpolarized beams. The measurements will study evolution of the dip structure with an increase in energy and will provide a unique opportunity to compare the results with pp at 63 and 546 GeV while LHC will explore the possibility or otherwise of the dip structure in going from 1.8 TeV to 14 TeV. Results from the Relativistic Heavy Ion Collider (RHIC) and Large Hadron Collider (LHC) will therefore open up new dimension to proton-proton physics.

In our brief review on the topic with special reference to the possibility of multiple dip structure, we take up various aspects in the light of Gemetrical picture and compare them with the predictions of other models. During the course of this study, many interesting questions arise which need to be addressed. In the next sections we will take up *Review of measurements, Predictions of Generalized Chou-Yang model, Comparison with other theoretical models and Conclusions.*

2. Review of measurements

An overview of the existing experimental data alongside the ongoing measurements at RHIC/LHC is now given [4]. Various physical parameters have been measured at CERN-ISR (pp and $p\bar{p}$), CERN-SPS ($p\bar{p}$), and

CP 998, Modern Trends in Physics Research
Third International Conference MTPR-08
edited by L. El Nadi

FERMILAB ($p\bar{p}$) [4]. Measurements from the Cosmic ray data corresponding to LHC energy have also been reported for pp elastic scattering [5]. Differential cross section measurements at these energies depict the following features:

- For both pp and $p\bar{p}$ elastic scattering, the local slope, B, of the diffraction peak increases with an increase in energy. Shape of the slope at ISR and SPS in the extreme forward region suggests a concave curvature. At FERMILAB this curvature seems to have disappeared. At RHIC, the PP2PP experiment will study pp elastic scattering from 60 GeV to 500 GeV in the Coulomb Nuclear Interference (CNI) region $0.0005 < -t < 0.12$ $(GeV/c)^2$. These measurements will provide us an opportunity to know more about disappearance of this curvature.

- Measurements of the differential cross section for pp and $p\bar{p}$ over large -t region at ISR have a rather complex behaviour, with a dip near - t = 1.4 $(GeV/c)^2$. Position of the dip moves towards -t = 0 with an increase in energy. This dip converts in to a shoulder at Collider and Tevatron energies $\sqrt{S} =$ 546 and 1800 GeV respectively for $p\bar{p}$ elastic scattering. The differential cross section at the dip first decreases and then increases at ISR energies. At Collider and Tevatron energies this dip turns into a shoulder and the differential cross section is two orders of magnitudes higher. At ISR, a secondary maximum is present beyond the dip, followed by a large -t regime, which can be described by a smaller slope. There is no evidence of a second diffraction minimum. At RHIC where measurements will be made from 60 to 500 GeV for pp elastic scattering, a study of the evolution of the dip structure with $\sqrt{S}$ is in progress in the medium -t region, -t < 1.5 $(GeV/c)^2$. This will provide us a comparison for both pp and $p\bar{p}$ in going from ISR (60 GeV) to SPS energy (500GeV).

- At LHC, measurements are planned at 14 TeV for pp elastic scattering (TOTEM experiments). LHC running at reduced c.m. energy of 1.8 TeV, will provide an opportunity to compare the results with FERMILAB for $p\bar{p}$. The main agenda of these experiments will include elastic scattering over a large range of -t in pp collisions at 14 TeV. This will provide us an opportunity to conclusively

observe the multiple dip structure besides other aspects.

3. Predictions of Generalized Chou-Yang model

We next take up the present data and future measurements in the light of impact picture and specifically in the light of generalized Chou-Yang model [6,7]. According to the generalized Chou-Yang model, hadrons consist of clusters of particles which, on collision, pass through each other, interacting in pairs i,j and scattering one another with invariant scattering amplitudes $f_{i,j}(t)$. The scattering amplitude $T\,(s,\,t)$ is then given by

$$T\left(s,t\right) = i\int b\,db\,J_0\left(b\sqrt{-t}\right)\left[1 - \exp\left(-\Omega\left(s,b\right)\right)\right]$$

where

$$\Omega\left(s,b\right) = K\left(1 - i\alpha\right)\int \sqrt{-t}\,d\sqrt{-t}\,J_0\left(b\sqrt{-t}\right)X$$
$$\left[f\left(t\right)/f\left(0\right)\right]G_A\left(t\right)G_B\left(t\right)$$

The function $\Omega\left(s,b\right)$ represents principally opacity effective for clusters passing with a relative impact parameter b, but is taken to be complex and thereby includes refractive as well as absorptive effects. $f(t)/f(0)$ takes account of the anisotropy of the parton-parton interaction. The parameter α is determined by the ratio of the real and imaginary parts of the scattering amplitude in the forward direction. $G_A\,(t)$ and $G_B\,(t)$ are form factors of colliding particles. For the isotropic scattering of the constituent partons, the model based on the multiple-diffraction theory reduces to the prestine Chou and Yang [6] model. Using the parameters as given in Ref. 7 for the proton form factor and the anisotropy function, a good fit was obtained for ISR, SPS and Tevatron data.

As reported earlier [7], predictions of our model for pp and $p\bar{p}$ at 53 GeV are shown in Fig. 1. The difference in the dip region is due to different values of ρ (0.077 $\pm$ 0.009 for pp and 0.106 $\pm$ 0.016 for $p\bar{p}$). RHIC measurements at 60 GeV will therefore throw more light on this.

For 546 GeV, our results [4,7] are plotted in Fig.2. It can be observed that the dip converts into a shoulder as the contribution of the real part increases. We thus observe that at RHIC, a higher value of ρ ($\approx$ 0.15) for pp at 500 GeV will mean a shoulder in the vicinity of 0.8 (GeV/c)^2. However, like ISR measurements if ρ for pp is lower than $\bar{p}p$, a more pronounced dip structure around this value (0.8 GeV/c)^2 at RHIC is more likely.

Predictions of our model for the differential cross-section at 1.8 TeV (Fig. 3) also give good agreement with the measured data while a dip in the differential cross section is exhibited at $-t = 0.58$ (GeV/c)^2. This fit is obtained by choosing a value of α which yields $\rho = 0.118$. This value of ρ is consistent with the measured value. This dip transforms into a shoulder if we choose a higher value of the ρ.

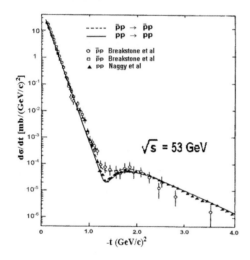

Fig. 1: Differential cross section for pp and $\bar{p}p$ elastic scattering at $\sqrt{S}$ = 53 GeV.

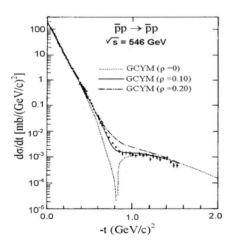

Fig. 2: Differential cross section for $\bar{p}p$ at $\sqrt{S}$=546GeV.

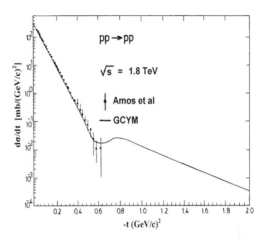

Fig. 3: Predicted Differential cross section for pp elastic scattering at $\sqrt{S}$ = 1.8 TeV.

At 14 TeV, our calculated results [7] obtained for a K value which yields ρ= 0.07 and a suggested total cross section of 115 mb, depict a multiple dip structure near $-t = 0.4$ and 1.5 (GeV/c)^2. Again, a higher value of ρ (≈ 0.15) as predicted by most models will fill up the first dip and turn the second dip into a shoulder/break. Thus we observe that in our model, the multiple structure (dip near 0.4 and a shoulder or break near 1.5 (GeV/c)^2 should appear at LHC energy of 14 TeV. Although most models predict a smoothening of the ρ value (~ 0.1) at TeV energies, higher value of this parameter as predicted by the models incorporating Odderon picture will mean disappearance of the dip structure.

4. Comparison with other theoretical models

We will now undertake a comparison of our predictions with some of the recent work, which takes into account the structure in the differential cross section. We would give special emphasis on the multiple dip structure at the RHIC and LHC.

The impact picture has also been applied to account for the dip structure incorporating the ideas of QCD by several authors, a detailed account of which is available in the proceedings of the recent conferences [4,6] and in excellent articles of various authors which are available on hep archives [8]. However, in order to have a feel of theoretical developments, we give a selective account of the work by some authors [9].

262

These models [4,9] incorporate semihard scattering of quarks and gluons or partons in the nucleus. The models mostly differ in their treatment of QCD and also whether or not they respect the constraints imposed by unitarity. Predictions of the Eikonal models describing the available elastic scattering data are generally similar and reproduce the experimental data. Their predictions differ at LHC energy of 14 TeV. In the last two decades, the picture is now fairly developed. In one such attempt, Kopeliovich et al [9] have given a soft QCD dynamics of elastic scattering in the impact parameter representation. They have calculated the elastic hadronic amplitude using the non-perturbative light-cone dipole representation for gluon bremsstrahlung. They argue that "data for large mass diffraction demand a two-scale structure of light hadrons: the gluon clouds of the valence quarks are much smaller (~ 0.3 fm) than the hadronic size". The presence of the two scales unavoidably leads to a specific form for the total hadronic cross section. To further test the model [9], they have analyzed the elastic pp and $\bar{p}p$ differential cross sections and extract the partial amplitudes in the impact parameter representation. The Pomeron trajectory as a function of the impact parameter is only slightly above one for central collisions, but steeply grows towards the periphery. The model predicts correctly the shape and energy dependence of the partial amplitude at all impact parameters.

Desgrolad et al [9] have also considered model based on Pomeron and Odderon contributions to high energy elastic pp and $\bar{p}p$ scattering. They have addressed the question about the behavior of the scattering amplitude at high energy, and how to fit all high energy elastic data. The relative virtues of Born amplitudes and of different kinds of eikonalizations were considered. They mention that an important point in this respect is that secondary structures are predicted in the differential cross-sections at increasing energies and the phenomena are directly related to the procedure of eikonalizing various Born amplitudes. Predictions of their computations at 0.5 and 14 TeV are given in Fig. 4. They predict a multiple structure at RHIC (500 GeV) and LHC (14 TeV).

Block et al [9] have incorporated QCD-inspired parameterization within the eikonal picture. Like other Eikonal models, their predictions fit the existing data

well. Predictions of their model for 14 TeV have been plotted together with predictions of our model. They predict a dip near 0.4 $(GeV/c)^2$ and a break near 2 $(GeV/c)^2$.

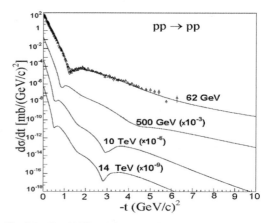

Fig. 4: Predicted Differential cross section for pp elastic scattering at LHC and RHIC using quasi-eikonalized and generalized eikonal procedure [9].

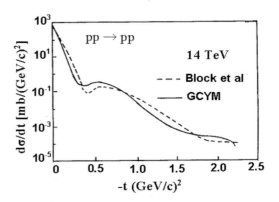

Fig. 5: A comparison of the predictions of our results [6] and Block et al [9] at LHC.

Petrov and A.V. Prokudin [4,9], motivated by the idea of multi-Pomeron structures, have undertaken soft physics in the light of three Pomerons. The model, which incorporates three Pomerons in to the eikonal picture, fits the available data well. In their model second structure (Fig. 6) begins to appear at 100 GeV in the vicinity of 6 $(GeV/c)^2$. This dip move towards – t = 0 as the energy increases. At 14 TeV this structure turns in to a dip at –t = 2 $(GeV/c)^2$. They predict a third dip at –t = 6 $(GeV/c)^2$. Their explanation, however, does not give an optimum in the number of "relevant" Pomerons.

Also the underlying physics of such a construction is missing.

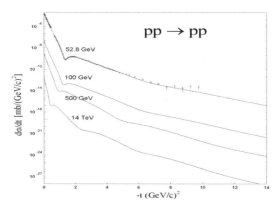

Fig. 6: Predictions of three-Pomeron model at RHIC and LHC [9].

Similar results have been predicted by Islam et al [11] in their recently reported work.

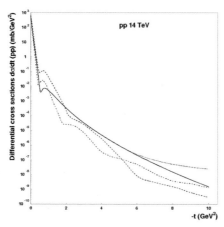

Fig. 7: Predictions of nucleon structure model [11].

5. Conclusions

We now give an overall picture of theoretical work versus measurements at LHC with special reference to shrinkage phenomena and multiple dip structure. In order to have a better understanding, we will also keep referring to RHIC measurements.

5.1 *Shrinkage of the diffraction peak*

The shape of the forward cross section appears to increase with an increasing energy. This is also called as *shrinking phenomenon*. We observe that:

- The shape of the slope at ISR and SPS energies in the extremely forward region is of slightly concave curvature.
- At Tevatron energy the curvature seems to have disappeared. Measurements at RHIC will therefore be very interesting from this point of view. In the simple Regge pole picture [10], t dependence of differential cross section is of a constant slope with no curvature. In the Eikonal picture this emerges as a natural consequence [9].
- Models based on impact picture predict a convex curvature at LHC and higher energies. This further enhances the need for measuring the forward scattering at extremely small angle.

5.2 *Multiple Dip Structure?*

Interesting picture emerges in the light of predictions of different theoretical models. It is found that:

- The dip structure observed at ISR which is moving toward - t = 0, seems to disappear with an increase in energy. At Tevatron it appears to have turned into a near shoulder [7,8].
- Simple Regge pole picture of soft Pomeron proves to be a good approximation in the forward elastic scattering from ISR to Tevatron. In order to account for dip structure, contribution of pole plus cut is needed [10].
- In the Eikonal picture, the dip mechanism emerges naturally. At RHIC, Tevatron, and LHC the first dip is predicted near - t = 0.8, 0.65 and 0.4 (GeV/c) ^{2} respectively [4]. As pointed out earlier, at RHIC and LHC with $\rho \approx$ 0.2, this dip will be filled and turn into a shoulder. Thus with an increasing contribution of the real part of the scattering amplitude, Eikonal models predict flattening/ disappearance of this dip.
- Eikonal models at LHC energies predict the appearance of another structure at large – t [4,7]. Similar structure is observed in the Regge like models. But the position of the dip

264

structure is different from those predicted by the Eikonal models. Thus, the position of the minima differ in different models for both RHIC and LHC energies. It will therefore be interesting to focus on the position of minimum (or minima in case of a multiple dip structure).

- Another important observation in the region of dip is the difference observed between pp and $\bar{p}p$ cross sections at ISR [7]. This difference is naturally taken account of in the Eikonal picture by the difference of the ratio ρ for two reactions. Will this difference persist or disappear at RHIC or LHC? This is another important observation to look for.

- Beyond the second maximum of the differential cross section, the interaction dynamic enters into a limit where pQCD can be applied [4,10]. In this region, the three gluon exchange multiple scattering interaction is shown to account for the t^{-8} dependence of the angular distribution at ISR energies. At LHC it (ggg exchange amplitude) predicts a continuous decrease of the differential cross section over large t region. This behaviour is different from the one predicted by the Eikonal picture and Regge type analysis.

- There exists an apparent dichotomy between the soft and hard Pomeron which poses problems for theorists and needs to be resolved [4]. Theorists are now beginning to develop formalisms which encompass the transition between hadron and quark and gluon degrees of freedom. QCD also predicts the existence of an Odderon based on three-gluon exchange, which through interference with the Pomeron exchange leads to exciting results. Measurements at TOTEM will help us understand the physics at this scale better.

- Substantial increase of data from the new measurements at LHC will help in a significant progress in the fundamental understanding of soft and hard diffraction.

NOTE

In view of the time/space constraint, many details have not been given. More details are available in the most recent conference proceedings on this topic as well as in excellent review articles of various authors which are available on hep archives [4,8]. We apologize to all those whose scholarly work have either been cited partially or could not be included due to representative selection of the literature. A detailed article, to be published separately on the same subject, will include all such contributions.

6. References

1. http://public.web.cern.ch/public/en/LHC/LHC-en.html.

2. http://totem.web.cern.ch/Totem/.

3. http://www.rhic.bnl.gov/pp2pp/.

4. D.E. Groom et al., Europhys. J., **C15,** 1 (2000). Proceedings of the "11th International Conference on Elastic and Diffractive Scattering: Towards High Energy Frontiers": The 20th Anniversary of the Blois Workshops, Chateau de Blois, Blois, France, 15-20 May 2005 (http://lpnhe-theorie.in2p3.fr/ EDS05 Accueil.html) ; 12th International Workshop on "Deep Inelastic Scattering" (DIS 2004), Strbske Pleso, Slovakia, 14-18 April 2004; Proceedings of "10th Blois Workshop On Elastic And Diffractive Scattering", 23-28 June 2003, Hanasaari, Helsinki, Finland; http://cerncourier.com/main/article/45/8/22. Fazal-e-Aleem et al., Int. J. Mod Phys. <u>A19,</u> 4455, (2004), Fazal-e-Aleem et al, *Elastic Scattering – Past, Present and Future*: **CP888** (American Institute of Physics), Modern Trends in Physics Research, Second International Conference on Modern Trends in Physics Research—MTPR-06, edited by L. El Nadi 241 (2007) , http://www.desy.de/~eds07/

5. S. Torii, 6th Blois Workshop France, 9 (1995) (Editions Frontieres); A.A. Arkhipov, hep-ph/ 0108118; http://www.cosmic-ray.org/ date/papers05.html.

6. T. T. Chou and C. N. Yang, Phys. Rev. 170, 1591 (1968); Phys. Rev. Lett. 20, 1213 (1968); L. Durand and R. Lipes, *ibid.* 20, 637 (1968).

7. Fazal-e-Aleem et al., J. Phys. **G16**, 269L, (1991); Phys. Rev. **D44**, 81 (1991); Fazal-e-Aleem and M Saleem, Monograph on "*Chou-Yang model and Elastic Reactions at high*

energies" Hadronic Press, FL, USA (1992) and references therin, Sohail Afzal Tahir, Ph. D. thesis, University of the Punjab, Lahore, Pakistan (Under review), M. Saleem, Fazal-e-Aleem and I.A. Azhar., Europhys. Lett. 6, 201 (1988).

8. http://www-spires.slac.stanford.edu/spires/ ; http://weblib.cern.ch/share/hepdoc/

9. L. Durand and H. Pi, Nucl. Phys. B Proc. Suppl. 12, 379 (1991); Phys. Rev. **D40** 1436 (1989); Phys. Rev. Lett. **58** 303 (1987); M. Block et al, Eur. Phys.J. **C23** 329 (2002); Phys. Rev. **D62** 077501 (2000); Phys. Rev. **D74** 117501 (2006); Phys. Rev. **D76**, 111503, (2007), [arXiv: 0705.3037]; B. Z. Kopeliovich et al, Phys.Rev. **D63** 054001 (2001); Phys. Rev. Lett. 85 507 (2000); Blois07, Forward physics and QCD 63-6; Nucl. Phys. **A782**: 24-32, (2007) [hep-ph/0607337]; Braz. J. Phys. **37**: 473-483 (2007) [hep-ph/0604097]; P. Desgrolard et al., hep-ph/9811384; Eur. Phys. J. **C16**: 499-511(2000) [hep-ph/0001149]; Pavel N. Bogolyubov (ed.), Lazslo L. Jenkovszky (ed.), Vladimir K. Magas (ed.) *"New trends in high-energy physics: Experiment, phenomenology, theory"* Proceedings, 21st Conference, Yalta, Crimea, Ukraine, September 15-22, (2007) Kiev, Ukraine: Bogolyubov Inst. Theor. Phys. (2007) 282; V.A. Petrov and A.V. Prokudin, hep-ph/0203162 (2002); [hep-ph/0701122] (2007); J. Hufner and B. Povh, Max Plank Institut fur Kernphysik preprint MPIH-V29 (1991); Phys. Lett. **B215**, 772 (1988) Phy. Rev. Lett **58**, 1612 (1987).

10. A. Donnachie et al, Eur. Phys. J. **C45**: 771-776, (2006) [hep-ph/05081 96]; Phys. Rept. 403-404: 281-301(2004); AIP Conf.Proc.717:797-806 (2004); DAMTP, Cambridge U. Preprint 96/66 (December 1996); Physics Lett. **B296**, 227 (1992); Nucl. Phys., **B348**, 297 (1991); Particle world, **2**, (1991); Nucl. Phys. **B267**, 657 (1986); **B231**, 189 (1984); A. Donnachie Cern Courier, 39, 29 (1999); Fazal-e-Aleem and Mohammad Saleem, Pramana, **31,** 99 (1988).

11. R.F. Avila et al, Braz. J. Phys. **37**: 675-678 (2007); Eur. Phys. J. **C54**:555-576, (2008) [arXiv:0712.3398] and references therin. M. M. Islam et al [arXiv:0804.0455] (2008); Blois07, Forward physics and QCD 267-27. [arXiv:0708.1156] and references therin.

12. B.A.Breakstone et al., Nucl. Phys. **B248**, 253 (1984); Phys. Rev. Lett. **54**, 2180 (1985).

13. E. Nagy et al., Nucl. Phys. **150B**, 221 (1979).

14. N.A. Amos et al., Phys. Rev. Lett. **68** 2433 (1992); and "Aspen 1993, **Multiparticle** dynamics" 395-399 In "Providence 1993, Elastic and diffractive scattering" 59-6.

SEARCH ON e$^+$ – e$^-$ PAIR AND OBSERVATION OF A NEW LIGHT NEUTRAL BOSON

M. S. EL–NAGDY[1], A. ABDELSALAM[2] AND B. M. BADAWY[3]

[1] *Physics Department, Faculty of Science, Helwan University, Helwan, Egypt*
[2] *Physics Department, Faculty of Science, Cairo University, Giza, Egypt*
[3] *Reactor Physics Department, Nuclear Research Center, Atomic Energy Authority, Egypt*
E. Mail: he_cairo@yahoo.com

Abstract

We present a unified description of e$^+$ e$^-$ dilepton production in heavy ion collisions at relativistic (3.7A GeV) and ultrarelativistic (200A GeV) energies. From the interactions of [^{12}C and ^{22}Ne] at 3.7A GeV and ^{32}S at 200A GeV with nuclear emulsion, 134 e$^+$ e$^-$ pairs are observed. The differential distribution of the energy asymmetry of pairs is compared to a background observation at Bristol. The events are consistent in a way that, they scale well in a single curve of exponential decay. This ensures correct asymptotics and provides a unified description of mesonic decay. The existence of light neutral boson of mass (1.55±0.14 MeV/c^2) is questionable, because the candidate peak is seen in the invariant mass spectrum of its decay into e$^+$ e$^-$ pairs. The data seem to suggest the production and subsequent decay of short lived neutral boson with lifetime of the order of 10^{-16} S.

Introduction

The key results on lepton pair production in ultra–relativistic nuclear collision had been starting from pp collisions in the seventies and are ending now at the RHC and LHC [1].

An understanding of dilepton production at excitation energies in the few–GeV range [2], will aid in the interpretation of dilepton production at ultra–relativistic energies [3 – 5] where it may be a probe of the quark gluon plasma [6].

An existence of a new pseudo scalar neutral boson was first suggested by Weinberg [7] and was given the name axion by Wilczek [8]. He suggested that the axion results from the breaking of the U(1) symmetry. The U(1) problem of the standard model was solved by t' Hooft [9] by demonstrating that the instanton should be taken seriously.

Previous experimental study conducted by El–Nadi and Badawy [2] at Cairo, has shown a production of e$^+$ e$^-$ pairs in the low mass region based on reorganization of dielectrons sources. They [2] suggested the existence of neutral boson with mass 1.51±0.14 MeV/c^2 or (2.95±0.27) m$_e$ and life time τ = 15 x 10^{-16} s. It was also noticed by de Boer and van Dantzig [10, 11] that, three bosons are probably present in data obtained in Cairo. moreover, El–Nadi et al [4] suggested the subsequent decay of four short–lived neutral bosons of average masses 1.8, 2.3, 2.6 and 12.2 MeV/c^2 with lifetime of order 10^{-16} s in the interactions of ^{32}S (200A GeV) ions with emulsion nuclei.

The previous experimental data [12] have been enhanced by a new statistical yield of pairs. We are interested here to explore new particles decaying into electron pairs at low mass region. This is the region where CERES experiment [13] had its

267

limitations with their detector and we want to use a different detector, which is ideal for this range.

Experimental Setup

This experiment (EMU03) was performed in a stack consisting of 24 Fuji films exposed horizontally to ^{32}S (200A GeV) ions at the CERN SPS. Details of the experimental setup, scanning of the pellicles, and event classification can be found in [3, 4]. The measuring microscope used is of type KSM1. The measurements were carried out for charge Z of projectile fragments using the lacunarity measurement method for Z = 2 fragments and δ–rays method in case of Z ≥ 3 fragments. The charge identification obeyed by δ–ray methods is explained in [14]. The observed interactions were carefully looked for the direct pairs in the forward cone of the primary beam under 100 X magnification. The following criteria must be verified during the measurements,

(a) The track length of neutral particle undergoing decay should be longer than 20 μm to overcome the uncertainty in the location of the pair origin. The electron pair tracks should come from the center of the interaction or located away from this center by no more than a distance L = 3 μm. This restriction [10, 11] is consistent with the Dalitz process taking into account the π° lifetime distribution.

(b) The ionization density for electron tracks should be less than or equal to the plateau value for relativistic singly charged–particle tracks ~ 30 grains per 100 μm.

(c) Coulomb–Scattering measurements on the electron tracks must be carried out with two different cell lengths to eliminate noise and spurious scattering. Other sources of errors are avoided following Refs. [15, 16].

Further checking on few selected pairs for energy – momentum balance, were used only in the calculations of the production cross section and excluded from the analysis of the data. This is due to the unstable physical conditions of the emulsion in the vicinity of the electron pairs and hence the energy determination might be affected for either one or both of the tracks of such pairs. The angles for all the analyzed electron pair tracks were measured with reference to the primary beam direction. The accuracy in the angle measurement was found to be about 0.05° when calculated according to ref. [3, 4 and references therein].

Results and Discussion

In this work we studied 134 fully measured electron pairs come from the interactions of ^{32}S (200A GeV) with emulsion nuclei as well as pairs from collisions of ^{12}C and ^{22}Ne at ~ 3.7A GeV [2]. We selected a clear examples of 8 direct electron pairs for which the charges of projectile fragments and its energy level have been carried out. Detailed information using kinematical analysis of the 8 pairs is obtained in Table (1).

On the basis of the discussion and comments given before by de Boer [10, 11] on the Cairo [2] and Bristol data [17] of electron pairs, we display in Fig. (1) the distributions of energy partition asymmetry Y, where Y = |E₊ – E₋| / (E₊ + E₋), for our data in comparison with Bristol results [17].

Table (1): Detailed information of the measured e^+e^- pairs

Star No.	P MeV/c	$\theta_p°$	$\theta_o°\pm0.1°$	M_c MeV/c²	$\tau .10^{-16}$ s	$S(\mu)\pm2\mu$	Y	E^* MeV
1	109.900±21.300 175.800±34.100	0.418±0.070	14.730	1.440±0.300	2.160±5.000	69	0.230	$^6Be^*$ 8.140±2.000
2	1st Pair 103.200±12.600 204.700±23.200	0.541±0.180	14.460	1.740±0.350	13.100±4.500	69	0.310	$^{12}C^*$ 10.300±0.500
	2nd Pair 163.550±16.230 85.270±10.400	0.147±0.070	23.750	1.150±0.080	7.700±1.600	50	0.330	$^{12}C^*$ 7.740±0.750
3	157.100±39.700 360.900±38.400	0.005±0.001	11.700	1.170±0.100	7.000±1.000	89	0.150	$^{12}C^*$ 10.300±0.500
4	149.140±42.160 121.340±13.460	0.181±0.070	8.260	1.120±0.120	9.500±3.540	115	0.100	
5	144.750±33.430 100.150±17.490	0.481±0.060	23.580	1.450±0.230	11.500±5.800	75	0.180	$^4He^*$ 17.120±2.080
6	218.340±20.310 147.090±12.760	0.330±0.070	8.870	1.460±0.300	10.400±3.400	78	0.190	$^8Be^*$ 5.600±1.000
7	1st Pair 288.500±32.320 225.250±24.750	2.200±0.200	17.310	9.820±1.520	31.200±9.600	49	0.120	$^8Be^*$ 18.100±1.000
	2nd Pair 111.220±28.140 504.770±127.700	0.310±0.050	12.350	1.490±0.650	10.900±6.200	108	0.640	$^{12}C^*$ 10.290±0.800
8	162.140±31.460 239.290±21.270	0.646±0.180	4.420	2.500±0.610	30.600±12.110	150	0.190	

P = Momentum of e^+ and e^-, θ_p = Opening angle of the pair in degrees, θ_o, M_c, τ, S: the emission angle, Mass, lifetime and decay time distance of the neutral particle, Y = Energy partition asymmetry of pair. E^*, is the energy of the excited level of the produced excited fragment.

Fig. (1) shows also the calculated distribution for Dalitz pairs [18] and the distribution of [~ 300 MeV γ–ray [19] phase space transition by two body decay].

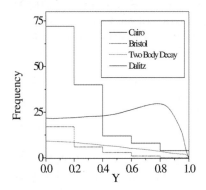

Fig. (1): The distributions of energy partition asymmetry of electron pairs observed in the experiment of Cairo, Bristol [17] compared with Dalitz pairs [18] and those due to two body phase transition [19].

In Fig. (1) one can observe similarity between Cairo and Bristol results showing forward peaking and deviating largely from both the calculated distribution for Dalitz pairs as well as for the phase space transition by two body decay.

As predicted by Koba, Nielsen, and Olsen [20] in high energy hadron–hadron collisions the multiplicity distributions obey a scaling law (KNO scaling), with energy independent function ψ. Now assuming the isotropic scenario in measurements for the two experiments of Cairo and Bristol [17], KNO scaling can display Y – distributions for the two experiments. Therefore, the scaling property of Y – distributions is seen in Fig. (2).

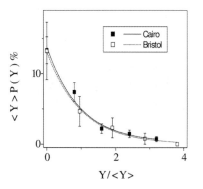

Fig. (2): The KNO scaling of the Y – distributions for Cairo and Bristol [17] data.

Data points in Fig. (2) of the two experiments fall on a unique curve. The figure beautifully illustrates a corresponding scaling property of the two distributions reproduced well by the universal behavior of exponential decay law of the form, $\psi = \alpha e^{-z/\beta}$. ψ is the scaling function and z is the scaling parameter. The fitting parameters $\alpha \approx$ 13.6 and $\beta \approx 1$ for the two curves in Fig. (2). Therefore, the two experiments of Cairo and Bristol [17] have the same behavior, where the associated pairs are originating from other source different from Dalitz pairs as discussed by de Boer [10, 11] and confirmed in [3, 4, 12].

Fig. (3) displays the invariant mass spectra of the selected sample of dielectrons associating the interactions of ^{32}S (200A GeV) with emulsion nuclei.

270

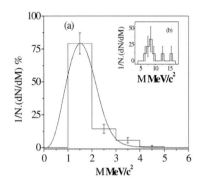

Fig. (3): The invariant mass spectra of dielectrons associated with the interactions of ^{32}S (200A GeV) with nuclear emulsion.

Clearly, most of electron pairs are locating in the region on lower invariant mass (M ≤ 5 MeV/c^2) as shown in Fig. (3a). The rest of electron pairs which forms ~ (6.72 %) electron pairs has mass distribution at (M > 5 MeV/c^2) as in Fig. (3b). The clear appearance of two peaks, one in the distribution over the invariant mass ≤ 5 MeV/c^2 and the other in the distribution over the invariant mass > 5 MeV/c^2, is observed. The two peaks are corresponding to average masses of (1.55±0.14 MeV/c^2) and (9.46±3.15 MeV/c^2) respectively. These masses are above the threshold of γ internal conversion. In Fig. (3a), the (1.55±0.14 MeV/c^2) line observed is found to be quantitavily consistent with the peak of Maxwell – Boltzman spectrum having the form,

$$P(E)dE = \frac{E-V}{T^2} e^{-\frac{E-V}{T}} dE$$

The potential V = 0 for particles of Z = 1. T is the temperature of the decay reactions. From fitting parameters T is found to be ~ 0.4 MeV. In Fig. (3b), it can be shown that, the distribution peaks at the lower M values and after

M = 10 MeV/c^2, the probability is almost constant.

The data indicates the production of clusters "A" and "C" obtained before by de Boer and van Dantzig [10, 11], discussion of the Cairo [2] and Bristol data [17]. The data characterizing the two clusters are included in Table (2).

Table (2): The data characterizing the two peaks of the invariant mass spectra of the pairs associating the interactions of ^{32}S (200A GeV) with nuclear emulsion.

Cluster A	Cluster C
Invariant Mass ≤ 5 MeV/c^2	Invariant Mass > 5 MeV/c^2
< M > = 1.55±0.14 MeV/c^2	< M > = 9.46±3.15 MeV/c^2
< τ > = 38.67±3.46 X 10^{-16} S	< τ > = 11.71±3.90 X 10^{-16} S
N = 125 Pairs	N = 9 Pairs

Table (3) gives the detailed data of masses and lifetimes of the observed pairs.

Fig. (4) shows a clear example of the production of two ^{12}C track fragments, the first is in the excited 10.30 MeV (0$^+$, 0) level which then decays to its g. s. (0$^+$, 0) emitting a neutral boson which decays after 69 μm into an (e$^+$ e$^-$ pairs). The path length of this excited ^{12}C–fragment is undetected due to the short lifetime of the 10.30 MeV level (Γ = 3000 KeV). The second ^{12}C fragment is produced at "A" probably in the 7.66 MeV (0$^+$, 0) level which decays after a detectable distance δ ≈ 5 μm at the point "B" into its g. s. emitting a neutral boson decaying into an (e$^+$ e$^-$ pair) after 50 μm. The δ distance corresponds to the 0.05 fem S lifetime of the 7.66 MeV level [21]. The probability for such pairs to be due to Dalitz decay $\pi° \rightarrow \gamma + e^+ + e^-$ is kinematically improbable for ranges higher than 49 μm in Table (1).

One can conclude that new light neutral bosons of average masses 1.55±0.14 and 9.46±3.15 MeV/c^2 and life times of orders 10^{-16} S are produced in ^{32}S (200A GeV) collisions in emulsion.

Table (3): the masses and lifetimes of electron pairs associated with the interactions of ^{32}S (200A GeV) with nuclear emulsion.

Mass Ranges	Frequency	$<M>$ MeV/c^2	$<\tau>$. 10^{-16} sec
$0 \leq M \leq 1$	–	–	–
$1 \leq M \leq 2$	99	1.24± 0.12	36.43± 3.66
$2 \leq M \leq 3$	18	2.36± 0.56	47.90± 11.30
$3 \leq M \leq 4$	8	3.31± 1.17	52.72± 18.63
$4 \leq M \leq 5$	1	4.24± 1.10	55.32± 16.71
$5 \leq M \leq 6$	–	–	–
$6 \leq M \leq 7$	1	6.65± 0.51	13.70± 4.89
$7 \leq M \leq 8$	2	7.30± 5.18	12.89± 9.14
$8 \leq M \leq 9$	3	8.64± 6.99	5.98± 3.46
$M \geq 9$	3	12.66± 7.32	16.00± 9.25
Total Sample	134	2.08± 0.18	37.21± 3.21

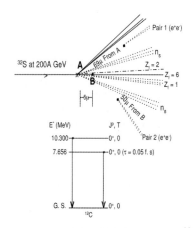

Fig. (4): A schematic diagram of a ^{32}S (200A GeV) interaction with nuclear emulsion accompanied by the emission of electron pairs.

Conclusion

From the interactions ^{32}S (200A GeV) with nuclear emulsion, the emission of neutral bosons of average masse 1.55±0.14 and 9.46±3.15 MeV/c^2 and life time of the order of 10^{-16} S is suggested. This is indicated through the tow narrow peaks found in the invariant mass spectrum of the emitted electron pairs. The mentioned neutral boson is expected to be emitted through the de–excitation of excited fragments ^{4}He, ^{8}Be, and ^{12}C produced in ^{32}S fragmentation. From the analysis of an example for these excited nuclei, two visible ^{12}C tracks fragmented from ^{32}S projectile are observed. One is corresponding to an excited state ~ 10.3 MeV (0^+, 0) of short lifetime, which then decays into its g. s. (0^+, 0) emitting a neutral boson after 69 μm. The other carbon fragment is emitted at 7.656 MeV (0^+, 0) state, which decays after a detectable distance ($\delta \approx$ 5 μm) into its g. s. emitting neutral boson which then decays into electron pair after 50 μm. The 5 μm distance corresponds to 0.05 fem s lifetime of 7.656 MeV state, which is kinematically improbable to be due to Dalitz decay.

Acknowledgement

The authors would like to dedicate this work to the souls of Prof. Dr. M. El–Nadi and Prof. Dr. O. E. Badawy of Cairo University, Egypt.

We are pleased to acknowledge the kind help of the CERN authorities for irradiation of photographic plates.

References

[1] Hans J. Specht, Nucl. Phys. A **805**, 338 (2008).

[2] M. El–Nadi and O. E. Badawy, Phys. Rev. Lett. **61**, 1271 (1988).

[3] S. Kamel, Phys. Lett. B **368**, 291 (1996).

[4] M. El–Nadi, S. Kamel, A. Hussein, Z. Abou–Moussa, E. A. Shaat, H. Salama, and W. Osman, IL NUOVO CIMENTO A **109**, 1517 (1996).

[5] P. L. Jain and G. Singh, J. Phys. G **34**, 129 (2007).

[6] DLS Collaboration, Nucl. Phys. A **525**, 299 (1991).

[7] S. Weinberg, Phys. Rev. Lett. **40**, 223 (1978).

[8] F. Wilczek, Phys. Rev. Lett. **40**, 279 (1978).

[9] G. 't Hooft, Phys. Reports **142**, 357 (1986).

[10] F. W. N. de Boer and R. van Dantzig, Phys. Rev. Lett. **61**, 1274 (1988).

[11] F. W. N. de Boer and R. van Dantzig, Phys. Rev. Lett. **62**, 2639 (1989).

[12] M. S. El–Nagdy, A. Abdelsalam, and B. M. Badawy, AIP Conference Proceeding **888**, 249 (2007).

[13] CERN Collaboration, Nucl. Phys. A **661**, 23c (1999).

[14] M. S. El–Nagdy, A. Abdelsalam, E. A. Shaat, B. M. Badawy, and E. M. Khashaba, Romanian Journal of Physics **53**, 487 (2008).

[15] H. Barkas, Nuclear Research Emulsion, Vol. **I**, Technique and Theory Academic Press Inc. (1963).

[16] C. F. Powell, F. H. Fowler and D. H. Perkins, The Study of Elementary Particles by the Photographic Method, Pergamon Press. London, New York, Paris, Los Angles, 474 (1958).

[17] B. M. Anand, Proc. Roy. Soc. (London) A **220**, 183 (1953).

[18] N. P. Samios, Phys. Rev. 121, 275 (1961).

[19] F. Ajzenberg–Selove and T. Lauritsen, Nucl. Phys. A **114**, 1 (1968).

[20] Z. Koba, H. B. Nielsen, and P. Olesen, Nucl. Phys. B **40**, 317 (1972).

[21] Table of Isotopes, 7th Edition, Edited by C. Michael Lederer and Virginia S. Shirley, Lawrence Berkeley Laboratory University of California, Berkeley USA(1989).

THE MOVING QUASARS

SHAHINAZ YOUSEF[1], M. KAMAL[1], H. MANSOUR[2], SAYEDA, M. AMIN[3],
KHADIGA ABDER RAHMAN[1], AL HASSAN ABD AL MONEM[4]

[1]*Astronomy & Meteorology Dept.,* [2]*Physics Dept., Faculty of Science*
[3]*National Research Institute of Astronomy and Geophysics, NRIAG, Helwan, Egypt*
[4]*Faculty of Engineering, Cairo University*

Abstract

Quasars, those very far objects with very high proper motions implying that their transverse velocities far exceed the velocity of light. Their space velocities are superluminal and are within a fraction of a degree from their transverse velocities. We propose that QSOs form a cloud that envelopes the cosmos. We refer to this cloud as Al Tareq Cloud. The jets although mildly superluminal may force the QSO forward opposite to the jet ejection direction. Jets also help in changing the moving direction of the QSO. In other words, jets help in speeding up and maneuvering the QSO. There are tw belts of Quasars; The inner Quasar Belt at Z= (0.25-0.4) and the Outer Quasar Belt at Z = (1.8-2.25). From the present Quasar study, there is an indication that the preliminary age of the universe is of the order 46-49 Gyrs..

Introduction

More than 100,000 quasars are known. All observed quasar spectra have redshifts between 0.06 and 6.4. Applying Hubble's law to these redshifts, it can be shown that they are between 780 million and 28 billion light-years away. Because of the great distances to the furthest quasars, we see them and their surrounding space as they existed in the very early universe. Most quasars are known to be farther than three billion light-years away.

Quasars are extremely luminous at all wavelengths and exhibit variability on timescales as little as hours, indicating that their enormous energy output originates in a very compact source. They are typically the size of the Solar System or smaller. Being so distant but still visible as faint stars implies that quasars give off more energy than 100 normal galaxies combined.

Although Quasars were first observed by radio telescopes, X-ray observations have been carried out e.g. Green et al. 2002.

The observations of superluminal velocities for the Quasars and their jets have been reported e.g. Cohen and Vermeulen (1977), Jorstad (2001), Rokaki (2003).

Quasar Distances

Many quasars have red shifts> 1; we get radial velocities greater than the velocity of light.

The radial velocity $V_r = (\Delta\lambda/\lambda) = Zc$ km/s

The distance in Mpc (one parsec = 3.26 light year) is given by:

$$d = V_r/H \qquad Mpc$$

The Hubble constant is in units of km/s/Mpc

Here we take H=50 and H=75 km/s/Mpc

Observation of quasars at large distances and their scarcity nearby implies that they were much more common in the early Universe. (Keel)

Because the faster an object appears to move, the farther away it is, at a redshift of 5.5, light traveling from the Stern's quasar has journeyed about 13 billion years to get here. That means the quasar existed at a time when the universe was less than 8 percent of its current age.

Proper Motion μ

The angular velocity of the star across the sky is how fast the star is moving across the sky in units of an angle. It is measured in milliarcseconds per year. The radial velocity, measured from Doppler shift and the proper motion are perpendicular to each other. The actual velocity of the star in three dimensional space is called the space velocity (Pasachoff 2002) (see Fig. 1).

The proper motion for faraway ordinary stars is small as seen in Fig 2(a). In the case of quasars which are almost on the border of the cosmos, they possess very large proper motions as seen in Table 1 and Fig 2(b). This is very surprising and can only be explained if the quasars are indeed moving with superluminal velocities.

Proper motions for 40 quasars are given in a table by (Varshni et al., 1995).

The corresponding distances have been calculated for Hubble constants of 50 and 75 km/s/Mpc. Table 1 (A&B) gives the redshift z, the distance of the quasar d in Mpc, the measured proper motion μ, the computed transverse velocity Vt in units of C and the space velocity Vs in units of C. The angle theta Θ which is the angle between the Vs and Vr.

Figure 3 illustrates the relation between the space velocity in C units and distance d in Mpc for H=50 and H= 75.

$$V_s = 446.6\,e^{0.0002\,d} \quad H=50 \quad km/s/Mpc$$
$$V_s = 280.77e^{0.0002\,d} \quad H=75 \quad km/s/Mpc$$

274

Proper Motion μ

The transverse velocity is given by;

Vt=4.74 μ d (km/s)

μ is in units of • (arcsec/year)

The Space Velocity is • given by

Vs = ((Vr)2 + (Vt)2)½ •

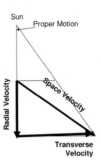

Fig 1. The velocity triangle.

Proper Motion For Far Away Stars Should Be Small

a

Proper Motions for Quasars Are Extremely Large

Although quasars and • QSAs are on the edge of the universe, their proper motion are extremely large. This implies that they are moving with superluminal velocities

b •

Fig 2. (a) The proper motion for nearby (wide angle) compared to far away objects (small angle for ordinary stars). (b) Is the proper motion for nearby stars compared to proper motion for Quasars with extremely large proper motion in spite of their very far distances. This contradiction can be explained. They have really moved very large distances within a year. They are moving with superluminal velocities.

Tables 1A&B: Computation for H=50&75 List of 40 Quasars, proper motion, redshift, distance, transverse & space velocities & revolution time T.

Name	μ	Z	d Distance H=50	Vt in C	V s in c	T Gyrs
LB 8956	60.8	1.896	11376 Mpc	10928.24064	10928.2408	21.31163
TON 202	52.6	0.362	2172	1805.10576	1805.105796	24.63397
LB 8991	50.5	1.013	6078	4849.6362	4849.636306	25.65835
PHL 1033	48.8	0.255	1530	1179.6912	1179.691228	26.55219
LB 9029	36.1	1.286	7716	4401.05208	4401.052268	35.89326
Br 337	35	0.3	1800	995.4	995.4000452	37.02134
PHL 1106	32.2	0.345	2070	1053.1332	1053.133257	40.24058
LB 9388	30.9	1.07	6420	3134.3724	3134.372583	41.93355
LB 9013	29.1	1.454	8724	4011.12072	4011.120984	44.52738
PHL 1072	28.6	0.615	3690	1667.4372	1667.437313	45.30583
PHL 1070	28.4	0.079	474	212.69328	212.6932947	45.62489
PHL 1194	26.2	0.299	1794	742.64424	742.6443002	49.45598
PHL 3632	25.1	1.479	8874	3519.25092	3519.251231	51.62338
PHL 8462	23.6	2.225	13350	4977.948	4977.948497	54.90452
PHL 1119	23.3	0.119	714	262.85196	262.8519869	55.61145
LB 9707	22.1	1.924	11544	4030.93392	4030.934379	58.63107
TON 616	21.9	0.268	1608	556.40016	556.4002245	59.16652
PHL 1049	21.8	0.147	882	303.79608	303.7961156	59.43792
PHL 1127	21.2	1.99	11940	3999.4224	3999.422895	61.12013
LB 9502	20.6	1.887	11322	3685.08456	3685.085043	62.90032
LB 9308	20.1	0.411	2466	783.15228	783.1523878	64.46501
PHL 1186	18.4	0.27	1620	470.9664	470.9664774	70.42101
PHL 828	17.9	0.624	3744	1058.87808	1058.878264	72.38808
LB 8741	17.9	0.568	3408	963.85056	963.8507274	72.38808
LB 8948	17.9	0.331	1986	561.68052	561.6806175	72.38808
PHL 1092	17.5	0.396	2376	656.964	656.9641193	74.04266
PHL 1027	16.6	0.363	2178	571.24584	571.2459553	78.05702
LB 8891	15.8	1.013	6078	1517.31192	1517.312258	82.00928
LB 8863	15.3	2.214	13284	3211.27416	3211.274923	84.68932
LB 9010	13	1.711	10266	2108.6364	2108.637094	99.6728
PHL 3424	12	1.847	11082	2101.1472	2101.148012	107.9789
RS 32	12	0.341	2046	387.9216	387.9217499	107.9789
LB 6158	11.3	2.052	12312	2198.18448	2198.185438	114.6678
RS 23	9.2	1.908	11448	1664.08128	1664.082374	140.842
PHL 3375	5.8	0.39	2340	214.4376	214.4379546	223.4043
LB 8775	5.1	1.926	11556	931.18248	931.1844718	254.0675
PHL 1222	4.2	1.923	11538	765.66168	765.6640949	308.5102
TON 621	4.2	2.011	12066	800.69976	800.7022854	308.5102
LB 19	4.1	2.043	12258	794.07324	794.0758681	316.0348
PHL 1226	1	0.404	2424	38.2992	38.30133074	1295.675

Name	μ	Z	d Distance H=75	Vt	V s	T Gyrs
LB 8956	60.8	1.896	7584 Mpc	6870.49728	6870.497542	22.5989
TON 202	52.6	0.362	1448	1134.85552	1134.855578	26.12193
LB 8991	50.5	1.013	4052	3048.9274	3048.927568	27.20819
PHL 1033	48.8	0.255	1020	741.6624	741.6624438	28.15601
LB 9029	36.1	1.286	5144	2766.90616	2766.906459	38.06131
Br 337	35	0.3	1200	625.8	625.8000719	39.25752
PHL 1106	32.2	0.345	1380	662.0964	662.0964899	42.67122
LB 9388	30.9	1.07	4280	1970.5548	1970.555091	44.46645
LB 9013	29.1	1.454	5816	2521.75944	2521.759859	47.21695
PHL 1072	28.6	0.615	2460	1048.3044	1048.30458	48.04242
PHL 1070	28.4	0.079	316	133.71856	133.7185833	48.38075
PHL 1194	26.2	0.299	1196	466.89448	466.8945757	52.44325
PHL 3632	25.1	1.479	5916	2212.52484	2212.525334	54.74156
PHL 8462	23.6	2.225	8900	3129.596	3129.596791	58.22089
PHL 1119	23.3	0.119	476	165.25292	165.2529628	58.97052
LB 9707	22.1	1.924	7696	2534.21584	2534.21657	62.17254
TON 616	21.9	0.268	1072	349.80432	349.8044227	62.74032
PHL 1049	21.8	0.147	588	190.99416	190.9942166	63.02812
PHL 1127	21.2	1.99	7960	2514.4048	2514.405587	64.81193
LB 9502	20.6	1.887	7548	2316.78312	2316.783888	66.69966
LB 9308	20.1	0.411	1644	492.36156	492.3617315	68.35885
PHL 1186	18.4	0.27	1080	296.0928	296.0929231	74.67461
PHL 828	17.9	0.624	2496	665.70816	665.7084525	76.76049
LB 8741	17.9	0.568	2272	605.96512	605.9653862	76.76049
LB 8948	17.9	0.331	1324	353.12404	353.1241951	76.76049
PHL 1092	17.5	0.396	1584	413.028	413.0281898	78.51502
PHL 1027	16.6	0.363	1452	359.13768	359.1378635	82.77185
LB 8891	15.8	1.013	4052	953.92184	953.9223779	86.96283
LB 8863	15.3	2.214	8856	2018.90232	2018.903534	89.80474
LB 9010	13	1.711	6844	1325.6828	1325.683904	105.6933
PHL 3424	12	1.847	7388	1320.9744	1320.975691	114.501
RS 32	12	0.341	1364	243.8832	243.8834384	114.501
LB 6158	11.3	2.052	8208	1381.98096	1381.982483	121.594
RS 23	9.2	1.908	7632	1046.19456	1046.1963	149.349
PHL 3375	5.8	0.39	1560	134.8152	134.8157641	236.8979
LB 8775	5.1	1.926	7704	585.42696	585.4301282	269.4129
PHL 1222	4.2	1.923	7692	481.36536	481.3692011	327.1434
TON 621	4.2	2.011	8044	503.39352	503.3975368	327.1434
LB 19	4.1	2.043	8172	499.22748	499.2316603	335.1224
PHL 1226	1	0.404	1616	24.0784	24.08178902	1373.82

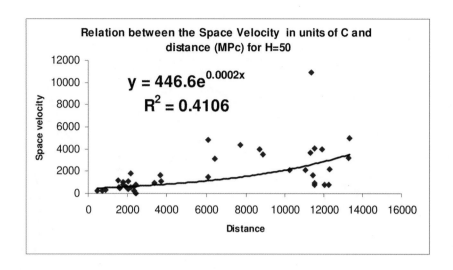

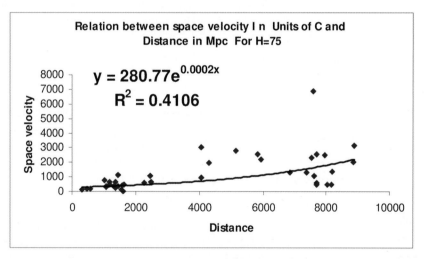

Fig 3. The variation of Quasars space velocity in units of C and their distances in Mpc. Above; computed for Hubble constant H =50 km∕s∕Mpc. Below for H= 75 km∕s∕Mpc. Note that Quasars space velocities are 10s of thousands of C.

For H = 50, with the exclusion of the first Quasar, the best fit is given in Fig. 4.

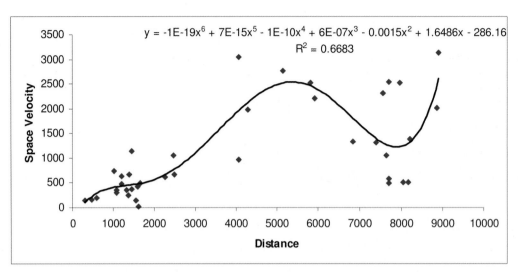

Fig 4. In the case of H=50. with the exclusion of the first quasar in the table, the best fit is polynomial f the 6th order.

On examination of Table 1 and Fig. 5, it is evident that Quasars space velocities and transverse velocities are almost overlapping. The angle separating them is only a fraction of a degree. This can be interpreted as the quasars being moving in orbits with such superluminal speeds almost perpendicular to the radius of the cosmos.

The Revolution Time Of Quasars

The revolution time for a quasar is given by

$T = 2\Pi d$ Mpc/ Vt in units of C

$= 6.28$ (d/Vt) 10^9 C $(3.26)/C = 20.4728$ (d/Vt) 10^9

The revolution times T for Quasars around the center of the cosmos have computed and tabulated in the last column of Tables 1 A&B.

Fig. 5 illustrates the linear relation between the revolution time T and Z for H = 50

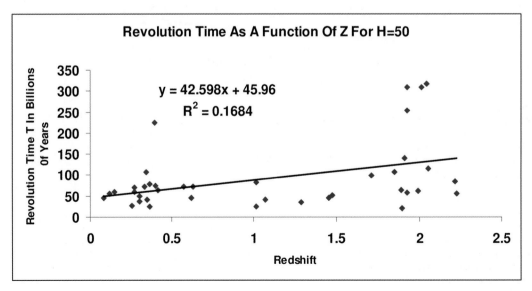

Fig 5. The linear relation between the revolution time of Quasar in billions of years and the redshift Z for H=50. Note the clustering of Quasars around Z= 2 and Z ≤ 0.4.

The following equations are found:

T (Billions of Years) = 42.598 Z + 45.96   For H= 50
T (Billions of Years) = 45.171Z + 48.736   For H= 75

For a Redshift Z=0, the revolution time for H=50 is 45.96 Gyrs, This value is an indication to the age of the universe. It is slightly larger in the case of H=75 (48.736 Gyrs).
The present list of Quasars is very short, better estimations can be found by future satellite measurements of proper motions

There seems to be two belts of Quasars:
The inner Quasar Belt at   Z= (0.25-0.4)
The Outer Quasar Belt at   Z = (1.8-2.25).
This outer belt is where most quasars are peaked in the Quasar Cloud.
In between these two belts, there are some scattered Quasars. Those clustering of Quasars are readily seen in figure 6.

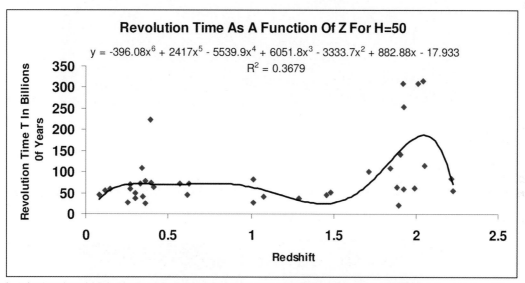

Fig 6. A 6 order polynomial fitting for the relation between the revolution time T in Billions of years and redshift Z.Note the clustering of Quasars in two belts.

Jets of Quasars
One very prominent feature of Quasars is the ejection of jets. These jets have been observed in the radio , X ray as well as Hubble Space Telescope (Pasachoff 2001). Chandra has made a survey of Quasar radio jets and found that X-ray emission is a common feature in Quasar jets (Gelbord et al. 2004). The jet of 3C 273 quasar extends 150,000 Ly from it. The most distant X ray jet in 2003 in the known universe (Astronomy Picture of the Day) is 100,000 light years in length, emerging from Quasar GB 1508+5714. An estimated 12 billion light years away,

The Unified Cosmos
There seems to be a unification of the cosmos from the minute micro- scale to the macro-scale. From the atomic system to the solar system for one. The merging of pairs of galaxies resulting in star birth for two. The solar system is enveloped by trillions and trillions of comets forming the Oort cloud as shown in Fig. 7(b). In analogy, we postulate the existence of a Quasar cloud that envelops the cosmos. We will refer to this cloud as Al Tareq Cloud. Quasars move within this cloud with superluminal velocities of thousands times the velocity of light. They can alter their directions by the ejection of jets of superluminal velocities of the order of several to perhaps 20C. These jets forces the QSO forward in the opposite direction of jet ejection. There is also an analogy between comets (head, coma and tail) and Quasar (black hole, disk and jet) as seen in fig 7a. Anther belt of Quasars exists redshift Z= (0.25-0.4)

Fig 7(a). Analogy between a quasar and a comet. The jet of the quasar resembles the tail of the comet. Better analogy is obtained if the comet looses its tail in space.

A Cloud of Quasars envelops the Cosmos similar to the Oort Cloud

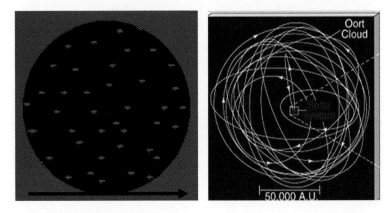

Diameter of the Cosmos

Fig 7(b). The solar system is all very well persevered by a shell of trillions and trillions of comets with orbits in every direction forming the Oort Cloud. In analogy, the whole cosmos is surrounded by a cloud of Quasars which we will call "Al Tareq Cloud" which envelops the cosmos and is composed of trillions of Quasars that moves in orbits with superluminal velocities of thousands time the present velocity of light but can change direction by ejecting jets.

Table 1 A 40 Quasars proper Motions, Redshift distances assuming H = 50, Transverse, Space Velocities And the Angle Between Them

Name	μ	Z	d Distance 50	Vt 50 in c	Space V 50 in c	Θ
LB 8956	60.8	1.896	11376	10928.24064	10928.2408	0.00994
TON 202	52.6	0.362	2172	1805.10576	1805.105796	0.01149
LB 8991	50.5	1.013	6078	4849.6362	4849.636306	0.01197
PHL 1033	48.8	0.255	1530	1179.6912	1179.691228	0.01238
LB 9029	36.1	1.286	7716	4401.05208	4401.052268	0.01674
Br 337	35	0.3	1800	995.4	995.4000452	0.01737
PHL 1106	32.2	0.345	2070	1053.1332	1053.133257	0.01877
LB 9388	30.9	1.07	6420	3134.3724	3134.372583	0.01956
LB 9013	29.1	1.454	8724	4011.12072	4011.120984	0.02077
PHL 1072	28.6	0.615	3690	1667.4372	1667.437313	0.02113
PHL 1070	28.4	0.079	474	212.69328	212.6932947	0.02128
PHL 1194	26.2	0.299	1794	742.64424	742.6443002	0.02307
PHL 3632	25.1	1.479	8874	3519.25092	3519.251231	0.02408
PHL 8462	23.6	2.225	13350	4977.948	4977.948497	0.02561
PHL 1119	23.3	0.119	714	262.85196	262.8519869	0.02594
LB 9707	22.1	1.924	11544	4030.93392	4030.934379	0.02735
TON 616	21.9	0.268	1608	556.40016	556.4002245	0.0276
PHL 1049	21.8	0.147	882	303.79608	303.7961156	0.02777
PHL 1127	21.2	1.99	11940	3999.4224	3999.422895	0.02851
LB 9502	20.6	1.887	11322	3685.08456	3685.085043	0.02934
LB 9308	20.1	0.411	2466	783.15228	783.1523878	0.03007
PHL 1186	18.4	0.27	1620	470.9664	470.9664774	0.03285
PHL 828	17.9	0.624	3744	1058.87808	1058.878264	0.03376
LB 8741	17.9	0.568	3408	963.85056	963.8507274	0.03376
LB 8948	17.9	0.331	1986	561.68052	561.6806175	0.03376
PHL 1092	17.5	0.396	2376	656.964	656.9641193	0.03454
PHL 1027	16.6	0.363	2178	571.24584	571.2459553	0.03641
LB 8891	15.8	1.013	6078	1517.31192	1517.312258	0.03825
LB 8863	15.3	2.214	13284	3211.27416	3211.274923	0.0395
LB 9010	13	1.711	10266	2108.6364	2108.637094	0.04649
PHL 3424	12	1.847	11082	2101.1472	2101.148012	0.05037
RS 32	12	0.341	2046	387.9216	387.9217499	0.05037
LB 6158	11.3	2.052	12312	2198.18448	2198.185438	0.05349
RS 23	9.2	1.908	11448	1664.08128	1664.082374	0.06569
PHL 3375	5.8	0.39	2340	214.4376	214.4379546	0.1042
LB 8775	5.1	1.926	11556	931.18248	931.1844718	0.11851
PHL 1222	4.2	1.923	11538	765.66168	765.6640949	0.1439
TON 621	4.2	2.011	12066	800.69976	800.7022854	0.1439
LB 19	4.1	2.043	12258	794.07324	794.0758681	0.14741
PHL 1226	1	0.404	2424	38.2992	38.30133074	0.60436

282

Conclusions

- Quasars are faint objects with peculiar emission lines. They are also known as Quasi stellar objects (QSOs). They are the most distant objects in the universe. They have very large red shift. Thus quasars must have 10 to 1000 the luminosity of large galaxies. Quasars fluctuate in brightness over few days. But cannot change its brightness appreciably in less time than it takes light to cross its diameter. The rapid fluctuations in quasars showed that they are small objects not more than few light days or light weeks in diameter(about the size of the solar system..
- Trillions of Quasars envelops the Cosmos in a cloud which we call All Tareq Cloud.
- Quasars have the capacity of moving in any direction in the cosmos.
- This is achieved by ejecting a jet or more and they thus move in the opposite direction of the jet or the resultant jets. The jets are used in the maneuverings of the quasars.
- There extraordinary large proper motions imply that they have superluminal velocities perhaps several thousand times the velocity of light.
- The space velocity is inclined by an angle within parts of a degree from the direction of the transverse velocity. There are two belts of Quasars; The inner Quasar Belt at   Z= (0.25-0.4)
  The Outer Quasar Belt at   Z = (1.8-2.25).
- There is an indication from Quasars that the age of the universe is between 46-49 Gyrs. A conclusion that needs to be confirmed from a larger sample of QSOS.

References

1. Astronomy Picture of the Day 2003 November 28.
2. Cohen, M. H. and Vermeulen R. C. (1977) Radio sources with superluminal velocities, Nature 268, 405-409 (04 August).
3. Cohen M. H., Kellermann K. I., Shaffer D. B., Linfield R. P., Moffet A. T., Romney J. D, Seielstad G. A., Pauliny-Toth I. I. K., Preuss E., Witzel A., Schilizzi R. T. and Geldzahler B. J. 1991. A statistical Study of Superluminal Velocities. Extragalactic Radio Sources. From Beams to Jets. Proceedings of the 7th. I.A.P. Meeting, held at the Institut d'Astrophysique de Paris, Paris, France, July 2-5, 1991. Editors, J. Roland, H. Sol, G. Pelletier. Publisher, Cambridge University Press (1992). p. 98–104. http://adsabs.harvard.edu/abs/1992ersf.meet...98C
4. Gelbord, J. M., Marshall H. L., Schwartz D. A., Worrall, D. M., Birkianshaw M., Lovell J. E. J., Janncey D. L., Perlman E. S. Murphy D. W. and Preston R. A. (2004). Continuing a Chandra Survey of Quasar Radio Jets. X-Ray and Radio Connections. Santa Fe NM, 3-6 February. http://aoc.nrao.edu/events/xradio
5. Green, P. J., Kochanek C., Siemiginowska A., Kim D.-W., Markevitch M., Verman J. S., Dosaj A., Jannuzi B. T. and Smith C. (2002). Chandra Observations of the QSO Pair Q2345+007: Binary or Massive Dark Lens. The AP. J. 571: 721-732 (2002).
6. Keel B. keel@bildad.astr.ua.edu
7. Pasachoff, J. M. (2002). Astronomy from the earth to the universe. Brooks/Cole Thomson Learning. Australia, Canada, UK and USA.
8. Svetlana G. Jorstad, Alan P. Marscher, John R. Mattox, Ann E. Wehrle, Steven D. Bloom and Alexei V. Yurchenko (2001). Multiepoch Very Long Baseline Array Observations of EGRET-detected Quasars and BL Lacertae Objects: Superluminal Motion of Gamma-Ray Bright Blazars, The Astrophysical Journal Supplement Series, 134:181–240, 2001 June.
9. Rokaki E., Alawrence A. Economou F, Mastichiadis A. (2003). Is there a disc in the superluminal quasars? Monthly Notices of the Royal Astronomical Society 340 (4), 1298–1308

III-2 INVITED LECTURE PAPERS

ELECTRON BEAM TECHNOLOGY – SOME RECENT DEVELOPMENTS

MUNAWAR IQBAL

National Institute of Laser and Optronics
P.O. Nilore, Islamabad, Pakistan

FAZAL-E-ALEEM

Theory Group, CHEP, University of the Punjab
Lahore-54590, Pakistan

Electron beam technology has been in focus since long due to wide variety of applications in research and industry. One of the important modes of e-beam production is through thermionic emission. Improvements and advancement in enhancing the capabilities of electron beam sources compatible with the task to be accomplished at a reduced cost are therefore necessary. We give an update of the recently developed and reported e-guns which are easy to fabricate, assemble and more efficient. Besides being cost effective, these guns are user friendly.

1. Introduction

Electron beam technology which was first developed in the middle of eighteenth century has gone through a wide variety of innovations based on its applications in fundamental research and industrial applications [1-26]. In the present work, we give an update on various aspects of development and applications of high power *thermionic* electron beam sources, partially reported earlier [3]. A common factor in these improved electron guns is the choice of thermionic filaments as a source of electrons. Our novel e-beam guns are easy to fabricate/assemble and user friendly besides being cost effective [3].

In view of vast variety of applications of this technology in research and industry, we have included some new aspects in the last section [1-3].

2. Sources of e-beam guns

Information about the sources of electron beam is available in several good books/review articles which give a detailed account of the beam assembly [1-3]. Electron beams can be produced from solids, liquids and gases. In case of solids, beam emission is by providing energy from an outside source. If electrons are emitted by proving heat energy, then the process is called *thermionic emission.* If electrons are emitted by photons, the emission is called *photo-emission.* In some of the cases, the beam may be emitted by applying a high voltage to the solid, resulting in what is called as - *field emission or cold emission.* In case of gases, a gas

discharge called plasma is also a source of electrons. In order to obtain good quality results, emission must be controlled [3].

3. Thermionic e-gun

Both diode and triode configurations are used for the purpose of assembling an electron beam source details of which are available in vast literature [1-26]. While diode is a simpler configuration, triode is used for an extra degree of freedom. We briefly give a general configuration of a thermionic triode gun [3]. It has three electrodes: one is electron source, called cathode, the other anode (beam extracting element) and third one a beam focusing or controlling element called the Whenelt electrode. Electrons emitted from the cathode due to heating, are attracted by the anode with a kinetic energy equal to the potential difference applied between these two electrodes. Cathode or filament is generally heated with 220 volt AC power supply, until it reaches its emission temperature that varies from metal to metal. A stream of electrons is produced that is accelerated by the positive potential towards anode. A negative electrical potential is applied to the Whenelt cap. If the extraction field is sufficient enough to extract all electrons emitted by the cathode, the gun emission will be limited by the melting temperature of the metal and the emission will be temperature limited. If the extraction field is not high enough then a space charge will be created around the cathode which will limit further emission of electrons. The emission is then said to be 'space charge limited'. Principal parameters on which a gun is based are

286

perveance, working pressure, acceleration voltage, beam current, focusing field, electrode's shape and spacing and minimum spot size [1, 3]. There are axial and transverse guns which basically depend on the mode of the trajectory being followed by the electron beam.

4. Significant features of our e-beam guns

In our novel e-beam assemblies, three different sources namely: Strip source gun, Line source gun and Hairpin source gun have been developed [3]. Details of fabrication, development and results have already been reported [3] and were partially presented these at MTPR-04. Work reported posed extreme demands on the assembly with respect to high temperature, high emission current and homogeneous emission from the cathode surface [3].

In our developments of directly and indirectly heated electron beam guns [3], we used the plane electrodes. However, we did not use the Pierce geometry in the design of the guns. Through our novel designs, we were able to obtain a high emission current with a nicely collimated and parallel beam of electrons. We also did not use any biasing to the focusing electrodes. This reduced the effective area at the cathode surface and decreased the emission current consequently. We also kept the focusing electrode at the same potential as of the cathode to keep the guns in a simple diode structure. Vacuum in these sources was obtained by an oil diffusion pump fitted with a liquid nitrogen trap and a rotary oil pump connected to the backing line. Normal vacuum close to the test assemblies was of the order of 10¡5mbar. The general profile of the current density distribution [3] of these cathodes is shown in Figure 1.

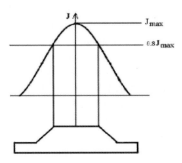

Figure 1: Emission current density variation across the cathode.

The life expectancy of these sources is expected to be more than other sources of the same standard. Mechanical and electronic features of these sources are very good. First phase of our work has been accomplished and significant features of the same are detailed bellow together with future goals.

❖ In our design, a diode configuration has been employed with different beam trajectories and heating modes [3]. Performance of these sources has been studied at temperature and space charge limited modes. The focal spot parameters at the work site of these sources are in good agreement with the designed dimensions ensuring correctness of the design geometry. High emission current, low thermal efficiency and high power density at the work-site have been recorded with an extended life expectancy for several hours. This was possible through novel designing techniques for these sources by protecting them from arcing and shortening even at hard vacuum. These electron beam guns have been utilized for penetration depth up to 5mm and evaporation rates up to the order of 100kg/h for refractory metals, with an acceleration voltage of 10kV. Besides scientific applications, these guns can also be used in large-scale vacuum metallurgy plants for melting, welding, coating and heat treatment [3].

❖ All the three sources are *thermionic* in nature and without any biasing and Pierce geometry for focusing of the beam.

❖ These Guns are very simple in construction, economical to manufacture and have high reliability and performance in operation.

❖ The present work [3] will add to the existing innovative pool of knowledge of the electron beam technology. It will enhance industrial capabilities of the technology especially in the area of welding, evaporation, machining and reprocessing of materials including refractory metals.

Future Directions

More work on this field is being perused and the future directions will include:

➢ Computer simulations of beam generation, will be used for further improvements in electron beam guidance with due consideration of the space charge limited operation and magnetic focusing. This will be accomplished by using the available established software. This will help us increase the beam power with the enhancement in the acceleration voltages particularly for electron beam for use in welding and other similar processes.

➢ The emission current densities will be enhanced further even at very low input heating power by using the pulse power supplies to obtain the electron beam in bunches.

➢ Coating of cathode material with the material of low work function like Cesium will be tried to obtain better emission current.

➢ Development of new (wide band gap) materials like SiC with physical vapor deposition technique by Line Source Electron Gun are on cards.

➢ Through enhancement in the beam power (especially) for the line source gun, attempts will be made to use the improved gun(s) for large scale vacuum metallurgy e.g., material melting and reprocessing, heat treatment of refractory metals as well as the deposition of high temperature metals on the larger substrate surfaces with faster evaporation rates.

5. Applications in Research and Industry

There is a diverse nature of applications of electron beam technology in different dimensions of research and industry. Unique properties, which one can develop by using different configurations, have been one of the strongest driving forces for this multi-dimensional technology [3]. Some of the applications which can be of interest to our audience are listed below:

• Fundamental research on various aspects of particle physics is ongoing through experiments at different accelerators. Higher and higher energies are used to discover new physics. For the past several decades, accelerator technology has improved sufficiently [27, 30]. We have now entered a new era with the commissioning of a Large Hadron Collider (LHC). All of the new electron-positron proposals are for linear accelerators because of the large energy loss due to synchrotron radiation in circular accelerators [1, 2]. Beam generation is the key to the experimental research.

• History of industrial scale electron beam devices picked up from the days of the World War II. One of the important applications of electron beams 'in terms of revenue' is CRTs, X-ray systems and electron microscopes. All these equipment use a focused electron beam [2]. The technology has improved through continuous experimentation and thus found new industrial and scientific uses.

• One of the widely used applications of e-beam technology is in *deep welding* and is in industrial use for several decades. When an electron beam with a high power density is bombarded on a metallic work-piece, the temperature rises and an intense evaporation of material occurs resulting in quick melting and re-solidification of the material along the beam path [17, 18]. On the whole, electron beam welding has many advantages over other welding techniques including high depth-to-width ratio weldments (30:1), high-energy efficiency and low distortion.

• Another innovation namely *Ebrazing* has opened an extra dimension to this field. The combined use of 'electron beam welding' with 'brazing' (Ebrazing), is in use of car industry. Similarly, EB welding has been established as a quality joining tool for spacecraft and nuclear industry [3].

• "Electron beams are widely established as a universal tool for evaporating materials at high rates with freedom from contamination, precise control, excellent economy and high thermal efficiency [3,17,18]". This technology can coat large area substrates (plastic webs, sheets, or metal films), with faster evaporation rates. The process is called 'Electron Beam Physical Vapor Deposition' (EBPVD).

• Another area, in which EBPVD has been successfully established, is called the 'Thermal Barrier Coatings' (TBC). TBC provides thermal protection to the parts of the turbine. It produces a thermal gradient between the hot gas stream and the cooler metal substrate [3,17, 18]. Multilayered thermal barrier coating is recommended for super alloy components in gas turbine engines. Multi-layer coating consists of a bond coat, a 'Thermally Grown Oxide' (TGO) film on top of the bond coat, and a low thermal conductivity top coat [3,17].

- Another application is in *industrial scale melting*: refining and purifying the metals. Two processes (drip and cold hearth melting) are used for this purpose. In drip melting, a metal piece is melted so that it drips in a water cooled ingot mold. It is particularly useful for 'melting and refining' [3,17] of metals having high melting points. In cold hearth melting [3,18], the molten metal is first allowed to flow in a water cooled hearth before it enters a water cooled ingot.

- *Surface modification:* When an electron beam of energy $60kV$ hits the surface of a matter, the kinetic energy of electrons is converted into heat that react with matter to transform their physical properties. The desired changes in the material's properties occur by phase transitions. The beam is impinged upon the target to the desired temperature and then stabilizing this temperature for a calculated time. This improves material properties, like strength, hardness, and corrosion resistance [3,17].

- *Lithography* is the key technology for fabrication of large integrated circuits (ICs), electronic devices, photonics and future nano-engineering at a smaller scale [3,18]. Electron beam lithography has now become a standard tool for creating patterns on semiconductors.

- In electron beam *irradiation*, accelerated electrons deposit energy in matter (gaseous, liquid or solid) that initiate two phenomenon; 'excitation and ionization'. These excited/ionized molecules interact with other molecules of the matter. Properties of the material are changed as a result of chemical reaction that varies from material to material [3, 18].

- Electron beam is also used for treatment and diagnostic purposes [2,3]. The first practical application of electron beams for medical diagnosis is of X-rays. The latest diagnostic technique is Electron Beam Computed Tomography (EBCT) which has emerged out of 'formerly existing Computed Tomography' (CT). Efforts are ongoing to use EBCT as a predictive tool to detect coronary heart disease and is the subject of intense debate in medical journals these days [2,3].

NOTE

In view of the diversity of subject many details which are available in literature and partially cited in references 1-3 have not been given. More details are available in books and recent conference proceedings. We apologize to all those whose scholarly work have either been cited partially or could not be included due to representative selection of the literature. A detailed article, to be published separately on the same subject, will include all such contributions.

References

1. S. I. Molokovsky and A. D. Sushkov, *Intense Electron and Ion Beams*, Springer-Verlag, Germany, (2005); J. Rosenzweig, G. Travish and L. Serafini, *The Physics and Applications of High Brightness Electron Beam*, World Scientific Publishing Co. Pte. Ltd., Singapore, (2003); J. Orloff, *Handbook of charged Particle Optics*, CRC Press, New York, (1997); M. Razor, *Theory & Design of Charged Particle Beams,* John Wily & Sons, Inc., New York, (1994); P. Hawkes, E. Kasper, *Principles of Electron Optics.* New York: Academic Press, (1994); H. Schultz, *Electron Beam Welding,* Woodhead Publishing Ltd., Cambridge, England, (1993); M. E. Perkin *Physics & Experiments with Linear Colliders,* World Scientific Singapore, (1992); S. Humphires, Jr., *Charged Particle Beams*, John Wiley & Sons, (1990); Oxford Paperback Reference, *A Concise Dictionary of Chemistry*, New York, (1990); S. Schiller, U. Heisig, and S. Panzer, *Electron Beam Technology*, John Wiley & Sons, New York, (1982); E. A. Kareh, E. C. Kareh, *Electron Beam Lenses & Optics*, Academic Press, New York, (1970); R. Bakish, *Introduction to Electron Beam Technology*, John Wiley & Sons, New York, (1962); J. H. Fewkes, *Electricity and Magnetism,* Vol, 2, University Tutorial Press, London, (1962); J. R. Pierce, *Theory and*

Design of Electron Beams, Van Nostrand, Princeton, NJ, (1949); A, Eastman, *Fundamentals of Vacuum Tubes*, McGraw-Hill Publishing, New York, (1941).

2. *Proc. of IEEE International Vacuum Electronics Conference*, IVEC/IVESC, Portola Plaza Hotel, Monterey, CA, USA April 25-27,(2006); *Seventh Int. Charged Particle Optics,* Trinity College, Cambridge, 24-28 July, (2006); *e-beam2004, Int. Conf on High Power Electron Beam Technology*, Reno, Nevada, October 18-19, (2004); *Proc. of the 7th Int. Conf. on Electron Beam Technology*, Varna, Bulgaria, (2003); S. W. Shultz, *Proc. of e-beam2002. Int. Conf. on High Power Electron Beam Technology*, USA, (2002), and references therein; *The 9th Symposium on Accelerator Science & Technology*, Tsukuba, Japan, (1993).

3. Munawar Iqbal, *Development and Applications of High power Electron Beam Sources,* Ph. D., Thesis, CHEP, University of the Punjab, Lahore, Pakistan (2007) and references therein; Rev. Sci. Instr., **77** (9) (2006), 1; Vacuum **77**, (2004), 19; Rev. Sci. Instr. **74**(11), (2003), 4616; Rev. Sci. Instr., **74**(3), (2003), 1196; partially reported at MTPR-04.

4. M. E. Read et al., *AIP Conference Proceedings*, **691**, (2003), 107.

5. H. Bluem, *AIP Conference Proceedings,* **680**(1), (2003), 1026.

6. H. Padamsee, in R. Bakish (ed.), *Electron Beam Melting and Refining, State of the Art 2000, Millennium Conference,* Englewood, (2000), 43.

7. A. C. Marshell, *AIP Conference Proceedings*, **504**, (2000), 1319.

8. M. Twada et al., *Proc. of 5th Workshop on JLC,* (1995), 31.

9. D. Schultz et al., SLAC-PUB-6606, in *17th Int. Linear Accelerator Conference (LINAC94)*, Tsukuba, Japan, Agusut 21-26, (1994).

10. T. Tiwari et al., *IETE - Technical-Review*, **9**, (1992), 41.

11. S. Ohsawa, et al., *Proceedings of Linac Conference* Ottawa, ON (Canada), (1992), 91.

12. Jialin Xie et al., *Particle Accelerator Conference* (1991), 2020.

13. M. Akemoto et al., *Proceedings of Linac Conference,* Albuquerque, NM (United States), (1990), 644.

14. O. W. Richardson, Phil. Mag. **28**(5), (1914), 633.

15. W. P. Dyke and F. M. Carbonnier, Adv. Electron Tube Tech, **2** (1962), 199.

16. N. K. Mitra, J. Phys. D: Appl. Phys., **4**, (1971), 39.

17. G. Mattausch et al., Proc. of ebeam2002. Int. Conf. on High Power Electron Beam Technology, USA, **1** (2002) ; P. Seserko et al., Proc. of ebeam2002, Int. Conf. on High Power Electron Beam Technology, USA, **6**, (2002); D. D. Hass, K. Dharmasena, and H. N. G. Wadley, Proc. of ebeam2002, Int. Conf. on High Power Electron Beam Technology, USA, **8**(2002), 1; D. von Dobeneck, Proc. of the 7*th* Int. Conf. on Electron Beam Technology, Varna, Bulgaria, (2003), 185.

18. R. J. Hill, 31st Technical Conference Proceedings, Society of Vacuum Coaters, **29**, (1988), 275; D. von Dobeneck, Proc. of the 7*th* Int. Conf. on Electron Beam Technology, Varna, Bulgaria, (2003), 185; K. Vutova et al., Proc. of the 7*th* Int. Conf. on Electron Beam Technology, Varna, Bulgaria,(2003), 469; E. Koleva and G. Mladenove, Proc. of the 7*th* Int. Conf. on Electron Beam Technology, Varna, Bulgaria,(2003), 514.

19. A. V. Aleksandrov et al., Nucl. Instr. Meth. Phys. Res. **A 340**, (1994), 114.

20. A. V. Crew, Rev. Sci. Instr., **34**(4), (1967), 576.

21. I. Langmuir and K. T Compton. Rev. Mod. Phys, **3**, (1931), 191.

22. L. Martina et al., Rev. Sci. Instr. **73**(7), (2002), 2552.

23. I. Osipov and N. Rempe, Rev. Sci. Instr., **71**(4), (2000), 1638.

24. D. M. Goebel and R. M. Watkins, Rev. Sci. Instr. **71**(2), (2000), 388.

25. E. M. Oks and P. M. Schanin, Phys. Plasmas **5**, (1999), 1649.

26. E. M. Oks, Plasma Sources Sci. Technol. **1**, (1992), 249.

STUDY OF TARGET FRAGMENTATION IN HEAVY ION INTERACTIONS AT 3.7A GeV

A. ABDELSALAM[1], N. RASHED[2], B. M. BADAWY[3], AND E. EL–FALAKY[4]

[1] Physics Department, Faculty of science, Cairo University, Giza, Egypt
[2] Physics Department, Faculty of science, Fayoum University, Fayoum, Egypt
[3] Reactor Physics Department, Nuclear Research Center, Atomic Energy Authority, Egypt
[4] Faculty of Education, Suez Canal University, El–Suez, Egypt
email: he_cairo@yahoo.com

Abstract

We report the experimental measurements of the multiplicity and angular distributions of the target associated particles (grey and black) produced in the interactions of (p, ^{3}He, ^{4}He, ^{6}Li, ^{12}C, ^{22}Ne, and ^{28}Si) with emulsion nuclei at nearly the same incident energy (3.7A GeV). The average values of the emitted grey and black particles increase with increasing target size. They are found to be dependent on the impact parameter of the interaction. The multiplicity and angular distributions reach asymptotic behavior with anisotropy factor $(F/B)_{g,b}$. This factor seems to be independent of the projectile and target masses as well as the impact parameter. The experimental angular distributions are analyzed in the framework of the modified Maxwell – Boltzman distributions. The results yield quite interesting information regarding the mechanism of target fragments production in the backward hemisphere. The $(F/B)_{g,b}$ ratios are good parameters to calculate the temperature of the systems emitting grey and black particles.

Introduction

The use of heavy nuclei as projectiles and targets in relativistic – particle collisions enables scientists to study the hadronic production mechanism.

Therefore, in this work we present some experimental data on the interactions of (p, ^{3}He, ^{4}He, ^{6}Li, ^{12}C, ^{22}Ne, and ^{28}Si) with emulsion nuclei at ~ (3.7A GeV).

At such high energies the target and projectile fragmentation regions are well separated in the rapidity. The physics of the two regions are believed to be similar [1]. We shall, therefore confine ourselves to the study of the target fragmentation region. This region is further classified according to the emulsion nomenclature into two main groups of particles namely the

black (E $\leq$ 26 MeV for protons) and grey particles (26 < E $\leq$ 400 MeV for protons). The fast (grey) and slow (black) target associated particles produced in relativistic heavy – ion reactions are a quantitative probe of the cascading processes in the spectator parts of the target nucleus [2]. These spectators of the reaction are excited primary fragments which then decay into the final fragments by a sequence of evaporation steps [3].

On the other hand, in free nucleon – nucleon collisions, the hadron emission in the backward hemisphere of the interactions (BHS) is kinematically restricted. The study of hadron emission beyond the kinematic limits ($\theta_{Lab} \geq$ 90°) in nucleus – nucleus collisions reveals signatures for a collective mechanism recognizing such emission. Therefore, it was concluded that [4, 5], the backward particle production is a consequence of a decay of a highly excited target system after the forward particle emission. This backward production is mainly dependent on the target.

Experimental Technique

The Dubna energy (a few GeV/A) is a special energy, at which the nuclear limiting fragmentation is applied initially [6]. Hence, the NIKFI–BR2 nuclear emulsion stacks measured in this experiment were irradiated by p, ^{3}He, ^{4}He, ^{6}Li, ^{12}C, ^{22}Ne, and ^{28}Si beams at the Synchrophasotron (JINR), Dubna, Russia. The beams energy is ~ 3.7A GeV. Each emulsion pellicle size is 10 cm x 20 cm x 600 μm.

The chemical composition of this emulsion type and the common emulsion terminology are given in [7].

Starting close to the entrance of the beam into the emulsion (0.5 cm) along–the–track, double scanning; fast in the forward and slow in the backward direction; was carried out. Oil immersion objective lens with magnification 100X was used for scanning the emulsion plates.

From the total scanned lengths of the primary beams, the reaction mean free paths for the used projectiles were displayed and given previously in details [8 – 12].

Depending on ionization, all tracks emitted from the interaction vertices were classified according to the commonly accepted emulsion experiment terminology [13] as:

1) Shower particles – tracks with $g \leq 1.5g_p$. They mainly, consist of pions having energy above 70 MeV and singly charged particles with energy above 400 MeV. They have relative velocity $\beta \geq 0.7$. Their multiplicity is denoted as n_s.

2) Grey particles – tracks with range > 3 mm and $1.5g_p < g < 4.5g_p$; they consist mainly of protons knocked – out from the target nucleus during the collision with kinetic energy ranging from 26 up to 400 MeV. They have a few percent admixture of π mesons. Their multiplicity is denoted as N_g. The multiplicity of grey particles emitted in the forward hemisphere (FHS) ($\theta_{Lab} < 90°$) is denoted as N_g^f. The multiplicity of grey particles emitted in the backward hemisphere (BHS) ($\theta_{Lab} \geq 90°$) is denoted as N_g^b.

3) Black particles – tracks are those having short range in emulsion $\leq$ 3 mm with $g > 4.5g_p$, mostly protons with kinetic energy < 26 MeV. Their multiplicity is denoted as N_b. Their multiplicity is denoted as N_g. The multiplicity of black particles emitted in the forward hemisphere (FHS) ($\theta_{Lab} < 90°$) is denoted as N_b^f. The multiplicity of black particles emitted in the backward hemisphere (BHS) ($\theta_{Lab} \geq 90°$) is denoted as N_b^b.

Here, g is the measured grain density and g_p corresponds to the grain density of a minimum ionizing track. Grey and black tracks amount the group of heavily ionizing tracks $N_h = N_g + N_b$.

4) Fragments of projectile nucleus with $Z \geq 1$, the projectile fragments PF's essentially travel with the same speed as

292

that of the parent beam nucleus, so the energy of the produced PF's is high enough to distinguish them easily from the target fragments. All PF's are emitted in a very narrow forward direction ($\theta_{Lab} \leq 3°$) within an angle given by the Fermi momentum. PF's with Z = 1 and 2 are identified by visual inspection of tracks where, their ionizations are similar to those of shower and grey particles respectively.

Results and Discussion

The number of heavily ionizing particles N_h emitted per event assists in determining the size of the collided nucleus in emulsion. Therefore, it is reliable to use N_h as a sensitive experimental parameter for the production mechanism.

Figures (1 and 2) show the multiplicity correlations for grey and black particles emitted in FHS and BHS as a function of N_h in the interactions of p, ^{3}He, ^{4}He, and ^{6}Li with emulsion nuclei at 3.7A GeV.

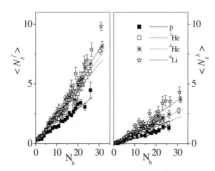

Fig. (1): The dependence of the average grey particle multiplicity $< N_g^f >$ in FHS and those in BHS, $< N_g^b >$ on the target fragments multiplicity parameter N_h through the interactions of p, ^{3}He, ^{4}He, and ^{6}Li with nuclear emulsion at 3.7A GeV, together with the linear fitting (lines).

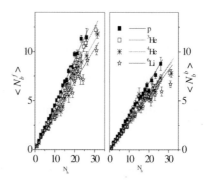

Fig. (2): The dependence of the average black particle multiplicity $< N_b^f >$ in FHS and those in BHS, $< N_b^b >$ on the target fragments multiplicity parameter N_h through the interactions of p, ^{3}He, ^{4}He, and ^{6}Li with nuclear emulsion at 3.7A GeV, together with the linear fitting (lines).

These correlations in Fig. (1 and 2) can be well fitted by positive linear dependences. The values of the slope parameters of the fitting are presented in Fig. (3) against the projectile mass numbers A_p.

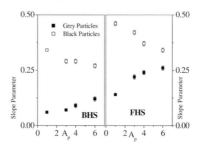

Fig. (3): The slope parameters of the linear relations correlating the average multiplicities of grey and black particles with N_h.

It can be observed that the dependences of black particles on N_h are stronger than grey particles. The slope parameters for grey particles increase with A_p in both hemispheres, while they decrease for black particles with A_p. However, the changes in the BHS with A_p are slow and tend to be constant. Therefore, the number of grey and black particles emitted in

BHS is mainly dependent on the target size and is roughly independent of the projectile size.

Now, Fig. (4) presents the angular distribution of the emitted grey particles from the interactions of ^{12}C with nuclear emulsion at 3.7A GeV (histogram).

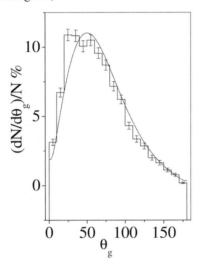

Fig. (4): The angular distribution of the grey particles emitted from the interactions of ^{12}C with emulsion nuclei at 3.7A GeV (histogram), together with the predictions of the modified Maxwell – Boltzman distribution (smooth curve).

On the other hand, Heckman et al [14] had modified Maxwellian distribution for the momentum P, of the emitted fragments in the form,

$$\frac{dN^2}{dPd\mu} \propto \exp\left[-\frac{P^2 - 2M\beta_{\parallel}P\mu}{P_0^2}\right] \quad (1)$$

$\mu = Cos\ \theta$, where θ is the laboratory space angle between the momentum of the fragment of mass M (nucleon mass) and the momentum of the incident projectile. $\beta_{\parallel}$ is normally the longitudinal velocity of the emitting system of particle. $\beta_{\parallel} = Tanh\ y$, where y is the emitting system rapidity. $P_0 = \mu\ \beta_0 = \sqrt{2ME_0}$, where E_0 is the

characteristic energy per particle in the hypothetical moving system and β_0 is the characteristic spectral velocity of the system. Eq. (1), could be modified to give the final form of the angular distribution according to the statistical model [14] as,

$$\frac{dN}{d\theta} \approx Sin\theta \cdot (F/B)^{Cos\theta} \quad (2)$$

(F/B) is the ratio of the forward – to – backward emitted particles.

$$(F/B) \approx e^{\frac{4}{\sqrt{\pi}}\cdot(\beta_{\parallel}/\beta_0)} \quad (3)$$

The mean temperature of the system T could be defined as,

$$T = \frac{M\beta_0^2}{2} \quad (4)$$

Applying the distribution of Eq. (2) on the system emitting grey particles, the smooth curve in Fig. (4) corresponds to such distribution.

From Fig. (4) one can notice that, the experimental histogram agrees with the calculated curve.

Fig. (5) shows the angular distributions of the grey particles emitted from the interactions of ^{12}C with nuclear emulsion at 3.7A GeV, categorized into two different target sizes, ($N_h < 9$, the region of light target nuclei) and ($N_h > 9$, the region of heavy target size). The experimental histograms are fitted by the calculated curves, using Eq. (2). Good agreements between the experimental results and the predictions of the statistical model are seen for the different target sizes. It is also seen that, the angular distributions is independent on the variation of the target size. While, the peaking behavior is observed in the angular distributions at $\theta < 90°$, the decay behavior is observed at $\theta \geq 90°$.

Table (1) displays the $(F/B)_{g,b}$ ratios for the system emitting grey and black particles from the interactions of p, ^{3}He, ^{4}He, ^{6}Li, ^{12}C, ^{22}Ne, and ^{28}Si with nuclear emulsion at ~ 3.7A GeV. Table (1) shows that, the anisotropy factor $(F/B)_g$ has a constant value ~ 3 for all projectiles interactions, while $(F/B)_b$ ~ 1.3. Therefore, the temperature of the emitting system is also independent of the projectile masses. The man temperature of the system emitting grey particles is predicted according to Eq. (4) and found to be of the order of 60 MeV, while that of the system emitting black particles ~ 6 MeV. This temperature has the same order of magnitude of the binding energy / nucleon in nuclei and is also compatible with projectile fragmentation [14, 15]. This result confirms the limiting fragmentation hypothesis [16], and its onset at such incident energy / nucleon.

Table (1): The anisotropy factor $(F/B)_{g,b}$ for the system emitting grey and black particles from different projectiles interactions with nuclear emulsion at 3.7A GeV.

Projectile	$(F/B)_g$	$(F/B)_b$
p	2.92±0.11	1.33±0.03
^{3}He	3.12±0.16	1.45±0.05
^{4}He	2.86±0.19	1.33±0.07
^{6}Li	2.18±0.13	1.30±0.06
^{12}C	3.28±0.50	1.27±0.10
^{22}Ne	3.38±0.50	
^{28}Si	3.51±0.46	

Conclusions

- The average multiplicities of grey and black particles depend strongly on the target size (N_h) either in FHS or in BHS.
- The anisotropy factor (F/B) has constant value for different projectiles, $(F/B)_g$ ~ 3, $(F/B)_b$ ~ 1.3.
- The angular distributions of grey particles show a good agreement with the predictions of the statistical model parameterized on the basis of Maxwell – Boltzman statistics. These predictions are factorized in terms of the anisotropy factor as, $\frac{dN}{d\theta} = Sin\theta(F/B)^{Cos\theta}$.

The temperature of the system emitting grey particles is predicted in the light of the proposed statistical model to be T ~ 60 MeV (hot), while that of black particles is T ~ 6 MeV (cold). The same angular behavior is observed for different regions of impact parameters. The angular distributions show peaking behavior at θ_g < 90° (FHS) while in BHS ($\theta_g \geq 90°$) the distributions show decay shape. This implies that the emission in BHS will occur at a letter

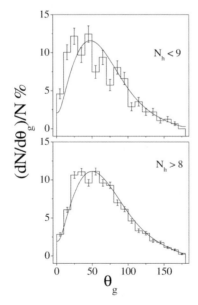

Fig. (5): The angular distributions of grey particles emitted from ^{12}C interactions with nuclear emulsion at 3.7A GeV depending on the two different regions of target size (histograms), together with the predictions of the statistical model (smooth curves).

stage of a decay system after the emission of the forward particles.

– The angular distribution of the system emitting grey particles is independent of the target size.

Acknowledgement

The authors thank the authorities and the staff of JINR, Dubna, Russia for their cooperation in supplying us the photographic emulsion plates used in this experiment.

References

[1] A. Abdelsalam, Physica Scripta **47**, 505 (1993).

[2] M. El–Nadi, A. Abdelsalam, M. M. Sherif, N. Ali–Moussa, and M. S. El–Nagdy, IL NUOVO CIMENTO **107**, 31 (1994).

[3] K. Sümmerer, W. Brüchle, D. J. Morrissey, M. Schädel, B. Szweryn, and Yang Weifan, Phys. Rev. C **42**, 2546 (1990).

[4] A. Abdelsalam, E. A. Shaat, N. Ali–Mossa, Z. Abou–Mousa, O. M. Osman, N. Rashed, W. Osman, B. M. Badawy, and E. El–Falaky, J. Phys. G **28**, 1375 (2002).

[5] A. Abdelsalam, B. M. Badawy, and E. El–Falaky, Can. J. Phys., **85**, 837 (2007).

[6] Fu–Hu Liu, Chinese Journal of Physics **40**, 159 (2002).

[7] C. F. Powell, F. H. Fowler and D. H. Perkins, The Study of Elementary Particles by the Photographic Method, Pergamon Press. London, New York, Paris, Los Angles, 474 (1958).

[8] A. Abdelsalam, M. S. El–Nagdy, N. Rashed, and B. M. Badawy, Chinese Journal of Physics **45**, 351 (2007).

[9] A. Abdelsalam, M. S. El–Nagdy, E. A. Shaat, N. Ali–Mossa, O. M. Osman, Z. Abou–Moussa, S. Kamel, N. Rashed, W. Osman, M. E. Hafiz, B. M. Badawy, and S. Magd Eldin, FIZIKA B **15**, 9 (2006).

[10] M. El–Nadi, A. Abdelsalam, N. Ali–Mossa, Z. Abou–Moussa, S. Kamel, Kh. Abdel–Waged, W. Osman, and B. M. Badawy, Eur. Phys. J. A **3**, 183 (1998).

[11] M. El–Nadi, A. Abdelsalam, N. Ali–Mossa, Z. Abou–Moussa, Kh. Abdel–Waged, W. Osman, and B. M. Badawy, IL NUOVO CIMENTO A **111**, 1243 (1998).

[12] M. El–Nadi, A. Abdelsalam, and N. Ali–Moussa, IL NUOVO CIMENTO A **110**, 1255 (1998).

[13] H. Barkas, Nuclear Research Emulsion, Vol. **I**, Technique and Theory Academic Press Inc., (1963).

[14] H. H. Heckman, H. J. Crawford, D. E. Greiner, P. J. Lindstrom, and Lance W. Wilson, Phys. Rev. C **17**, 1651 (1978).

[15] J. V. Geaga, S. A. Chessin, J. Y. Grossiord, J. W. Harris, D. L. Hendrie, L. S. Schroeder, R. N. Treuhaft, and K. Van Bibber, Phys. Rev. Lett., **45**, 1993 (1980).

[16] J. Benecke, T. T. Chov, C. N. Yang, and E. Yen, Phys. Rev. **188**, 2159 (1969).

SOLAR INDUCED CLIMATE CHANGES AND COOLING OF THE EARTH

SHAHINAZ M. YOUSEF

Astronomy & Meteorology Dept.,
Faculty of Science, Cairo University, Cairo, Egypt
e-mail: habibat_arrahman@yahoo.com

Abstract. Evidences are given for the cooling effect induced by solar weak cycles. It is forecasted that the coming solar cycle number 24, which has started on January 2008, would be very weak. This cycle would be followed by several weak cycles. Its very start on January 2008 have induced a climate change that forced global cooling, Indeed all global temperature monitors have shown temperature drops. The GISS monitor showed a 0.75°C drop between January 2007 and January 2008. This sharp temperature drop characterizes cooling induced by weak cycles as was evident by historical temperature records. It also happened in the right exact timing of the start of cycle 24. This cooling is real and could last for some time. The cooling well width is location dependant. Last January cooling left many countries in deep freeze. Cooling is very serious and can destroy crops and cause famines.

This cooling is instrumentally recorded. This is an appeal to scientists to consider the present cooling seriously, after all the truth ought to be followed. Alert is also given to the responsible authorities to work promptly to choose the proper crops that can tolerate the cold otherwise it would be a disaster worldwide.

INTRODUCTION

The Earth's weather machine is an exquisitely complex affair, in which many processes are simultaneously at work. Clearly however the weather system is principally driven by the sun (Roberts 1976).

Solar radiation is an unfailing source of energy for the Earth. Without visible and infrared (IR) radiation from the sun, Earth's surface temperature would be too cold to support life. Nor would there be energy to fuel photosynthesis or power the circulation of the lower atmosphere and oceans that profoundly influence living organisms. Lack in solar ultraviolet (UV) radiative inputs, Earth's middle atmosphere would be devoid of ozone and its upper atmosphere cold and unionized. Living things would be exposed to damaging high –energy solar photons (Lean 1997).

Solar induced climate fluctuations are crucial to life in earth. When weak solar cycles are predominant on the sun, cooling occurs on the earth.

In the 1970s, the CIA encouraged research in climatic fluctuations because of its implication on food production and political instabilities. It was found that optimum climatic conditions between 1930-60 was ideal for green revolution in world food production and that there is a likelihood to return to the 19th century climate characterized by marked drop in global temperature and wide spread drought-flood hazards that led to many revolutions and political instabilities. It is estimated that a one degree drop in European temperature will lead to a drop of the number of people sustained by one hectare from three individuals to two and to the possibilities that one fifth of the world population will be subjected to hunger(extracted from Abu El Ezz (1980) short translation of CIA book "The Weather Conspiracy: The Coming of the New Ice Age").

The present solar cycle number 24 is the first of a weak cycle series on the sun. This eventually will lead to drops in the solar energy budget received by the earth including drops in solar irradiance, UV radiation which plays a vital role in photosynthesis and IR radiation.

Biological effects will happen on the earth. It is found that fish population is very much controlled by the Solar (80-120 year) Wolf-Gleissberg cycles (Yousef 2006a).

Solar induced climate changes can be attributed to the drop of solar activity level and the appearance of a weak solar cycles causing acceleration of spinning of the sun in the photosphere.

It is the purpose of the present paper to study the effect of weak solar cycles on cooling using primarily the Central England Temperature as it extends back to 1659 given by the Hadley Center in England. Other data set used is the Greenland temperature records (Vinther et al. 2006).

Temperature reflection into the future is also given using forecast of weak solar cycles.

WEAK SOLAR CYCLES

The magnetic solar cycles are variable in strength and durations. Figure 1 illustrates such variability since 1700. Strong and moderate solar cycles have duration 10 and 11 year, whilst the weak solar cycles mostly have 12 years duration Nesme-Ribes et al (1994)

The Wolf-Gleissberg solar cycle (WGC) is the moderate range cycle (80-120 year cycle) which stands up on smoothing the sunspot cycles shown in Fig 1.

The main purpose of this paper is to understand temperature variations in the light of solar stimulation (Yousef 2004a). Fig. 2 gives the monthly mean for January for central England.

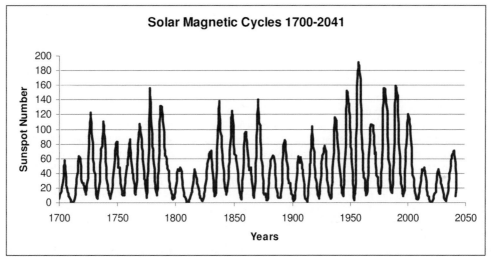

FIGURE 1. Time series of sunspot number illustrating the existence of weak solar cycle's series around 1700s, 1800s, 1900s and 2000s.The prediction of cycles 24, 25 and 26 is given at the extreme right of Fig. 1 as a repetition of cycles 4, 5 and 6 around 1800s.

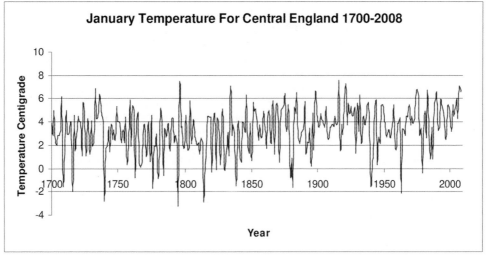

FIGURE 2. Mean monthly temperature time series for Central England. Notice the 13 cooling wells that dropped below zero degree.

HOW ARE THE WEAK CYCLES GENERATED BY THE SOLAR DYNAMO?

Tachocline and Variability of Solar Spin Rate

The solar differential rotation persists to the bottom of the connective zone. The rotation velocity becomes uniform from pole to pole nearly 1/3 of the way to the core Lower down the rotation remains independent of latitude acting as if the Sun were a solid body. Shearing motion along this interface may be the dynamo source of magnetism. The top of the radiative zone which rotates at one speed meet the overlying convective zone which spins faster in its equatorial middle. The roughly 20 million meter wide layer where these very different zones meet and shear against one another, called the tachocline, is the likely site of the solar dynamo, the source of sun's magnetism. The hot circulating gases, which are good conductors of electricity, generate electrical currents that create magnetic fields, these fields in turn sustain the generation of electricity. The dynamo amplifies and regenerates the sun's magnetic field. The strong fields eventually rise through the connective zone and emerge at the photosphere (Lang 2001 and references therein)

There is a temporal variation in the rotation of the Sun in the base of the connective zone and close to the probable site of the solar dynamo in equatorial regions, the rotation speeds up, slows down and speeds up again with a period of 1.3 year. There are observed differences of the rotation rate from the mean at 0.72 R and 0.63 R. When the lower gas speeds up, the upper gas slows down and vise versa. The anticipated 11-year signal is not observed. However, the amplitude of those variations seems to decrease at the solar maximum. It is intriguing that similar 1.3-year are observed in the surface sunspot data (Kosovichev 2003 after his references).

If we tie up those variations of rotation speed near the tachocline with those observed at the photosphere, it is thus evident that faster rotations of weak solar cycles means slower rotation at the tachocline, i.e. of the solar dynamo and this implies the decline of solar magnetism. Vice versa, for stronger more active solar cycles the equatorial photosphere is observed to rotate slower, implying that the solar dynamo spins faster and that the hot circulating ionized gases generate stronger electrical currents and thus create stronger magnetic fields.

The solar spin rate is generally faster during the start of weak cycle series as evident from Fig 3a for the weak cycles earlier than 1900 and for cycle 20 in the 1960s (Hoyt and Schatten1997after their reference). This faster rotation rate leads to a reduction in solar energy budget emitted from the sun in all wavelengths as well as solar wind flux, density, velocity and interplanetary magnetic field.

Solar induced climate changes do occur at the turning points of those WGCs whenever there is a change in the solar spin rate. Solar induced climate changes do occur at the start, end of those weak cycles and following the WGC maximums (Yousef 2006a). In addition, solar induced climate changes can occur within the weak cycle's series following the end of 22-year Hale magnetic cycle.

SOLAR SPIN RATE, SOLAR WIND, INTER-PLANETARY MAGNETIC FIELD AND TEMPERATURE ON EARTH

Svalgaard et al. (2003) have succeeded in working out the near-earth interplanetary magnetic field since 1600. They have also reconstructed solar wind speed since 1880 (Fig. 3c). The number of 3-hour intervals per year where the aa-index was at its lowest possible values as shown in Fig. (3b). We study the IMF during the previous weak solar cycles series (cycles 5-7) and (12-14) around 1800 and 1900. In response to the drop of the level activity, the magnitude of IMF and the solar wind speed dropped. The IMF for cycles 5 and 6 is in the range of 4.5-6.5 nT compared to 8.5 nT for cycle 19 which is the maximum of the Wolf-Gleissberg cycle.

Fig. 3 shows that the years 1900-1902 seem to be in a class of their own with very low solar wind speed (275-300 km/s). In order to understand what happened in 1900-1902 let us put the pieces together, compare Figs. (3 a, b, c and d) starting from the sun down to earth

a)   At those years solar equatorial rotation rate abruptly slowed down (Fig. 3a), i.e. the Tachocline spins faster than the convection zone leading to stronger magnetic cycles.
b)   The solar wind speeds up following its drop (Fig. 3c).
c)   Before those years, it was exceptionally quiet with more than half the time there was no measurable geomagnetic activity at all (Fig. 3b).
d)   There was a sudden rise in the temperature of the small town Soria in Spain Fig. 3d (Donaire 2000).

The 1900 year slowing down of the equatorial rotation means that the Tachocline rotation speeds up. It is at those years that a solar induced climate change happened ending the previous one at 1878. This 1878 earlier solar induced climate change occurred with the start of the weak cycle number 12 leading to about 5 degrees sudden drop of little town Soria in Spain. Several examples of temperature drops in southern Mediterranean Stations are found in Aesawy and Hasanean (1998). In between 1878 and 1900 the IMF dropped down and it was a geomagnetic quiet period. The sun was spinning much faster for a period of 22 years Hale magnetic cycle. Less irradiance reached the earth in all wave lengths of spectrum and this lead to a global cooling. So what counts here is the spin of the sun.

EFFECT OF WEAK SOLAR CYCLES ON COOLING OF THE EARTH

As seen in Fig. 1, there are three intervals of weak solar cycle series namely
a) cycles 4, 5 and 6 (1797-1823)
b) cycles 12, 13, 14,15 and possibly 16 (1877-1933)
c) cycles 23, 24, 25 and perhaps 26 (1997-2032)

In addition there are one or two weak cycles following the maximum of each the Wolf-Gleissberg cycle namely
a) 1744-1766
b) cycle 10 (1844-1856)
c) cycle 20 following the most active cycle number 19 (1965-1976)
The temperature variation will be studied in terms of solar activity.

I-Abrupt Temperature Anomalies Around 1800

The weak solar cycles 4, 5 and 6 occurred during the period (1797-1823). Figure 4 shows the weak solar cycles around 1800 and their forcing on January temperature for Central England. Cycle 4 started on 1797. Just prior to its start in 1793 the temperature was 2.8°C but dropped by 5.9°C to -3.1°C abruptly but suddenly rose by 10.2°C to 7.3°C on January the following year. This abrupt change indicates a solar induced climate change.

The following abrupt temperature anomaly started in 1812 when the temperature was 2.6°C then dropped within a couple of years by 5.5°C to -2.9°C but rose by 7.4°C in 1817 to 4.5°C. The end of this induced climate change occurred around the maximum of cycle 7 by another abrupt anomaly as the temperature dropped from 7.1°C in 1834 to -1.5°C in 1838 and then rose to 4.1°C in 1840. Thirty eight years have elapsed between the two warmest January months while 43 years have passed between the first and last temperature drops. Details of these temperature anomalies are given in Table 1.

Several biannual temperature oscillations occurred in-between the second and last temperature anomalies stimulated by the weak solar cycles. Similar observations of biannual oscillations are reported by Burroughs (1992) for the winter temperature in the central United States for 11 years in the 1870s and 1880s (Yousef 2004b).

The temperature records of Greenland for 13 stations starting 1784 indicate that the 1810s (1811-1820) with an annual temperature of -4.4°C is the coldest decade in record (Vinther et al. (2006),

Table 1: Abrupt Temperature Anomalies

Year	Sunspot	January Temp	Year	Sunspot Number	January Temp	Year	Sunspot Number	January Temp
1793	46.9	**2.8**	1811	1.4	1.2	1834	13.2	<u>**7.1**</u>
1794	41	1.8	1812	5 Start	**2.6**	1835	56.9	2.9
1795	21.3	**-3.1**	1813	12.2	1.9	1836	121.5	3.7
1796	16	<u>**7.3**</u>	1814	13.9	**-2.9**	1837	138.3 max	2.7
			1815	35.4	0.3	1838	103.2	**-1.5**
			1816	45.8	2.7	1839	85.7	2.8
			1817	41.1	**4.5**	1840	64.6	**4.1**
			1818	30.1	4.4			
			1819	23.9	4.4			
			1820	15.6	-0.3			

300

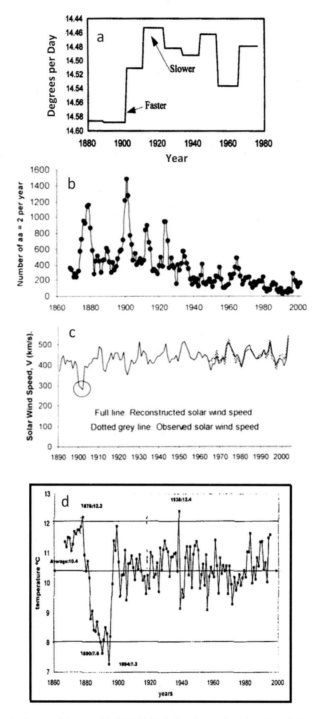

FIGURE 3. The variation of spin rate of the sun (a) 1880-1900 during the weak cycles 12 and 13 as stimuli to the consequences of interplanetary and terrestrial events. (b) the number of magnetically quiet days show sharp decreases between 1878 and 1900 at the start and end of 22 year Hale magnetic cycle. (c) drop of solar wind speed prior 1900-1902 to 275-300 km/s, Svalgaard and Le Sager (2003) d) Sudden 5 degrees drop in the temperature of Soria (a small town in Spain in 1878 followed by 5°C abrupt rise in 1900 showing end of 1878 solar induced climate change (Donaire 2000).

301

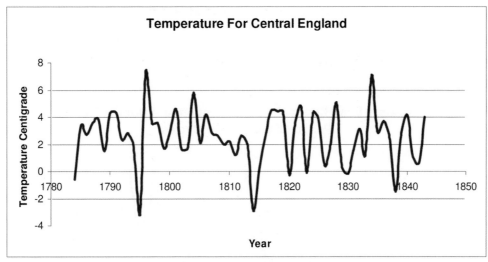

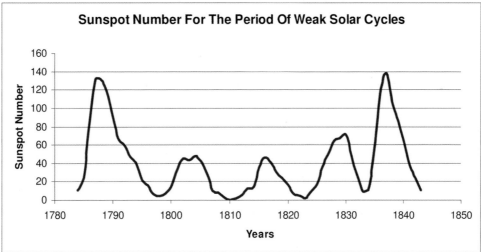

FIGURE 4. Temperature anomalies during the period of weak cycles around 1800. Notice the narrow cooling wells and the biannual oscillations of temperature between about 1819 and 1831.

2-Abrupt Temperature Anomalies Around 1900

Another set of weak solar cycles occurred around 1900s. Fig. 5 shows the response of the January average anomalies of the temperature of Central England to the weak solar cycle stimuli. The anomalies started with the decline of cycle 11 on the extreme left of Fig. 5 - bottom.

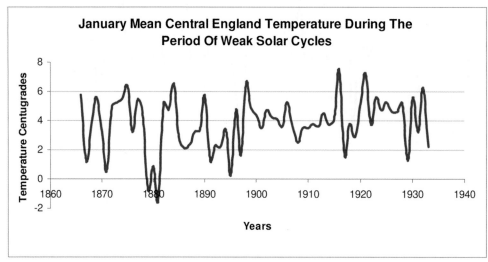

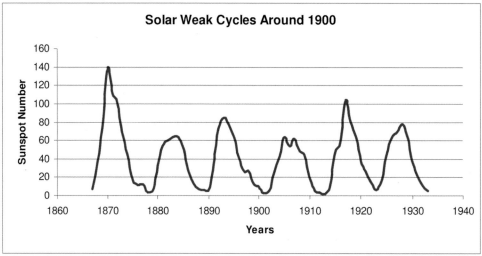

FIGURE 5. Solar forcing of the weak solar cycles around 1900 on the January temperature for Central England.

The cooling well in response to the weak cycle 12 which started in 1878 is shown in Fig. 6 for Central England. Exactly in 1878 the temperature dropped. The average January temperature dropped from 4.6°C on 1878 to -0.7°C the following year and to -1.5°C in 1881 and then rose to 5.2°C in 1882 and 6.5°C in 1884.

Fig. 7 shows the cooling wells for Edinburgh, Wakefield, and Greenwich in Great Britain shown with Wolf sunspot numbers as well as the cooling well for the small town of Soria in Spain. Both of the cooling wells started promptly with the start of cycle 12 in 1878. However, in Spain the well was closed in 1902, while in Britain's case it lasted till around 1940.

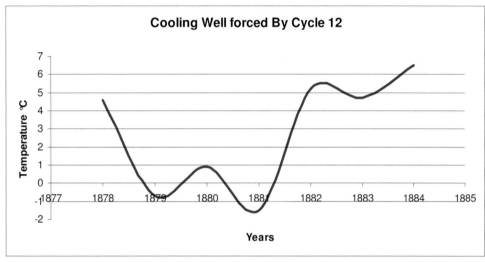

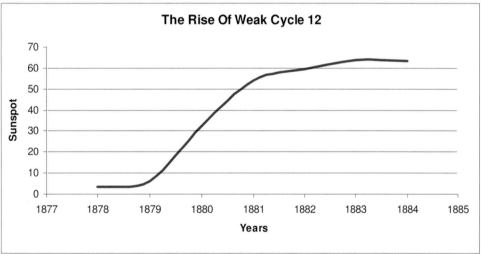

FIGURE 6. The s cooling well in response to cycle 12 for the January temperature for Central England.

EFFECT OF WEAK CYCLE NUMBER 20 FOLLOWING THE MAXIMUM OF WOLF-GLEISSBERG CYCLES

Cycle number 19 is the most active solar cycle since records of sunspot number. It was followed by the weak cycle number 20 which started in 1964 as seen in Fig. 8. By the end of cycle 19 a serious induced climate change happened (Yousef 2000). The January temperature for Central England dropped from 4.3°C on 1962 to -2.1°C in 1963 (a 6.4°C change) and then rose to 3.4°C on 1964. This drop was contemporary with sudden increase in the equatorial spin rate of the sun as seen in Fig 3 a. The decline of temperate zone westerlies and increased frequency of blocking in high latitudes have been associated with anomalies (or changes) of temperature and rainfall regime that are having serious effects in many parts of the world (Lamb (1966).

304

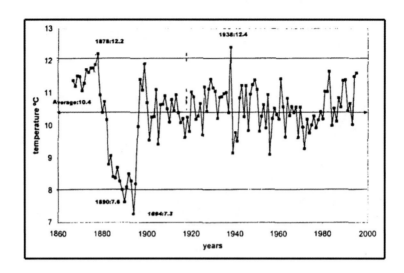

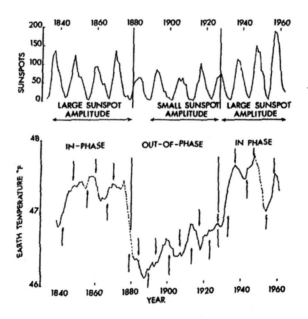

FIGURE 7. Cooling wells in response to the weak solar cycles around 1900. Above, is the cooling well for the town of Soria in Spain (Donaire 2000). Below, is the cooling well of the air surface temperatures for Edinburgh, Wakefield, and Greenwich in Great Britain shown with Wolf sunspot Numbers. Temperature appears to be out of phase with solar activity from 1880 to 1930, but in phase for other years. (Adapted from Hoyt and Schatten 1997 and references therein.)

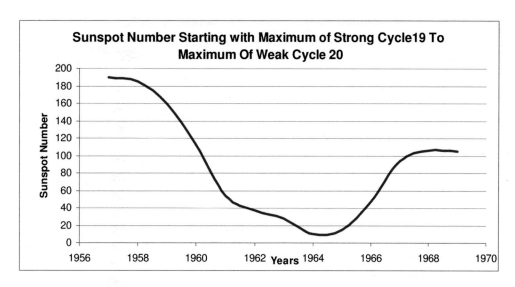

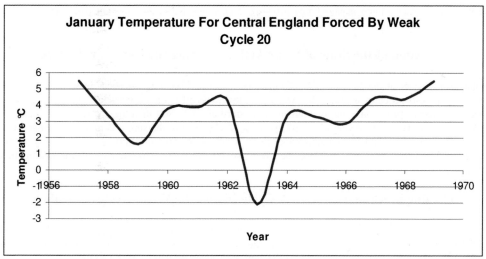

FIGURE 8. The cooling well due to the decline of cycle 19 and rise of the weak cycle 20.

THE PRESENT COOLING OF THE EARTH IN RESPONSE TO WEAK CYCLE 24

Contrary to the expectations of Global warming, a sudden cooling of the earth occurred between January 2007 and January 2008. Such temperature drops were recorded by the four international centers monitors namely, GISS, HadCRUT, UAH MSU and RSS MSUh. The GISS showed a temperature drop of 0.75°C within one year, big enough to wipe out global warming for decades Fig. 9.

306

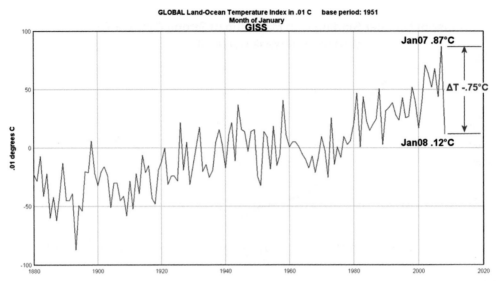

FIGURE 9. GISS global temperature monitor for the period 1880 till present. Notice the present drop of temperature.

Prediction of the State of Solar Activity During the Next Few Decades

Weak solar cycles occur at the bottom of Wolf-Gleissberg cycles. They tend to occur in series of 3-4 cycles. A single weak cycle also occur in between the two maximums of Wolf-Gleissberg cycle. Since the last weak solar cycles occurred around 1900 while the previous ones occurred around 1800 then the newly started cycle 24 should be a weak solar cycle. However, owing to the 200-years de Verie cycle of the sun, it is more likely that the status of the coming solar activity would be something like those weak cycles around 1800 as shown in Fig 1. Svalgaard (2005) also predicted that cycle 24 would be the lowest so far in the past 100 years with the maximum sunspot number around 75.

Cooling in Response to Weak Cycle 24 and the Weak Cycles to Follow

As it was shown earlier, weak cycles do induce cooling on the earth. The harsh winter of 2007-2008 and the deep freeze that had resulted in the worst snowfalls for decades swept across the globe – almost everywhere across much of the northern hemisphere, from Greece and Iran to China and Japan, Afghanistan, Kashmir.

In Greece and Turkey, the temperatures dropped to as low as minus 31 degrees Celsius. Even Saudi Arabia had snow fall. Following last year's freak snowfalls in such southern cities as Buenos Aires and Sydney, satellite observations from the other end of the world have this winter shown ice cover round the Antarctic at easily its greatest extent for this time of year since data began in 1979, 30 per cent above average (Booker 2008).

This cooling is real and as was expected earlier by the author it happened by the end of cycle 23 and start of cycle 24. On the 6th of January 2008, the first sunspot **of cycle 24** has been observed and the NOAA Confirms Start of New Sunspot Cycle.

Will the global cooling last? And for how long?

There are two points of view in my opinion; either the cooling well be as broad as it was around 1900 or narrow cooling wells followed at the end of the last weak cycle by biannual oscillations. Perhaps the second opinion is more likely. In addition to global cooling of the air, the sea surface temperature of all oceans cools in response to weak solar cycles. This phenomenon was observed around 1900 and is expected to happen in response to the coming era of weak solar cycles (Yousef 2006b). This will initiate several La Nina events and induce several drought-flood hazards.

FIGURE 10. Frozen water falls in China

CONCLUSION

The observed concept of the sudden cooling of the earth between January 2007 and January 2008 may be a surprise for researchers working on global Warming. However according to my expectation, sudden cooling would arise with the start of cycle 24.

People may argue that one year drop is not to be trusted, but as we have seen solar induced cooling is abrupt.

Solar induced cooling due to weak cycles is already in the way. How long it would last is location dependant. It is most likely that what is to be expected is something like what happened around 1800 rather than the cooling that occurred around 1900 owing to the 200 years solar cycle.

My sincere advice to meteorologists is to study previous cooling wells in different locations in their own countries as the cooling wells vary much from place to place.

Cooling is harmful to crops, to the wheat cultivating zone etc. One evidence for this statement is the spoil of the Irish potatoes crop in 1878, the year of sudden drop of temperature, this cooling caused agriculture disaster led to the flee of a lot of the Irish people to America.

The implications of weak cycles on the earth are numerous, they include, the sudden rises and falls of lakes and closed seas (Yousef 2004b), fisheries, North Atlantic Oscillation etc. (Yousef 2006a).

One important point, is that the cycle just preceding the low sunspot cycle number cycles which are of moderate sunspot maximum are peculiar. Those cycles are numbers 3, 11 and 23.

Solar forcing of cycle 23 for example caused lake Victoria to rise by 1.3m.

On analyzing Greenland temperature data Vinther et al. (2006) found that 1811-1820 was the coldest decade in data set and incorrectly attributed this coldness to 1809 "unidentified" volcanic eruption and the eruption of Tambora in 1815 – unusual geologic events that defined the climate. However, this coldness surely was induced by weak solar cycles forcing of the earth rather than an unidentified volcano.

Investigators are encouraged to apply computer models to study the effect of cooling in different countries on the yield of agriculture and work out the suitable crops to be cultivated and perhaps cultivation of the present crops ought to be shifted to lower warmer latitudes.

The Egyptian government is urged to put forward a revised strategy taking into account the expected cooling effect. This is a call for the Egyptian authorities to provide historical temperature records for researchers for the benefit of the economy of the country.

Since, it is more likely that the cooling well due to the weak cycle 24 and its series of weak cycles would be some how similar to that around 1800, it is thus recommended to find reconstructions of temperature from proxy data like tree ring widths.

ACKNOWLEDGEMENTS

My gratitude is due to my mother Mrs Ikram El Attar and to My Father Mr. Moustafa Ali Yousef. Sincere thanks are also due to Prof. Lotfia EL Nadi.

REFERENCES

1. Abu El Ezz, M. S., World Climatic Fluctuations, Kuwait. In Arabic, 1980.
2. Aesawy, A. M. and Hasanean, H. M., Theor. Appl. Climatol. 61, 55-68 (1998).
3. Booker, C., Global Cooling: Amazing Pictures of Countries joining Britain in the Big Freeze. Daily Mail, 21 February (2008). http:// dailymail.com.uk
4. Burroughs, W. J., *Weather Cycles, Real or Imaginary*, Second Edition, Cambridge University Press, 2003, p. 9.
5. Donaire, J. J. S., Bulletin of the Egyptian Geographical Society, 73, 127-144 (2000).
6. Hoyt , D.V. and Schatten, K., *The Role of the Sun in Climate Change*, Oxford University Press, 1997, p. 193.
7. Kosovichev, A. G., What Helioseismology Teaches Us about the Sun, Proceedings of ISCS Solar Variability as an Input to the Earth's Environment, 23-28 June 2003, Tatranska Lomnica, Slovak Republic, Europe Space Agency. ESA. Edited by A. Wilson 795-806 (2003).
8. Lamb, H. H., Climate of the 1960s, changes in world wind circulation reflected in prevailing temperatures, rainfall patterns and the level of African lakes. The Geographical Journal, 132 (1994).
9. Lang, K. R., *The Cambridge Encyclopedia of the Sun*, Cambridge University Press, 2001.
10. Lean, J. The Sun's Variable Radiation and Its Relevance for Earth, Annual. Rev. Astron. Astrophys. 35.33-67 (1997).
11. Nesme-Ribes E., Sokoloff D., Ribes, J. C. and Kremliovsky, M., The Maunder minimum and the solar dynamo. NATO ASI series 25. The solar engine and its influence on terrestrial atmosphere and climate. Edited by Nesme-Ribes, 527 (1994).
12. Richardson, J. D., Paularena, K., Belcher, J., Solar wind oscillations with a 1. 3 year period ,Geophysical Research Letters 0094-8276, (1994).
13. Roberts, W. O., Science *Technology and the Modern Navy. The Thirtieth Anniversary 1946-1976.* Department of Navy, Office of Naval research, Arlington, Virginia, 371-368 (1976).
14. Svalgaard, L., Cliver, E. W., LeSager, P., Determination of Interplanetary Magnetic Field Strength, Solar Wind Speed and EUV Irradiance, 1890-Present, Proceedings of ISCS, European Space Agency ESA. SP-535. Solar variability as an input to the Earth's environment, 23-28 June , Tatranska Lomnica, Slovak Republic, pp15-24 (2003).
15. Svalgaard, L., Cliver, E.W., Kamide, Y., Sunspot cycle 24; smallest cycle in 100 years. Geophysical Res. Lett. Vol 32, L01104 (2005).
16. Yousef, S. M., Proceedings of the Third SOLTIP Symposium, Oct 1996. Beijing, 569-575 (1998).
17. Yousef, S.M., "The Solar Wolf-Gleissberg Cycle and Its Influence on the Earth" in the Proceedings of the International Conference on the Environmental Hazards Mitigation, Cairo, Sept. 2000. Conferences of Virtual Academia, www.virtuacademia.com/conferences
18. Yousef, S, M. Expected cooling of the earth and its implications on food security., Bulletn De L'Insttut D'Egypte, Tom LXXX. 53-82 (2004a).
19. Yousef, S, M. Deciphering the Fall and Rise of the Dead Sea in Relation to Solar Forcing. In Proceedings of Modern Trends in Physics Research. Lotfia EL Nadi edt., American Instittute of Physics, 441-449 (2004b).
20. Yousef, S, M., 80-120-year Long Term Solar Induced Effects on the Earth, Past and Predictions, Physics and Chemistry of the Earth. 31 pp. 113-122 (2006a). Elsevier, available online www.sciencedirect.com
21. Yousef, S, M., El Nino and La Nina During 2200-1200 BC and 622-1467 AD as Deduced from Nile Records as Indicators of Past and Future Solar Induced Climate Changes, Proceedings of 8th International Conference on the Geology of the Arab World, GAW8 Vol. 1 pp 383-388 (2006b).
22. Vinther, B. M., Andersen, K. K., Jones, P.D., Briffa, K. R., Cappelen, J., Extending Greenland temperature records into the late eighteenth century. Journal of Geophysical Research, 111, 10.1029/2005JD006810 (2006).

III-3 CONTRIBUTING PAPERS

III.3 CONTRIBUTING PAPERS

CROSS-SECTIONS CALCULATIONS FOR PRODUCING PAIR HIGGS BOSONS AND NEUTRALINO VIA THREE NEUTRAL HIGGS BOSON PROPAGATORS

M. M. AHMED

Physics Department, Helwan University, Cairo, Egypt
mkader4@yahoo.com

Abstract: The production cross-sections calculations for the process $e^- + e^+ \to H_k^o \to H_i^+ + H_j^- + \tilde{\chi}_l^o$. Where H_k^o ($h^o.H^o, A^o$) represent the three neutral Higgs bosons, are calculated through all the allowed modes of interactions. Which are (1296) situations are consider in two different groups of Feynman diagrams are taken into account.
(1) Production of $H_i^+ + H_j^- + \tilde{\chi}_l^o$ from H_k^o and Z^o propagators exchange (from 1-864 Feynman diagrams).
(2) Production of $H_i^+ + H_j^- + \tilde{\chi}_l^o$ from H_k^o propagators and the neutralino is leg (from 865-1296 Feynman diagrams).
where ($i, j = 1,2,3$, and $\ell = 1,2,3,4$)
The values of the cross sections σ are taken as a function of the incident center of mass energy S. The values of the cross-sections have greatest values at S ranges from 3000 to 3400 GeV., and the against cross section is from rang 2.5 x 10^{-10}- 4 x 10^{-10}, and the best calculations have numbers from 865-1080, where the production mechanisms can be detected in mode $H_k^o(p_2 + p_4) \to H_i^+(P_2)H_j^-(P_4)$ when the neutralino is leg from positron.

Keywords: Higgs bosons, neutralinos, MSSM model.

1. Introduction:

In two-Higgs-doublet models of electroweak symmetry breaking, such as the minimal supersymmetric extension of the standard model (MSSM), there are five physical Higgs bosons: two neutral CP-even scalars, h^o, H^o and A^o being the heavier one; a neutral CP-odd state A; and two charged states H^+ and H^-. The coupling of the A boson to the down quarks, such as the b quark, is enhanced by a factor of $\tan \beta$ compared to the Standard Model (SM) one, where $\tan \beta$ is defend as the ratio of the vacuum expectation values of the two Higgs doublets. At high $\tan \beta$, this is also true for either h^o or H^o.

In particle physics, the neutralino is a hypothetical particle, part of the doubling of the in supersymetry models, all Standard Model particles have patner particles with the same quantum numbers but spin differing by 1/2. Since the superpartners of the Z boson (zino), the (photino) and the neutral Higgs (Higgion) have the same quntum numbers, they can mix to from four eigenststes of the mass operator called "neutralions". In many models the lightest of the four neutralinos turns out to be the lightest supersymetric particle [1] (LSP), though other particles may also take on this role. The standard symbol for neutralinos is $\tilde{\chi}_l^o$, where l runs from 1 to 4.

CP 998, Modern Trends in Physics Research
Third International Conference MTPR-08
edited by L. El Nadi
Copyright @ 2011 by World Scientific Publishing Co. 978-981-4317-50-4 / 981-4317-50-0

We analyze cross sections in the regions of the MSSM parameter space for the process $e^- + e^+ \rightarrow H_k^o \rightarrow H_i^+ + H_j^- + \tilde{\chi}_l^o$. as a function of center of mass energy S, all the possible processes has been calculated for this reaction [3] Where $(i, j = 1, 2, 3,$ $l = 1,2,3,4,$ and $k = 1, 2, 3$ for h^o, H^o, A^o respectively) [2,4] The neutralino sector depends on four parameters: the U (1) and SU(2) gaugino masses m_1, and m_2, the higgsino mass parameter μ, and the ratio of the vacuum expectation values (vev) of the Higgs fields, with tan β obviously[4] restricted to 0.7 $\langle \tan \beta \rangle$ 1.9. [5]

$$tan\beta = {}^{v_2}/_{v_.}$$

The MSSM model (minimal supper symmetric model) has two Higgs doublets and additional constraints [6, 7].

$$m_3^2 + m_\pm^2 = m_3^2 + M_w^2,$$
$$m_3^2 + M_z^2 = m_1^2 + m_2^2,$$
$$m_\pm^2 = m_3^2 + M_w^2,$$

From these constrains, it also follows that

$$m_2 \leq m_3 \leq m_1, m_\pm \geq (M_w, m_3)$$

$$m_{1,2}^2 = 1/2 \left\{ m_3^2 + m_z^2 \pm \left[(m_3^2 + m_z^2)^2 - 4m_z^2 m_3^2 \cos^2 2\beta \right]^{1/2} \right\}$$

where $m_1, m_2, m_3,$ and $m_\pm$ are the masses of Higgs particles ($h^o, H^o, A^o,$ and $H^\pm$) respectively. The two angles β and α and are fixed in terms of the Higgs boson masses [9]. and θ_w is the standard weak mixing angle. Importantly, these SUSY relations guarantee that the neutral Higgs particle exists H^o with a mass less than that of the m_z [4,8]

The two angles β and α are fixed in terms of the Higgs boson masses [9].

$$\cos 2\alpha = -\cos 2\beta [(m_3^2 - M_z^2)/(m_1^2 - M_z^2)]$$
$$\sin 2\alpha = -\sin 2\beta [(m_1^2 + m_2^2)/(m_1^2 - m_2^2)]$$
$$\tan 2\alpha = \tan 2\beta [(m_3^2 - M_z^2)/(m_3^2 - M_z^2)]$$

The angel α can be taken to lie in the interval ${}^{-\pi}/_2 \leq \alpha \leq 0$. And the angel β lie in the interval $0 \leq \beta \leq {}^{\pi}/_2$ [3].

FEYNMAN DIAGRMS FOR THE PROCESS:

$$e^- + e^+ \rightarrow H_k^o \rightarrow H_i^+ + H_j^- + \tilde{\chi}_l^o$$

2. Production via different propagator:

2.1. H_k^o and Z^o *are the propagators exchange*

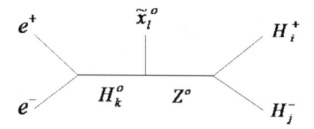

Fig. 1. Feynman diagrams for the process $Z^o(p_2+p_4) \to H_i^+(P_2) + H_J^-(P_4)$ via neutral Higgs bosons H_k^o and Z^o boson propagators exchange. There are (1-432) diagrams.

2.1.1. *The matrix elements for a (1-432) are:*

$$M_{a(1-432)} = \frac{-ig^3\, m_e\, Cos2\,\theta_w}{8M_W Cos^3\theta_W}\, \bar{V}_{e^+}(P_1)\; B_x\; U_{e^-}(P_3)\,(S^2 - m_{H_k^o}^2)^{-1}(S+P_5)_\lambda(\sigma^2 - M_{Z^o}^2)^{-1}\,(p_2+p_4)_\kappa\, \bar{U}_{\tilde{\chi}_l^o}(P_5)$$

where m_e is the electron mass, for B_x (x=1, 2, 3) are

$$B_1 = \frac{Cos\beta}{Cos\alpha} \quad , \quad B_2 = \frac{Sin\beta}{Cos\alpha} \quad , \quad B_3 = \gamma_5\, tan\alpha$$

$$P_1 + P_3 = P_2 + P_4 + P_5$$
$$S = \sigma + P_5$$

2.1.2. *Cross section calculations from situations (1-432)*

In this work we have 3-body final states with momentum P_2, P_4, P_5 and the initial states have momentum P_1, P_3. In general, the cross section for the process

$$e^+(p_1) + e^-(p_3) \to H_i^+(P_2) + H_J^-(P_4) + \tilde{\chi}_\ell^o(P_5)$$

Can be written in the form [1]

$$\sigma = \int \Pi^2 |M|^2\, \frac{dx\; dy\; d\sigma^2}{\Lambda(S, m_1, m_2)\, \Lambda(S, \sigma, m_5)}$$

where M is the matrix element previously mentioned, the integration is performed using a simple approximation obtained by an improved Weizsacker-Williamson procedure [3,10]. Where

314

$$\Lambda(x,y,z) = [x^4 + y^4 + z^4 - 2x^2y^2 - 2x^2z^2 - 2y^2z^2]^{1/2}$$

The limit of integration is given as follows:

$$x_{\pm} = \frac{1}{4S^2}[(S^2 + m_1^2 - m_2^2)(S^2 - \sigma^2 + m_5^2) \pm \Lambda(S, m_1, m_2)\Lambda(S, \sigma, m_5)]$$

$$y_{\pm} = \frac{1}{4\sigma^2}[(\sigma^2 + m_3^2 - m_4^2)(S^2 - \sigma^2 + m_5^2) \pm \Lambda(\sigma, m_3, m_4)\Lambda(S, \sigma, m_5)]$$

$$(m_2 + m_4)^2 \leq \sigma^2 \leq (S^2 - m_5^2)^2$$

In all our calculations, we assume the following values for vector-boson masses suggested by [4, 10] recent collier runs:

$$M_W = 81 Gev$$

$$M_Z = 100 Gev$$

$$M_{H_k^o} = 600, 700, 800 Gev.$$

$$M_{H^{\pm}} = 500, 600, 700 Gev$$

$$m_{\tilde{\chi}_1^o} = 700, 800, 900, 1000 Gev.$$

For: $m_{\tilde{\chi}_1^o} = 700 Gev$

$$M_{h^o} = 600 Gev.$$

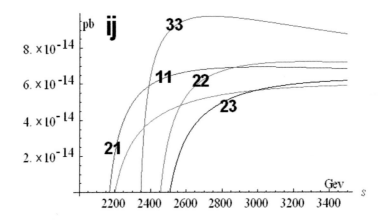

Fig. 2. The relation between the cross sections (pb) and the center of mass energy (Gev) for mode $Z^o(p_2 + p_4) \rightarrow H_i^+(P_2) + H_j^-(P_4)$, when H_k^o and Z^o are propagators exchange, from situations (1-144), $m_{\tilde{\chi}_1^o} = 700 Gev$, and $M_{h^o} = 600 Gev$.

315

$M_{H^o} = 700 Gev.$

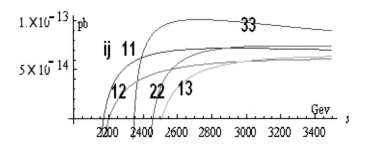

Fig. 3. The relation between the cross sections (pb) and the center of mass energy (Gev) for mode $Z^o(p_2 + p_4) \to \mathrm{H}_i^+(P_2) + \mathrm{H}_J^-(P_4)$, when H_k^o and Z^o are propagators exchange, from situations (145-288), $m_{\tilde{\chi}_1^o} = 700 Gev$, and $M_{H^o} = 700 Gev.$

$M_{A^o} = 800 Gev.$

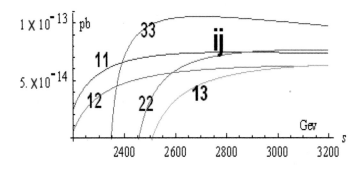

Fig. 4. The relation between the cross sections (pb) and the center of mass energy (Gev) for mode $Z^o(p_2 + p_4) \to \mathrm{H}_i^+(P_2) + \mathrm{H}_J^-(P_4)$, when H_k^o and Z^o are propagators exchange, from situations (288-432), $m_{\tilde{\chi}_1^o} = 700 Gev$, and $M_{A^o} = 800 Gev.$

Results: The results of the calculated cross sections for the process $Z^o(p_2 + p_4) \to \mathrm{H}_i^+(P_2) + \mathrm{H}_J^-(P_4)$ via H_k^o and Z^o Bosons are propagators exchange is presented in the above figures, as a function of the incident energy (S). We found that: At S increase from 2000 to 2800 the cross-section σ increase.

2.2. Z^o and H_k^o *are the propagators exchange*

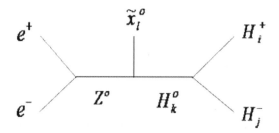

Fig. 5. Feynman diagrams for the process $H_k^o\,(p_2+p_4)\to H_i^+(P_2)+H_j^-(P_4)$ via Z^o and H_k^o bosons propagators exchange. There are (433-864) diagrams.

2.2.1. *The matrix element for a (433-864) are*

$$M_{a(433-864)} = \frac{-ig^3}{8\,Cos^2\theta_W}\,\bar{V}_{e^+}(P_1)\,\gamma_\mu\,C\,U_{e^-}(P_3)\,(S^2-M_Z^2)^{-1}(P_5+S)_\lambda\bar{U}_{\chi^0}(P_5)(\sigma^2-m_{H_y^0}^2)^{-1}$$

$$\left[M_W\cos(\beta-\alpha)-\frac{M_z}{2\cos\theta_w}\cos 2\beta\cos(\beta+\alpha)\right]$$

where $C = 4\,Sin^2\theta_W + \gamma_5 - 1$

2.2.2. *Cross section calculations*

For: $m_{\tilde{\chi}_1^o} = 700Gev$

$M_{h^o} = 600Gev.$

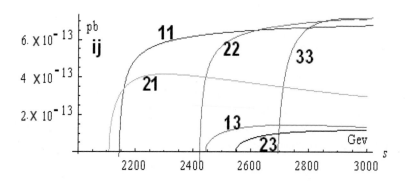

Fig. 6. The relation between the cross sections (pb) and the center of mass energy (Gev) for mode $H_k^o\,(p_2+p_4)\to H_i^+(P_2)+H_j^-(P_4)$, when Z^o and H_k^o are propagators exchange, from situations (433-576), $m_{\tilde{\chi}_1^o} = 700Gev$, and $M_{h^o} = 600Gev.$

$M_{H^o} = 700 Gev.$

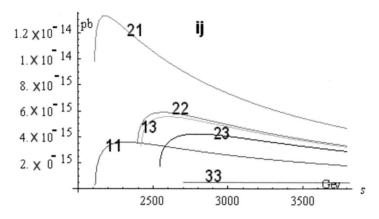

Fig. 7. The relation between the cross sections (pb) and the center of mass energy (Gev) for mode $H_k^o \left(p_2 + p_4 \right) \to H_i^{+} \left(P_2 \right) + H_J^{-} \left(P_4 \right)$, when Z^o and H_k^o are propagators exchange, from situations (577-720), $m_{\tilde{\chi}_i^o} = 700 Gev$, and $M_{H^o} = 700 Gev.$

$M_{A^o} = 800 Gev.$

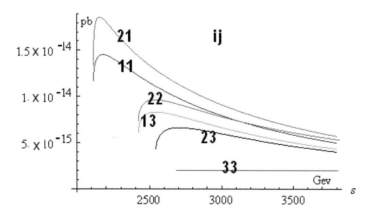

Fig. 8. The relation between the cross sections (pb) and the center of mass energy (Gev) for mode $H_k^o \left(p_2 + p_4 \right) \to H_i^{+} \left(P_2 \right) + H_J^{-} \left(P_4 \right)$, when Z^o and H_k^o are propagators exchange, from situations (721-864), $m_{\tilde{\chi}_i^o} = 700 Gev$, and $M_{A^o} = 800 Gev.$

Results: The results of the calculated cross sections for the process $H_k^o \left(p_2 + p_4 \right) \to H_i^{+} \left(P_2 \right) + H_J^{-} \left(P_4 \right)$ via Z^o and H_k^o bosons are propagators exchange is presented in the above figures, as a function of the incident energy (S). We found that: At S increase from 2000 to 2500 the cross-section σ increase and if S increase from 2500 to 2700 the cross-section σ decrease.

3. Production via H_k^o is a propagator and $\tilde{\chi}_l^o$ is leg:

3.1. $\tilde{\chi}_l^o$ is leg from positron

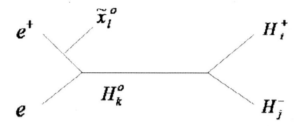

Fig. 9. Feynman diagrams for the process $H_k^o\left(p_2+p_4\right)\to H_i^+\left(P_2\right)+H_j^-\left(P_4\right)$ mediated by H_k^o propagator, and the neutralino $\tilde{\chi}_l^o$ is leg from positron. There are (865-1080) diagrams.

3.1.1. *The Matrix Elements from a (865-1080) are:*

$$M_{a(865-1080)}=\frac{g^2}{2}M_w\bar{V}_{e^+}+(P_1)(N+N^*\gamma_5)_\mu\bar{U}_{\chi^o}(P_5)\frac{(P_1-P_5)+m_e}{(P_1-P_5)^2+m_e^2}U_{e^-}(P_3)(\sigma^2-M_{H^o}^2)^{-1}$$

$$[M_w\cos(\beta-\alpha)-\frac{M_Z}{2\cos\theta_w}\cos 2\beta\cos(\beta-\alpha)]$$

3.1.2. *Cross section calculations*

For: $m_{\tilde{\chi}_1^o}=700Gev$

$M_{h^o}=600Gev.$

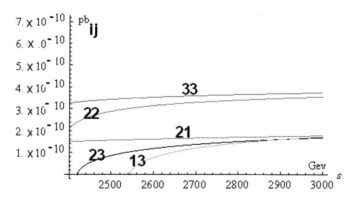

Fig. 10. The relation between the cross sections (pb) and the center of mass energy (Gev) for mode $H_k^o\left(p_2+p_4\right)\to H_i^+\left(P_2\right)+H_j^-\left(P_4\right)$, when H_k^o is propagator and neutralino $\tilde{\chi}_l^o$ is leg from positron, from situations (865-938), $m_{\tilde{\chi}_1^o}=700Gev$, and $M_{h^o}=600Gev.$

$M_{H^o} = 700Gev.$

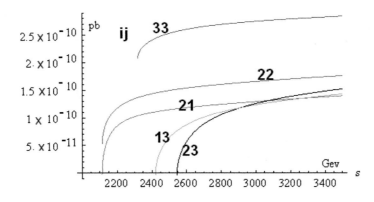

Fig. 11. The relation between the cross sections (pb) and the center of mass energy (Gev) for mode $H_k^o\left(p_2+p_4\right)\to H_i^+\left(P_2\right)+H_J^-\left(P_4\right)$, when H_k^o is propagator and neutralino $\tilde{\chi}_l^o$ is leg from positron, from situations (938-1008), $m_{\tilde{\chi}_l^o}=700Gev$, and $M_{H^o}=700Gev.$

$M_{A^o} = 800Gev.$

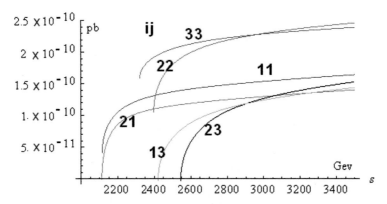

Fig. 12. The relation between the cross sections (pb) and the center of mass energy (Gev) for mode $H_k^o\left(p_2+p_4\right)\to H_i^+\left(P_2\right)+H_J^-\left(P_4\right)$, when H_k^o is propagator and neutralino $\tilde{\chi}_l^o$ is leg from positron, from situations (1009-1080), $m_{\tilde{\chi}_l^o}=700Gev$, and $M_{A^o}=800Gev.$

Results: The results of the calculated cross sections for the process $H_k^o\left(p_2+p_4\right)\to H_i^+\left(P_2\right)+H_J^-\left(P_4\right)$ When H_k^o is propagator and the neutralino is leg from the positron is presented in the above figures, as a function of the incident energy (S). We found that: At S increase the cross-section σ increase.

320

3.2. $\tilde{\chi}_l^o$ is leg from electron

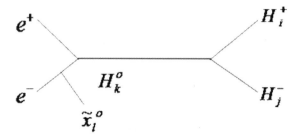

Fig. 13. Feynman diagrams for the process $H_k^o\left(p_2+p_4\right)\to H_i^+\left(P_2\right)+H_J^-\left(P_4\right)$ mediated by H_k^o propagator, and the neutralino is leg from electron. There are (1081-1296) diagrams.

3.2.1. The matrix elements for a (1081-1296) are:

$$M_{a(1081-1296)} = \frac{g^2}{2} M_w U_{e^-} + (P_1)(N+N^*\gamma_5)_\mu \bar{U}_{\chi^o}(P_5)\frac{(P_3-P_5)+m_e}{(P_3-P_5)^2+m_e^2}\bar{V}_{e^+}(P_3)$$

$$(\sigma^2 - M_{H^o}^2)^{-1}[M_w\cos(\beta-\alpha)-\frac{M_Z}{2\cos\theta_w}\cos 2\beta\cos(\beta-\alpha)]$$

3.2.2. Cross section calculations

$$\text{For: } m_{\tilde{\chi}_1^o} = 700Gev$$

$$M_{h^o} = 600Gev.$$

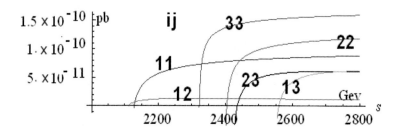

Fig. 14. The relation between the cross sections (pb) and the center of mass energy (Gev) for mode $H_k^o\left(p_2+p_4\right)\to H_i^+\left(P_2\right)+H_J^-\left(P_4\right)$ mediated by H_k^o propagator, and the neutralino is leg from electron., from situations (1081-1152), $m_{\tilde{\chi}_1^o} = 700Gev$, and $M_{h^o} = 600Gev$.

$M_{H^o} = 700 Gev.$

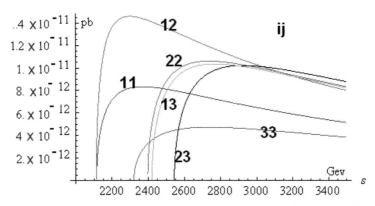

Fig. 15. The relation between the cross sections (pb) and the center of mass energy (Gev) for mode $H_k^o\left(p_2 + p_4\right) \to \mathrm{H}_i^+\left(P_2\right) + \mathrm{H}_J^-\left(P_4\right)$ mediated by H_k^o propagator, and the neutralino is leg from electron., from situations (1153-1224), $m_{\tilde{\chi}_1^o} = 700 Gev$, and $M_{H^o} = 700 Gev.$

$M_{A^o} = 800 Gev.$

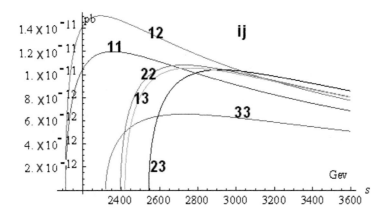

Fig. 16. The relation between the cross sections (pb) and the center of mass energy (Gev) for mode $H_k^o\left(p_2 + p_4\right) \to \mathrm{H}_i^+\left(P_2\right) + \mathrm{H}_J^-\left(P_4\right)$ mediated by H_k^o propagator, and the neutralino is leg from electron., from situations (1225-1296), $m_{\tilde{\chi}_1^o} = 700 Gev$, and $M_{A^o} = 800 Gev.$

Results: The results of the calculated cross sections for the process $H_k^o\left(p_2 + p_4\right) \to \mathrm{H}_i^+\left(P_2\right) + \mathrm{H}_J^-\left(P_4\right)$ When H_k^o is propagator and the neutralino is leg from the electron is presented in the above figures, as a function of the incident energy (S). We found that: At S increase from 2200 to 2600 Gev, the cross-section σ increase and if S increase from 2600 to 3400 the cross-section σ decrease.

4. Discussion:

Comparison the values of different cross-sections, it seems that the most probable mechanism for the reaction

$$e^+(p_1 - p_5) + e^-(p_2) \rightarrow H^o(p_3 + p_4) \rightarrow H_i^+(p_3) + H_j^-(p_4)$$

In which the cross-sections are in fig. (3.1) is approximately equal to that in fig. (3.2) and fig. (3.3), when $\tilde{\chi}_\ell^o$ is emitted from positron as a leg through H_k^o (neutral Higgs bosons; M_{h^o}, M_{H^o}, and M_{A^o}) propagator for all different values i & j.

At S increase from 2200 to 2800 Gev, we have the maximum values of the cross-sections σ various from 2.5 x10^{-10} pb. To 4 x 10^{-10} pb.

No expected significant difference in the values of calculated cross sections if $m_{\tilde{\chi}_l^o} = 700, 800, 900, 1000 Gev$.

References:

1. B.C. Allanach et al., Nuclear Physics, B **135** 107(2004).
2. Citation: S. Eidelman et al. (particle Data group), Phys. Lett. B **592** (2004) 1 and 2005 partial update for edition 2006 (URL: http://pdg.IbI.gov).
3. E. Witten, Nucl. Phys. B **188** 513 (1981): S. Bimopoulose and H. Georgei, Nucl. Phys. B **193** 150 (1981): N. Sakia, Z. Phys. C**11** 153(1981).
4. H. E. Haber and G. L. Kane. Nuclear Physics B **272** 1 (1986).
5. L. Randall and R. Sundrum. Out of this world supersymmetry breaking. Nucl. Phys. B **557**:79-118 (1999).
6. M. Boillargeon, F. Baud Jema, F. Cuypers, E. Gabrielli and B. Melle. Nucl. Phys. B **424** 343 (1994).
7. S. Dawson, C. Jackson, L. Reina and D. Wackeroth, Phys. Rev. D 69, 074027 (2004).
8. S. Dittmaier, M. Kramer and M. Spira, Phys. Rev. D 70, 074010 (2004).
9. W. Buchmuller, D. Wyler, Nucl. Phys. B **268** 621(1986); C. J. C. Burges and H. J. Schmitzer, Nucl. Phys. B **228** 424(1983).
10. W. Williamson, American J. of Physics, **33** 987(1965).

THE PRODUCTION CROSS-SECTIONS FOR THE PROCESS
$$e^+(P_1)+e^-(P_3) \to H_i^+(P_2)+H_i^-(P_4)+\tilde{\chi}_\ell^0(P_5)$$

M. M. AHMED, M. H. NOUS, SHERIF YEHIA AND ASMAA A. A.

Physics Department, Helwan University Cairo, Egypt
mkader4@yahoo.com

Abstract. The cross-section, in electron (e⁻) positron (e⁺) collision, are calculated over range of center of mass energy for the process.

$$e^+(P_1)+e^-(P_2) \to H_{\pm}^+(P_2)+H_-^-(P_4)+\tilde{\chi}_j^0(P_5)$$

Four different group of Feynman diagrams are taken into consideration depending on the type of the propagator:

i - Production of χ^o and $H^{\pm}$ when z^o and H^o are the propagators exchange.

ii - Production of χ^o and $H^{\pm}$ when χ^o is a leg from electron or positron, and γ is the propagator.

iii - Production of χ^o and $H^{\pm}$ when χ^o is a leg from electron and positron, and Z^o is the propagator.

iv - Production of χ^o and $H^{\pm}$ when χ^o is a leg from electron or positron, and H^o is the propagator.

Where $(i, j = 1,2,3,$ and $= 1,2,3,4)$

The cross section for each group is calculated according to a carefully selected set of parameters. These different possible (1155) situations are graphed and tabulated.

The production mechanisms can be detected as:

$$e^+(P_1)+e^-(P_3-P_5) \to \gamma^0(P_2+P_4) \to H_1^+(P_2)+H_i^-(P_4)$$

$$e^+(P_1)+e^-(P_3-P_5) \to Z^0(P_2+P_4) \to H_i^+(P_2)+H_i^-(P_4)$$

$$e^+(P_1)+e^-(P_3-P_5) \to H^0(P_2+P_4) \to H_i^+(P_2)+H_i^-(P_4)$$

$$e^+(P_1)+e^-(P_3) \to H_i^+(P_2)+H_i^-(P_4)$$

Keywords: Higgs bosons; neutralinos.

1. Introduction

One of the main open problems of particle physics is the understanding of the mechanism responsible for breakdown of the electroweak symmetry. The cross-sections for the production of two Higgs boson with neutralinos due to electron-positron annihilation, is calculated according to the reaction.

$$e^+(P_1)+e^-(P_3) \to H_i^+(P_2)+H_i^-(P_4)+\tilde{\chi}_j^0(P_5)$$

(where $i, j = 1, 2, 3,$ and $\ell = 1, 2, 3, 4$)

CP 998, Modern Trends in Physics Research
Third International Conference MTPR-08
edited by L. El Nadi

In the Standard Model (SM) [1] the SU (2)× U(1) group is assumed to be spontaneously broken and $W^{\pm}$ *and* Z^0 bosons acquire their masses through Higgs mechanism.

In the minimal version of the SM where only one Higgs doublet is present, the theory predicts one neutral scalar particle H^0 with an arbitrary mass. The existence of the Higgs boson remains the main missing ingredient for the complete consistency of the SM but it might just as well play a crucial role in the discovery of new physics.

Now, the low energy super gravity model [2] represents one of the most popular extensions of the SM super symmetry standard model is actually able to solve the problem of "naturalness" [3] for light scalar particles. In the general CP-conserving two-Higgs doublet model, three of the eight original scalar degrees of freedom become the longitudinal components of the $W^{\pm}$ *and* Z^0 via the Higgs mechanism. The five remaining physical degrees of freedom manifest themselves as three neutral Higgs H_1^0, H_2^0, H_3^0 and a pair of charged Higgs $H^{\pm}$. We assume the model to be CP-conserving, in which case H_1^0 *and* H_2^0 are CP- even while H_3^0 is CP-odd (pseudo scalar) with respect to coupling to the SM fermions.

In particle physics, a slepton is a sfermion which is hypothetical boson superpartner of a lepton whose existence is implied by Supersymmetry. Slepton have the same flavour and electric charge as corresponding leptons and their spin is zero. For example selectron $\tilde{e}_h$ is superpartner of electron

The MSSM (Minimal Supersymetric Model) contains four neutralinos $\tilde{\chi}_i^0$, which are due to the mixing of photino, Zion and neutral Higgsinos. The neutralino sector depends on four parameters: gaugino masses M and M' associated with the U(1) and SU(2) subgroups of standard model, the Higgs mass parameter μ, and the ratio of the vacuum expectation values (vev) of the Higgs fields.

$$tan\beta = {}^{v_2}/v_{\circ}$$

The MSSM model has two Higgs doublets and additional constraints [4,5].

$$m_3^2 + M_Z^2 = m_1^2 + m_2^2,$$
$$m_+^2 = m_3^2 + M_w^2,$$
$$0 \leq m_2 \leq M_Z \leq m_1.$$

From these constraints, it also follows that

$$m_2 \leq m_3 \leq m_1,$$
$$m_{1,2}^2 = 1/2 \left\{ m_3^2 + m_z^2 \pm [(m_3^2 + m_z^2)^2 - 4m_z^2 m_3^2 \cos^2 2\beta]^{1/2} \right\}$$

m_1, m_2, m_3, m_+ are the masses of the Higgs particles $H_1^0, H_2^0, H_3^0, H^{\pm}$ respectively and θ_w is the standard weak mixing angle. Importantly, these SUSY relations guarantee that the neutral Higgs particle H_2^0 exists with a mass less than that of the Z.

The two angles β *and* α and are fixed in terms of the Higgs boson masses [6].

$$\cos 2\alpha = -\cos 2\beta[(m_3^2 - M_z^2)/(m_1^2 - M_z^2)]$$
$$\sin 2\alpha = -\sin 2\beta[(m_1^2 + m_2^2)/(m_1^2 - m_2^2)]$$
$$\tan 2\alpha = \tan 2\beta[(m_3^2 - M_z^2)/(m_3^2 - M_z^2)]$$

The angel α can be taken to lie in the interval ${}^{-\pi}/_2 \leq \alpha \leq 0$. And the angel β lie in the interval $0 \leq \beta \leq {}^{\pi}/_2$

2. Production via Z^0 and H^0 Propagators Exchange

2.1. *Feynman Diagrams*

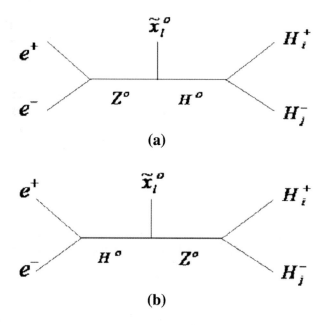

Fig.1.1. Feynman diagrams for the process $e^+(P_1) + e^-(P_3) \rightarrow H_i^+(P_2) + H_j^-(P_4) + \tilde{\chi}_l^0(P_5)$ via Z and Higgs boson exchange. There are (1-288) diagrams.

2.2. *The Matrix Elements*

2.2.1. *For a (1-144) are:*

$$M_{a(1-144)} = \frac{-ig^3 m_e \cos 2\theta_w}{8M_w \cos^2\theta_w} \bar{V}_{e^+}(P_1) B_x U_{e^-}(P_3) \left(S^2 - m_{H^0}^2\right)^{-1} (S + P_5)_\lambda (\sigma^2 - M_Z^2)^{-1}(P_2 + P_4)_K \bar{U}_{\tilde{\chi}^0}(P_5)$$

where m_e is the electron mass, for B_x (x = 1, 2, 3) are

$$B_1 = \frac{\cos\beta}{\cos\alpha} \ , B_2 = \frac{\sin\beta}{\cos\alpha} \ , B_3 = \gamma_5 \tan\alpha$$

The Feynman rules for $e^- e^+ H^0$ vertices are [7,8]

$$P_1 + P_3 = P_2 + P_4 + P_5$$

$$S = \sigma + P_5$$

2.2.2. *For b (145 - 288) are:*

$$M_{b(145-288)} = \frac{-ig^3}{8\cos^2\theta_w} \bar{V}_{e^+}(P_1)\gamma_\mu C U_{e^-}(P_3)(S^2 - M_{\tilde{Z}}^2)^{-1}(S + P_5)_\lambda \bar{U}_{\tilde{\chi}^0}(P_5)(\sigma^2 -$$
$$m_{H^0}^2)^{-1}\left[M_w\cos(\beta - \alpha) - \frac{M_Z}{2\cos\theta_w}\cos2\beta\cos(\beta + \alpha)\right]$$

where

$$C = 4\sin^2\theta_w + \gamma_5 - 1$$

2.3. *Cross section calculations*

In this work we have 3-body final states with momentum P_2, P_4, P_5 and the initial states have momentum P_1, P_2. In general, the cross section for the process

$e^+(P_1) + e^-(P_2) \rightarrow H_i^+(P_2) + H_i^-(P_4) + \tilde{\chi}_i^0(P_5)$ Can be written in the form

$$\sigma = \int \pi^2 |M|^2 \frac{dx\,dy\,d\sigma^2}{\Lambda(S,m_1,m_2)\,\Lambda(S,\sigma,m_5)}$$

where M is the matrix element previously mentioned, the integration is performed using a simple approximation obtained by an improved Weizsacker-Williamson procedure [9,10]. Where

$$\Lambda(x,y,z) = [x^4 + y^4 + z^4 - 2x^2y^2 - 2x^2z^2 - 2y^2z^2]^{1/2}$$

The limit of integration is given as follows:

$$x_\pm = \frac{1}{4S^2}[(S^2 + m_1^2 - m_2^2)(S^2 - \sigma^2 + m_5^2) \pm \Lambda(S,m_1,m_2)\Lambda(S,\sigma,m_5)]$$

$$y_\pm = \frac{1}{4\sigma^2}[(\sigma^2 + m_3^2 - m_4^2)(S^2 - \sigma^2 + m_5^2) \pm \Lambda(\sigma,m_3,m_4)\Lambda(S,\sigma,m_5)]$$

$$(m_2 + m_4)^2 \le \sigma^2 \le (S^2 - m_5^2)^2$$

In all our calculations, we assume the following values for vector-boson masses suggested by [11, 12] recent collier runs:

$$M_W = 81 Gev$$

$$M_Z = 100 Gev$$

$$M_{H^0} = 600 Gev$$

$M_{H^{\pm}} = 500,600,700 Gev$

$m_{\tilde{\chi}_1^0} = 700 GeV, m_{\tilde{\chi}_2^0} = 800 GeV, m_{\tilde{\chi}_3^0} = 900 GeV, m_{\tilde{\chi}_4^0} = 1000 GeV$

m χ_1^0 = 700 Gev

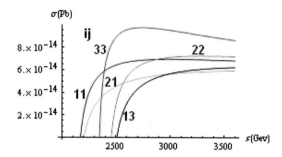

m χ_2^0 = 800 Gev

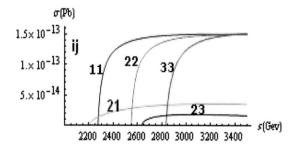

m χ_3^0 = 900 Gev

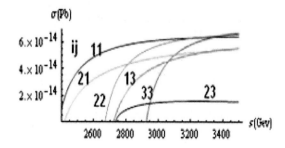

Fig.1.2. The cross section for a (1-144)

328

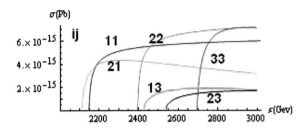

$m_{\chi_1^0} = 700\,\text{Gev}$

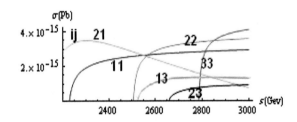

$m_{\chi_2^0} = 800\,\text{Gev}$

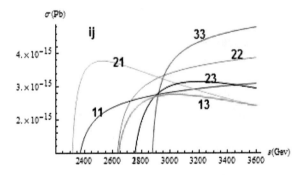

$m_{\chi_3^0} = 900\,\text{Gev}$

Fig.1.3. The cross section for b (145-288)

Results: the calculated cross section for the process $e^+(P_1) + e^-(P_3) \rightarrow H_i^+(P_2) + H_j^-(P_4) + \tilde{\chi}_\ell^0(P_5)$ via Higgs and Z propagators fig.(1.2) and (1.3) As a function of center of mass energy (S) we found that: At S increase from 2000 to 3600 we have different values for the cross-sections at different values of Higgs masses i&j and different value of neutralino mass.

3. Production via γ Photon and Neutralino is a Leg

3.1. *Feynman Diagrams*

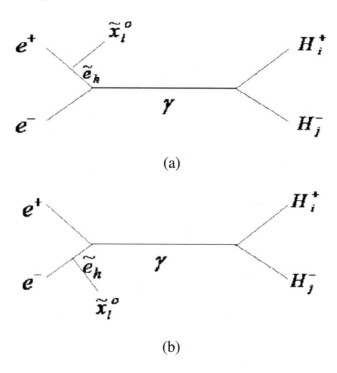

(a)

(b)

Fig.1.4. Feynman diagrams for the process $e^+(P_1) + e^-(P_3) \rightarrow H_i^+(P_2) + H_j^-(P_4) + \tilde{\chi}_\ell^0(P_5)$ mediated by γ photon propagator. There are 289 diagrams.

3.2. *The Matrix Elements*

3.2.1. *For a (289 – 433) are:*

$$M_{a\ (289-433)} = -\bar{V}_{e^+}(P_1)i(N + N^*\gamma_5)_\mu \bar{U}_{\tilde{\chi}^0}(P_5)i\frac{(P_1-P_5)+m_e}{(P_1-P_5)^2+m_e^2} g\cos\theta_W [(P_1 - P_5) + P_3]_\nu U_{e^-}(P_3)i\frac{g_{\nu\lambda}}{(P_1+P_3-P_5)^2} ie (P_2 + P_4)_K$$

330

3.2.2. *For b (434 – 577) are:*

$$M_{b\ (432\ -\ 577)} = -U_e-(P_1)\,i(N + N^*\gamma_5)_\mu\,\bar{U}_{\tilde{\chi}^0}(P_5)\,i\,\frac{(P_3-P_5)+m_e}{(P_3-P_5)^2+m_{\tilde{e}}^2}\,g\cos\theta_W\,[(P_3 - P_5) +$$

$$P_1]_\nu\ \bar{V}_{e^+}(P_2)\,i\,\frac{g_{\nu\lambda}}{(P_1+P_3-P_5)^2}\,ie\,(P_2 + P_4)_\kappa$$

where: N are the (4×4) matrices diagonalizing of the neutralino mass matrix. The Feynman rules for $e^-e^+\tilde{\chi}^0$ vertices [13,14,15]

3.3. *Cross Section Calculations*

The Cross section as a function of center of mass energy for the Feynman diagrams of Fig. (1.4a) are calculated and the result are given in Figs. (1.5) by interchanging the indices i & j and the mass of Neutralino and the result corresponding to Fig. (1.4b) are obtained from Fig. (1.6). It shows the cross-sections for the process as a function of (S) (Neutralino is emitted from the leg through γ photon exchange). If S increase from 2200 to 3600 Gev. We have maximum values for the cross-section,

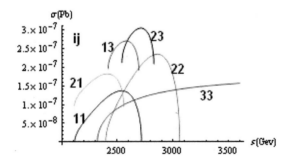

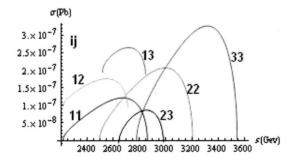

$m_{\tilde{\chi}_3^0} = 900\,\text{Gev}$

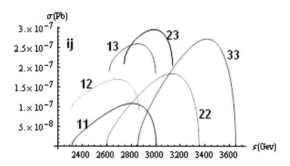

Fig.1.5. The cross section for a (289 – 433).

$m_{\tilde{\chi}_1^0} = 700\,\text{Gev}$

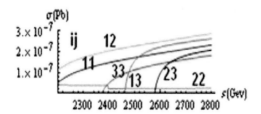

$m_{\tilde{\chi}_2^0} = 800\,\text{Gev}$

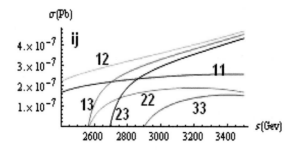

332

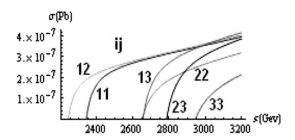

$$m_{\tilde{\chi}_3^0} = 900 \, \text{Gev}$$

Fig.1.6. The cross section for b (434 – 577)

4. Production via z^0 Boson Propagator

4.1. *Feynman Diagrams*

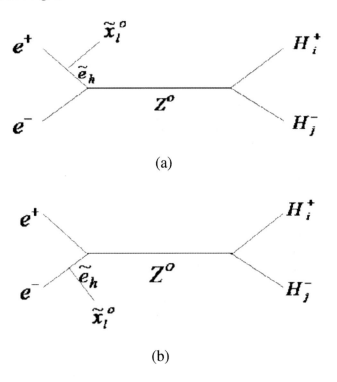

(a)

(b)

Fig.1.7. Feynman diagrams for the process $e^+(P_1) + e^-(P_2) \rightarrow H_i^+(P_2) + H_i^-(P_4) + \tilde{\chi}_j^0(P_8)$ mediated by z^0 boson propagator. There are 289 diagrams.

4.2. *The Matrix Elements*

4.2.1. *For a (578 - 722) are:*

$$M_{a(578 - 722)} = -\bar{V}_{e^+}(P_1)i(N + N^*\gamma_5)_\mu \bar{U}_{\chi^0}(P_5)i \frac{(P_1-P_5)+m_e}{(P_1-P_5)^2+m_e^2} g\cos\theta_W [(P_1 - P_5) +$$
$$P_3]_\nu U_{e^-}(P_2)(\sigma^2 - M_z^2)^{-1} \frac{g\cos 2\,\theta_W}{2\cos\theta_W} (P_2 + P_4)_\kappa$$

4.2.2. *For b (723 - 866) are:*

$$M_{b(723 - 866)} = -U_{e^-}(P_1)i(N + N^*\gamma_5)_\mu \bar{U}_{\chi^0}(P_5)i \frac{(P_3-P_5)+m_e}{(P_3-P_5)^2+m_e^2} g\cos\theta_W [(P_3 - P_5) +$$
$$P_1]_\nu \bar{V}_{e^+}(P_3)(\sigma^2 - M_z^2)^{-1} \frac{g\cos 2\,\theta_W}{2\cos\theta_W} (P_2 + P_4)_\kappa$$

4.3. *Cross Section Calculations*

The Cross section as a function of center of mass energy for the Feynman diagrams of Fig. (1.7a) are calculated and the result are given in Figs. (1.8) by interchanging the indices i & j and the mass of Neutralino and the result corresponding to Fig. (1.7b) are obtained from Fig. (1.9).

It shows the cross-sections for the process as a function of (S) (Neutralino is emitted from leg through Z^o boson exchange). At S increase from 2200 to 3600 Gev we have maximum value for the cross-section, the value of σ various from 1×10^{-8} to 5×10^{-8}

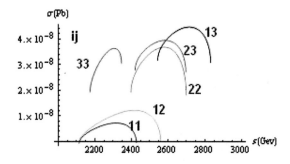

334

$m\,\tilde{\chi}_2^0 = 800\,\text{Gev}$

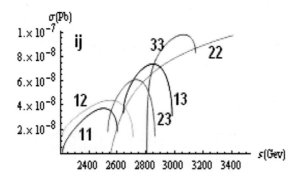

$m\,\tilde{\chi}_3^0 = 900\,\text{Gev}$

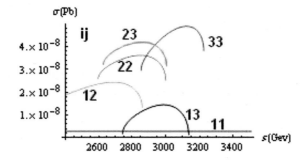

Fig.1.8. The cross section for a (578 - 722)

$m\,\tilde{\chi}_1^0 = 700\,\text{Gev}$

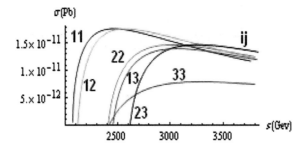

$m_{\tilde\chi_2^0} = 800\,\mathrm{Gev}$

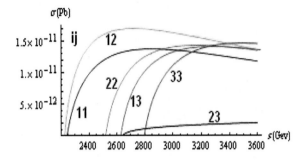

$m_{\tilde\chi_3^0} = 900\,\mathrm{Gev}$

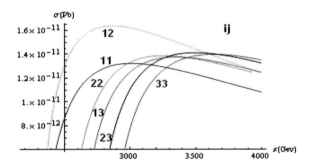

Fig.1.9. The cross section for b (723 - 866)

5. Production via Higgs Boson Propagator

5.1. *Feynman Diagrams*

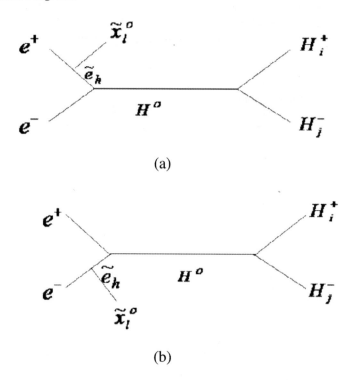

(a)

(b)

Fig.1.10. Feynman diagrams for the process $e^+(P_1) + e^-(P_3) \to H_i^+(P_2) + H_j^-(P_4) + \tilde{\chi}_l^0(P_5)$ mediated by Higgs boson propagator. There are 289 diagrams.

5.2. *The Matrix Elements*

5.2.1. *For a (867 - 1011) are:*

$$M_{a(867-1011)} = \frac{g^2}{2} M_W \bar{V}_{e^+}(P_1)(N + N^* \gamma_5)_\mu \bar{U}_{\tilde{\chi}^0}(P_5) \frac{(P_1 - P_5) + m_e}{(P_1 - P_5)^2 + m_{\tilde{e}}^2} U_{e^-}(P_3)\Big(\sigma^2 -$$
$$M_{H^0}^2\Big)^{-1}\Big[M_W cos(\beta - \alpha) - \frac{M_Z}{2cos\theta_w} cos2\beta cos(\beta + \alpha)\Big]$$

5.2.2. *For b (1012 -1155) are:*

$$M_{b(1012-1155)} = \frac{g^2}{2} M_W U_{e^-}(P_1)(N + N^* \gamma_5)_\mu \bar{U}_{\tilde{\chi}^0}(P_5) \frac{(P_3 - P_5) + m_e}{(P_3 - P_5)^2 + m_{\tilde{e}}^2} \bar{V}_{e^+}(P_3)\Big(\sigma^2 -$$
$$M_{H^0}^2\Big)^{-1}\Big[M_W cos(\beta - \alpha) - \frac{M_Z}{2cos\theta_w} cos2\beta cos(\beta + \alpha)\Big]$$

5.3. *Cross Section Calculations*

The cross section as a function of center of mass energy for the Feynman diagrams of Fig. (1.10a) are calculated and the result are given in Figs. (1.11) by interchanging the indices i & j and the mass of Neutralino and the result corresponding to Fig. (1.10b) are obtained from Fig. (1.12) it show the cross-sections for the process as a function of (S) (Neutralino is emitted from the leg through Higgs boson exchange). At S various from 2200 to 3200 Gev we have different values for the cross-section various.

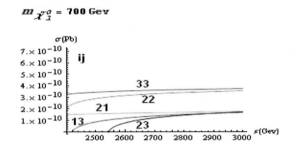

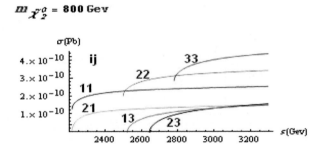

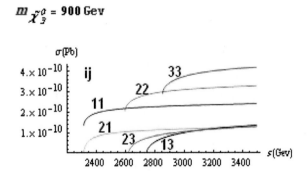

Fig. 1.11. The cross section for a (867 – 1011).

338

$m_{\tilde{\chi}_1^0} = 700\,\mathrm{Gev}$

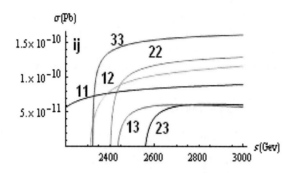

$m_{\tilde{\chi}_2^0} = 800\,\mathrm{Gev}$

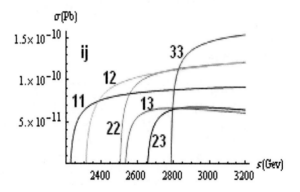

$m_{\tilde{\chi}_3^0} = 900\,\mathrm{Gev}$

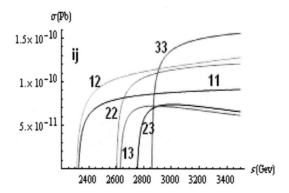

Fig. 1.12. The cross section for b (1012 - 1155)

6. Conclusion

Comparing the values of different cross-sections, it seems that the most probable mechanism for the reaction is in the following table:

Table (1) Shows the peak values of the cross sections of the interaction $e^+(P_1)+e^-(P_2) \to H_i^+(P_3)+H_j^-(P_4)+\tilde{\chi}_\ell^0(P_5)$ with different I & j values of Higgs Boson at different values of incident energies.

Group No.	$e^+ + e^- \to H_i^+ + H_j^- + \tilde{\chi}_\ell^0$	Fig. No.	$m_{\gamma q}$	i j	S(Gev)at max σ	σ(Pb)
i	Production via Z^0 and H^0 propagators exchange	1.2	700GeV	33	3500	8×10^{-14}
			800GeV	11 22 33	3400	1.5×10^{-13}
			900GeV	11 22 33	3400	6×10^{-14}
		1.3	700GeV	22 33	3000	6×10^{-15}
			800GeV	12	3000	4×10^{-15}
			900GeV	33	3600	5×10^{-15}
ii	Production via γ photon and Neutralino is a leg	1.5	700GeV	23	2700	3×10^{-7}
			800GeV	33	3300	3×10^{-7}
			900GeV	23	3400	3×10^{-7}
		1.6	700GeV	12	2800	3×10^{-7}
			800GeV	12	3400	4×10^{-7}
			900GeV	12	3200	4×10^{-7}

340

iii	Production via Z^0 boson and Neutralino is a leg	1.8	$700GeV$	13	2700	4×10^{-8}
			$800GeV$	33	3000	1×10^{-7}
				22	3400	
			$900GeV$	33	3100	4×10^{-8}
		1.9	$700GeV$	11	2300	1.5×10^{-11}
				12	2500	
			$800GeV$	12	2600	1.5×10^{-11}
			$900GeV$	12	2800	1.6×10^{-11}
iv	Production via H^0 and Neutralino is a leg	1.11	$700GeV$	33	3000	4×10^{-10}
			$800GeV$	33	3200	4×10^{-10}
			$900GeV$	33	3400	4×10^{-10}

From Table (1), it could be concluded that the reaction has the most probable of cross section ($\sigma = 4 \times 10^{-7}\ Pb$) over all interaction processes at center of mass energy at range 3200 to 3400 Gev so it consider the main mode of the reaction is via γ propagator.

The production mechanisms can be detected as:

$$e^+(P_1) + e^-(P_3 - P_5) \to \gamma^0(P_2 + P_4) \to H_i^+(P_2) + H_i^-(P_4)$$

$$e^+(P_1) + e^-(P_3 - P_5) \to Z^0(P_2 + P_4) \to H_i^+(P_2) + H_i^-(P_4)$$

$$e^+(P_1) + e^-(P_3 - P_5) \to H^0(P_2 + P_4) \to H_i^+(P_2) + H_i^-(P_4)$$

$$e^+(P_1) + e^-(P_3) \to H_i^+(P_2) + H_i^-(P_4)$$

References

1. B. C. Allanach et al., Beyond the Standard Model Working Group Colloboration, hep-ph/0402295; J. Rosiek, Phys. Rev. D **41**, 3464 (1990).
2. B. C. Allanach et al., Nuclear Physics, B **135** 107 (2004).
3. C. N. Leung, S. T. Love and S. Rao, Z. Phys. C **31** 433 (1986).
4. Debajyoti Choudhucy and Frank Cuypers, Nucl. Phys. B **429** 33 (1994).
5. E. Witten, Nucl. Phys. B **188** 513 (1981): S. Bimopoulose and H. Georgei, Nucl. Phys. B **193** 150 (1981): N. Sakia, Z. Phys. C **11** 153 (1981).
6. G. B´elanger, F. Boudjema and J. Fujimoto, Apr (2006) 87. hep-ph/0308080.
7. H. E. Haber and G. L. Kane, Nuclear Physics B **272** 1 (1986).
8. J. Ellis and G. G. Ross, Phys. Lett. B **117** 396 (1982).
9. J. F. Gunion and H. E., Haber Nucl. Phys. B **272** 1(1986); B **278** 449(1986).

10. J. F. Gunion and H. E. Haber, G. I. Kane and S. Dawson, The Higgs Hunter's Guide 1 (1990).
11. K. Desch, J. Kalinowski, G. Moortgat-Pick, M. M. Nojiri and G. Rolesello, JHEP0402(2004)035[hep-ph/03/2069].
12. M. Boillargeon, F. Baud Jema, F. Cuypers, E. Gabrielli and B. Melle, Nucl. Phys. B **424** 343 (1994).
13. S. M. Bilenky and J. Hosek, Phys. Rep., **90** 73 (1990).
14. W. Buchmuller, D. Wyler, Nucl. Phys. B **268** 621 (1986); C. J. C. Burges and H. J. Schmitzer, Nucl. Phys. B **228** 424 (1983).
15. W. Williamson, American J. of Physics, **33** 987 (1965).

AN EARLIER NATURAL MECHANISM PROPOSAL FOR THE CLOSURE OF THE OZONE HOLE AND THE PRESENT 30% CLOSURE

SHAHINAZ M. YOUSEF

Astronomy & Meteorology Dept., Faculty of Science, Cairo University, Egypt
email: habibat_arrahman@yahoo.com

SIHAM A. AL-KUHAIMI

Girls College of Education, Scientific Dept., Malaz Riyadh, Saudi Arabia

AISHA BEBARS

National Research Institute of Astronomy and Geophysics, NRIAG, Helwan, Egypt

Abstract. A prolonged period of reduced solar activity of the order of few decades is expected owing to the presence of weak solar cycles series like those around 1800 and 1900 AD. Reduced UV flux is forecasted.

The multitude of phytoplanktons in the Antarctic Ocean which are harmed by excessive UV passing through the ozone hole are expected to recover owing to the reduced solar UV doze even with the existence of ozone hole. An increase of only 10% of the phytoplankton would remove about 5 Gigatons of carbon dioxide from the atmosphere annually (which is equal to the amount of carbon dioxide emitted currently by fossil fuel utilization) and sink it into the ocean. Reduction of carbon dioxide from the atmosphere will lead to cooling of the troposphere and hence warming of Antarctic stratospheric clouds which are the sight of ozone destruction. Eventually, this procedure will hopefully lead to Antarctic ozone hole closure.

The paper also discuss the implication of the 1997 solar induced climate change on the appearance of the Arctic ozone hole and the reduction of the Antarctic ozone hole. Anther more serious solar induced climate change is currently on due to the end of the first weak solar cycle number 23 and the start of predicated second weaker solar cycle number 24. A climate change, which has already brought global cooling to the earth.

The Ozone hole has been closed by 30% in 2007 as predicted which a triumph is for the subject of sun-Earth connections. It is also predicted that further closures in the coming few years will occur due to solar induced climate changes. The forecast of the ozone hole closure was predicted in earlier papers.

INTRODUCTION

The ozone layer lies in the stratosphere. It absorbs most of the harmful solar UV. The small percent of UV radiation which reaches the earth is vital for photosynthesis and some other biological processes.

In the equatorial zone, the annual flux of solar radiation with $\lambda < 315$ nm (currently designed as UV-B) at sea level is 12.6 MJ/m^2. In the upper atmosphere, the UV-B amounts to 1.85 % of the total radiation but at the sea level it is only 0.1% at 0° N and .04% at 60° latitude, due to its absorption by ozone (Khrgian and Thang 1991 and references therein). The higher the latitude, the thicker the ozone layer is (Xanthakis et al. 1993).

The discovery of the Antarctic ozone hole has caused concern among scientists. The concentration of the stratospheric ozone varies in response to a number of competing factors given by Hood and McComrmack (1992) and references therein:

1. Interannual changes in internal stratospheric dynamics such as those associated with the quasi-biennial oscillation (QBO) and the El Nino Southern Oscillations (ENSO).

2. Interannual changes in the solar ultraviolet spectral irradiance and energetic particles precipitation.

3. Long term changes in the atmospheric content of anthropogenic trace gases such as chloroflourocarbons (CFCs).

CP 998, Modern Trends in Physics Research
Third International Conference MTPR-08
edited by L. El Nadi

4. Changes in the abundance of volcanically ejected trace gases and aerosols.

The CO_2 concentration in the earth's atmosphere is related to the ozone depletion (Shaaban 1995 and references therein).

It is the purpose of this paper to point out that a termination of Solar Wolf-Gleissberg cycle occurred in 1997 with the start of cycle 23 which is the first of a weak cycle series (Yousef 2003) and to explain the appearance of Arctic ozone hole as a manifestation of solar induced climate change. A mechanism for closure of ozone hole is proposed.

Such mechanism is capable of explaining the present reduction in ozone hole area.

Another more serious solar induced climate change occurred in response to the end of solar cycle 23 and start of cycle 24 of a predicted very weak second weak cycle. As a sequence of this present climate change, there is a switch from global warming to global cooling (Yousef 2004). Details of such a switch will be given else where. Another sequence is the closure of the ozone hole. The start of such closure is well in the way.

1997 Solar Induced Climate Change and Its Implication on the Appearance of Arctic and Contraction of the Antarctic Ozone Holes

Turning points of the Wolf-Gleissberg cycles

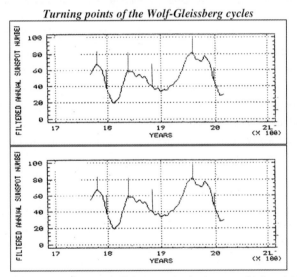

FIGURE 1. The solar magnetic Wolf-Gleissberg cycles showing some of their turning points

On smoothing time series of annual sunspot number, the Wolf-Gleissberg Cycles (WGC) become evident as seen in fig 1. They are magnetic solar cycles of the order 80-120 year. Each cycle has two maximums of normal cycles separated by one or two weak solar cycles. The minimum of those cycles consist of a series of weak cycles of faster solar rotation and lower solar irradiance, UV, X-rays, IR radio waves and weak interplanetary magnetic field, etc. It is found that solar induced climate change do occur at the turning points of those cycles, at the start and end of the weak cycles series and following the maximums (Yousef 2003 and 2006). In other words whenever there is a change in solar spin rate. Recently there are the 1997 with the start of the first weak cycles series and the more dramatic 2007-2008 solar induced climate change with the end of cycle 23 and start of the anticipated still weaker solar cycle.

1997 SOLAR INDUCED CLIMATE CHANGE

1997 marked a solar induced climate change as solar cycle number 23 is the first of 12 yr. weak solar cycles series (Yousef 2003). Several abrupt responses to solar stimuli occurred in 1997. The Mediterranean Sea, the little Aral Sea and Lake Victoria are few examples shown in Fig 2.

a) 1997 Rise of the Little Aral Sea

The Aral and the Dead Seas have been shrinking in coherence ever since the 1960 solar induced climate change following the first maximum of Wolf-Gleissberg cycle. Since the 1960s drying up, the shoreline has receded by more than 120 kilometers in places and the Aral Sea has since split into two basins, the Big Aral and the Little Aral, each supplied by one of the rivers. This shrinkage is misinterpreted as a result of the large amounts of water diverted by cotton farmers from the two rivers feeding into it, the Syr Darya and the Amu Darya. Oceanography satellites such as Topex/Poseidon measure sea level continuously over the long term and thus enable us to monitor variations in the Aral Sea.

Only the little Aral showed the 1997 sudden level rise of about 3 m. Maximum rise was attained early 1999 followed by a sudden drop of about 3 m. This rise must have been due to increased inflow over the source of its feeding river, the Syr Darya. Central Asia's two most important rivers, the Syr Darya and the Amu Darya, flow to the Aral Sea from the Tian Shan and Pamir mountains. The Tian Shan system is the highest mountain system north of Tibet. Pobeda Peak, the highest peak, rises 7,439 meters above sea level. Rivers flow north from Tian Shan into Kyrgyzstan and Kazakhstan, and south into China. The system includes some of the world's largest glaciers (World Book Encyclopedia 1997). Since only the little Aral Sea showed the 1997 sudden rise then, only the Tian Shan was affected by the 1997 solar induced climate change but not the Pamir mountains. This indicates space dependence of the solar induced climate change. If the Syr Darya river flooding is due to snow melting, then it is likely that a temperature increased happened over the Tian Shan mountains up till early 1999.

b) 1997 Rise in Mediterranean Sea Level and Temperature

Aviso/Altimetry - Applications of Mean Level Altimeters of little Aral Sea, Mediterranean Sea (Source: Legos, Toulouse, France - 2001), show sudden rise in 1997. Figure 2 illustrates the Mediterranean Sea rise in level and temperature. This large rise in Mediterranean Sea level is contemporary with the start of solar cycle number 23 and with increment of solar irradiance flux shown also in Fig 2. It is evident that the Mediterranean Sea level rise is due to solar forcing.

c) The 1997 Sharp Rise of Lake Victoria

Figure 2 shows the 1997 Lake Victoria level rise. It was unique as the lake level dropped steadily from June to October to 11.32 m then rose abruptly up to 12.87 m in May 1998. Around this time was the end of the previous solar cycle and the beginning of weak solar cycle 23. 1997-1988 was also a strong El Nino year. This rise was followed by a drop which is expected to continue up till about 2008-2009 when severe drought years are expected and then the Equatorial lakes levels will rise and fall in coherence with the following few solar weak cycles (Yousef and Amer 2003).

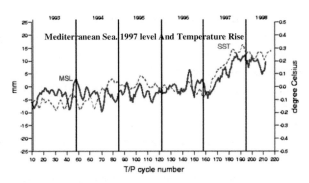

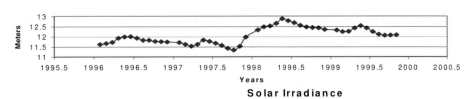

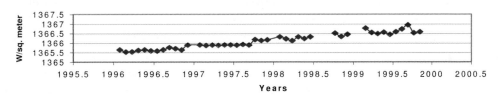

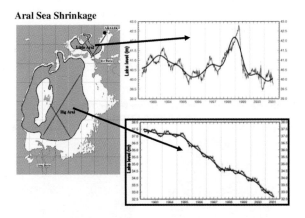

FIGURE 2. 1997 solar induced climate change

346

Comparison of the 1997-98 sharp rise with solar irradiance is shown for comparison in the same Fig 2. Solar irradiance increase of about 0.11% in the above period is well associated with an increase of about 1.55 meters in Lake level. The mechanism of this relationship is not currently well defined

This sudden rise must have been similar to the rise   which was reported earlier at 1877-78 at the end of a Wolf-Gleissberg cycle, however the recent rise is milder than the 1877 rise. 1997-98 marked the end of the last Wolf-Gleissberg cycle which started in 1913.

POLAR STRATOSPHERIC CLOUDS

(PSCs) are common in Antarctica, but a rare sight in the Arctic. They form when temperatures in the stratosphere become extremely cold – below -78°C. PSCs spell trouble for ozone; tiny ice crystals and droplets within the clouds provide surfaces where CFCs are converted into ozone-destroying molecules. Credit: Lamont Poole, NASA.

"Typically a wave will warm the polar region by 5 to 10°C," Newman continued. "A 'warm' polar stratosphere is typically in the temperature range -73°C to -63°C. Of course, as soon as the wave has dissipated, the polar region begins to cool down again."

THE 1997 ARCTIC OZONE HOLE APPEARANCE
Planetary Waves

High mountains and land-sea boundaries combine to generate vast undulations in the atmosphere called "planetary-scale waves," or "long waves," which act to heat polar air. Planetary-scale waves are so large that some of them wrap around the whole Earth!

Planetary waves displace air north and south as they travel around our planet. They form in the troposphere and propagate upward, transferring their energy to the stratosphere. Stronger planetary waves in the northern hemisphere warm the Arctic stratosphere and suppress ozone destruction (NASA News 2001). However in 1997, the effect of planetary waves reversed, they act in cooling the Arctic stratosphere thus arctic ozone hole came into being. However they warmed the Antarctic Stratosphere trying to close the Antarctic ozone Hole.

Figure 3 shows the 1997 Arctic ozone hole due to cooler stratospheric temperature in comparison with the 1984 warmer conditions which led to no Arctic ozone hole appearance.

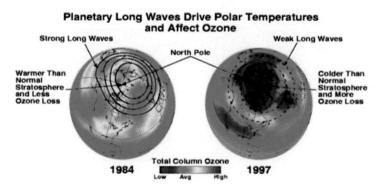

FIGURE 3. Demonstration of the appearance of the 1997 Arctic ozone hole (NASA News 2001)

Natural Mechanism for the Antarctic Ozone Hole Closure
DROP OF SOLAR UV FLUX

Computation of the flux of solar irradiance, IR and UV (200-250 nm), which is of relevance to ozone formation, by Lean (2000) show a considerable drop contemporary with the weak cycles series Since the present solar cycle number 23 is the first of weak cycles series (Yousef 2003), then we should expect a reduced UV-B drop of the order given in Fig 4.

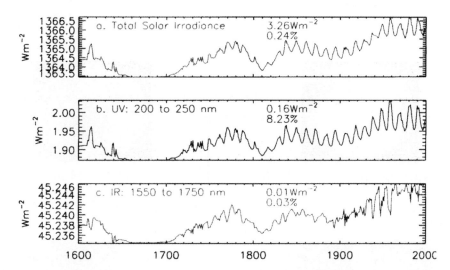

FIGURE 4. Model estimate of long-term variability in total solar irradiance and in two spectral bands in the UV and IR showing the Wolf-Gleissberg cycles (after Lean 2000)

PHYTOPLANKTONS REMOVE CO2 FROM ATMOSPHERE

Since experts (Huder, et al. 1991) indicate that UV-B penetrates 65 meters deep in clear Antarctic waters, consequently the UV-B irradiance in Antarctic waters drastically increased during the occurrence of the ozone hole causing considerable damage to Planktonic marine organisms. Those organisms account for over half of the total global amount of carbon fixed annually. Any reduction in this productivity will undoubtedly affect global food supply and global climate. The expected prolonged UV-B flux reduction will eventually lead to the revival of Phytoplankton and hence increased photosynthesis. Considerable reduction of carbon dioxide from the atmosphere will lead to cooling of the troposphere and hence warming of Antarctic stratospheric clouds which are the sight of ozone depletion. Generally speaking warming of stratosphere will lead to ozone hole disappearance as in the case of warming by planetary waves seen in Fig. 3.

348

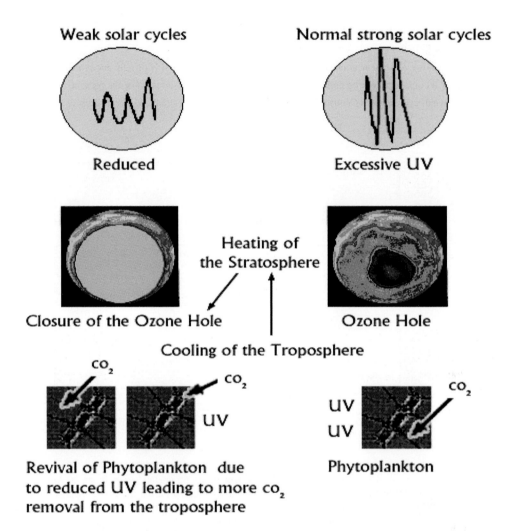

FIGURE 5. Mechanism for closure of ozone hole; starting with reduced solar activity less UV will be emitted from sun. The Phytoplankton underneath the Antarctic ozone hole will revive thus absorbing more CO_2 in the process of photosynthesis. Reduction of CO_2 from the troposphere will cool it and heat the stratosphere, hence reduce the size (or even eliminate) of the stratospheric clouds. Those clouds are the sites of ozone depletion via CLFCs. Cooling of the stratosphere will lead to ozone hole closure as seen in Fig 3.

THE 2007 SOLAR INDUCED CLIMATE CHANGE

With the end of the weak solar cycle number 23 and the start of cycle 24 the era of global warming has ended and we have entered into a new era of solar induced cooling. All four monitors of global temperatures have recorded a sharp decrease of about 0.75°C between January 2007 and January 2008 (Blog 2008). Further cooling is expected to last between 1.5 to 4 solar cycles location dependant similar to earlier cooling around 1800 and 1900 due to weak solar cycles. The present cooling was predicted (Yousef 2004). Details of the 2007 climate change will be given in a special paper.

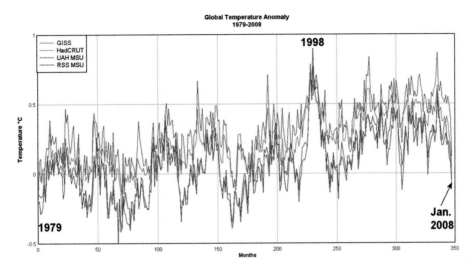

FIGURE 1. Instrumental measurements of global temperature anomaly 1979-2008 by four centers.

THE 2007 ANTACTIC OZONE HOLE 30% CLOSURE

As predicted earlier (Yousef 1998, Yousef, Al-Kuhaimi 1999 and Yousef et al. 2004) the ozone hole has started to close. It was announced in October 2007 that the ozone hole over Antarctica has shrunk 30 percent in 2007 as compared to 2006 record size. According to measurements made by ESA's Envisat satellite, the 2007 ozone loss peaked at 27.7 million tones, compared to the 2006 record ozone loss of 40 million tones (ESA News 3 October 2007). This partial closure is coincident with the present drop of global temperature (ref). The proposed model is capable of exploring such drop. In addition, the present global temperature drop of the troposphere must have also contributed to such partial closure. Complete ozone hole closure is expected within few years.

CONCLUSION

1997 was an extraordinary year worldwide; it was a year of solar induced climate change. There are 80-120 yr. solar cycles termed the Wolf-Gleissberg cycles. They are magnetic cycles evident on smoothing of the annual sunspot cycles. It is found that climate change occur at the turning points of those cycles. 1878 and 1997 were identical turning points arising from the start of the first of a 3 to 4 weak (mostly12 yr.) solar cycles 1997 solar forcing was great and induced several solar-terrestrial events. Among those events are the Arctic ozone hole appearance, the shrinkage of the Antarctic ozone hole, the rise of the little Aral sea, the rise of the Mediterranean Sea level and temperature, the excessive heat over Egypt, the sudden rise in Lake Victoria level and the shrinkage of the Antarctic ozone hole etc.

The polar ozone holes are affected by planetary waves that can wrap over the earth pushing air towards the poles causing heating on one of the stratospheric poles and cooling on the other. Cooling of the stratosphere polar region induce ozone hole while warming close it.

In 1997, it seems that there was a reverse of the effect of those planetary wave, inducing cooling in the Arctic stratosphere and warming in the Antarctic. It seems that those planetary waves are not only affected by the distribution of global mountains but are also controlled by solar forcing.

The natural proposed mechanism of ozone hole closure is summarized as follows:

The presence of weak solar cycles, already cycle 23 is the first of weak cycles series each of duration 12 yrs. Reduced emission of UV-B. Recovery of Phytoplankton. Increased carbon dioxide removal from atmosphere hence cooling of troposphere. Heating of stratosphere. Reduced stratospheric clouds, reduced destruction of ozone by CFC etc Gradual closure of ozone Hole.

UV-B-derived variation in phytoplankton biomass ought to be incorporated into global climate change models (Huder, et al. 1991).

The 2007-2008 solar induced climate change has stimulated wide spread terrestrial effects among them are the sudden cooling of the earth and the closure of the ozone hole. There is still the fear of wide spread drought flood hazards owing to the expected cooling of the oceans with special emphasis on the Pacific Ocean similar to what happened around 1900(Reid 2000). This Pacific cooling is expected to bring persistence La Nina cold currents for several successive years

ACKNOWLEDGMENTS

My sincere thanks are due to my parents

REFERENCES

1. Bolg, M.A., Blog Science Daily Technology. Temperature Monitors Report Widescale Global Cooling (2008).
   http://www.dailytech.com/Temperature+Monitors+Report+Worldwide+Global+Cooling/article10866.htm
2. ESA Observing the Earth, 2007 ozone hole smaller than usual. ESA News (3 October 2007). www.esa.int/esaCP/SEM6MD7H07F_index_1.html
3. Hood, H.H. and McCommrack, J.P., Geophys. Res. Lett. 19, 2309-2313 (1992). Huder, D.P., Worrest, R.C. and Kumar, H.D., Aquatic Ecosystems, Chapter 4 in *Environmental Effects of Qzone Depletion*, Nairobi: UN Environment Programme, 1991.
4. Khrgian, A. Kh. and Thang N.V., Soviet Meteorology and Hydrology 1, 23-31 (1991).
5. Lean, J., Space Sci. Rev. 94, 39-51 (2000).
6. Legos, Toulouse, France - 2001
7. www.jason.oceanobs.com/html/applications/niveau/tchad_uk.html
8. NASA Science News, Oct 11 2001 internet site.
9. Reid, G.C., Solar Variability and the Earth's Climate: Introduction and Review, *Solar Variability and Climate*; Edited by Friis-Christensen, E., Fröhlich, C., Haigh, J.D., Schüssler, M. and Von Steiger, R. Space Sci.Rev. 94, No1-2 (2000)
10. Shaaban, S., Al-Qafila 43, No 12, 15-19 (1995).
11. Xanthakis, J., Poulakos, C. and Zerefos, C., Earth Moon and Planets, 60,109-125 (1993).
12. World Book Multimedia Encyclopedia, IBM 1997.

13. Yousef, S.M., A Warning of Solar Inactivity During the Next Few Decades and Its Influence on IMF and Cosmic Rays. Proceedings of the Third SOLTIP Symposium, Beijing, 569-575 (1996).

14. Yousef, S.M., Cycle 23, The First of Weak Solar Cycles Series and the Serious Implications on Some Sun-Earth Connections, in Proceedings of ISCS, European Space Agency ESA. SP-535. Solar Variability as an Input to the Earth's Environment, 23-28 June, Tatranska Lomnica, Slovak Republic, 2003, 177-180.

15. Yousef, S.M., Expected cooling of the earth and its implications on food security, Bulletin De L'Institut D'Egypte, Tom LXXX. pp 53-82 (2004).

16. Yousef, S.M., 80-120 yr. Long Term Solar Induced Effects on the Earth, Past and Predictions", Physics and Chemistry of the Earth, 31 pp.113-122 (2006). Elsevier, available online www.sciencedirect.com

17. Yousef, S.M. and Al-Kuhaimi, S.A., Expected Drop in Solar Ultraviolet Flux During the Coming Three Decades and Ozone Hole Recovery, Journal of Environmental Sciences Vol. 18, 149-164 (1999).

18. Yousef, S. and Amer, M., The Sharp Rise of Lake Victoria, A Positive Indicator to Wolf-Gleissberg Cycle Turning Points, in Proceedings of ISCS, European Space Agency ESA. SP-535. Solar variability as an input to the Earth's environment, 23-28 June, Tatranska Lomnica, Slovak Republic, 2003, pp. 397-400.

19. Yousef, S.M., Al-Kuhaimi, S.A. and Bebars, A., Modern Trends in Physics Research Conference. Cairo, Egypt, Lotfia EL Nadi, Abstract Book (2008).

AUTHOR INDEX